COLUMBIA

FINAL VOYAGE

THE LAST FLIGHT OF NASA'S FIRST SPACE SHUTTLE

PHILIP CHIEN

T0222401

COPERNICUS BOOKS
An Imprint of Springer Science+Business Media

in Association with

PRAXIS PUBLISHING, LTD.

All photos are courtesy of NASA and the Columbia Accident Investigation Board (CAIB) unless otherwise noted.

an imprint of Springer Science+Business Media.

Copernicus Books
Springer Science+Business Media
233 Spring Street
New York, NY 10013
www.springer.com

Additional information is available at the author's website—http://www.sts107.info

Manufactured in the United States of America.
Printed on acid-free paper.

9 8 7 6 5 4 3 2 1

e-ISBN 0-387-27149-X

ISBN 978-1-4419-2092-8   e-ISBN 978-0-387-27149-1

FOREWORD

JONATHAN B. CLARK

Columbia—Final Voyage is the most comprehensive book about the final mission of Columbia STS-107 to date. I have known the author, Phil Chien, since I started working shuttle missions as a flight surgeon in 1998. He was always there asking the shuttle crew questions when they were at the Cape for their Terminal Count Down Tests and press conferences just prior to launch. I was always impressed with his very insightful questions and deep grasp of technical issues. As I have come to know him more, I am profoundly impressed with his encyclopedic knowledge of human spaceflight. I can think of no more dedicated journalist to write this book. He had come to know the STS-107 crew very well in the long period prior to launch. He was one of a handful of journalists who was at the Shuttle Landing Facility at Kennedy Space Center on February 1, 2003, and like all of us felt the visceral emotion as the clock ticked down to landing time and then started counting up. This book has captured the essence of human spaceflight and the intricacies of space shuttle operations. A companion CD-ROM [available for purchase separately] will provide additional historical references for countless space aficionados, including myself.

Phil has been able to explain the myriad of science experiments on this flight so that even I can understand them. The book also presents a very well documented analysis of the Columbia Accident Investigation Board. The author has also crafted the story behind the story concerning how politics is intertwined with NASA. More importantly, he has captured the essence of the human spirit—of the crew, the families and friends, and the people who really make it happen behind the scenes, in training, science support, and launch and Mission Control. As we all come to deal with the aftermath of the triumph and tragedy of Columbia's final flight, this wonderful book will serve as a guidebook for us to learn from the past and enable the future.

INTRODUCTION

BUZZ ALDRIN

About 400 people have had the opportunity to fly in space over the past 45 years. Only a few are famous or household names—primarily the ones who have flown on historic missions, and the ones who died. It's regrettable that we remember the Apollo 1, Challenger, and Columbia crews primarily because of their deaths. In contrast, most astronauts remain relatively anonymous people. In this book Philip Chien takes a passionate look at the people who flew on Columbia's last mission. Not just as astronauts—but as people. Most reporters considered STS-107 to be a "boring non-newsworthy mission." But Philip Chien went to the effort to get to know the astronauts and their mission in detail because they were important as human beings and their mission was important even if it wasn't as historic as some other flights.

Rick, Willie, Dave, Kalpana, Laurel, Mike, and Ilan were a varied group of people—they included military pilots, like the early astronauts, but also medical doctors and engineers.

In 1969, Neil, Mike, and I flew to the moon in a ship named "Columbia." Eleven years after our mission, another ship named Columbia launched from the very same launch pad. But it was drastically different. Instead of just space for three people in a small living space, the Columbia space shuttle was massive—with room to carry many more people and lots of cargo. And a varied cargo it did carry—commercial satellites, military payloads, scientific satellites, and plenty of scientific laboratories. On Columbia's 28th and last mission it was carrying one of these scientific laboratories—filled to the brim with science which could only be performed in space. Many shuttle critics have claimed that it was a "make work" mission that didn't have to be flown. But judge for yourself after reading this book's chapters about each of the experiments. STS-107 certainly wasn't as historic as going to the moon or building a space station, but it was important in the long run—extending our knowledge about what we can do in space and even extending the capabilities of scientific laboratories on Earth.

This book will tell you about the real people who flew on STS-107 and their mission.

CONTENTS

PREFACE

February 1, 2003—Space Shuttle Columbia was on its 28th mission, completing a routine microgravity science mission. Like over 100 shuttle missions before, it fired its engines to reduce its speed enough to drop out of orbit and reenter the Earth's atmosphere. But unlike every other time a shuttle returned from space, Columbia didn't make it. Instead of the signature twin sonic booms normally heard a couple of minutes before landing, there was only an eerie silence. In east Texas people saw Columbia come apart and saw its pieces fall out of the sky. A relatively plain spaceflight had suddenly become major international news.

Every shuttle flight is important, but certainly some are more glamorous than others. The STS-107 mission was not an extraordinary shuttle flight—while there were some minor "firsts," it was basically another flight for fundamental scientific research. In no way was STS-107 a "historic" mission—it didn't accomplish the first servicing of a satellite in orbit, make the first docking with a Russian space station, assemble the first components of the International Space Station, or even collect radar mapping data of the entire world. Had Columbia landed safely, it would be just one of 112 shuttle flights that accomplished its goals and returned excellent results for its scientists (the only exception being the Challenger accident). But because the shuttle was destroyed and the seven-person crew perished, STS-107 entered the history books.

What caused the Columbia accident is simple—a 1.67 pound chunk of foam fell off of the External Tank 81 seconds after launch. The foam hit the front of the left wing, and as it descended into the atmosphere, the hot reentry gasses flowed into the damaged wing, leading to Columbia's destruction. Why the foam fell is a book in itself—the compromises in the original space shuttle design, how the engineers decided to use foam as an insulator, and how NASA treated the issue of falling foam throughout the shuttle program. But this book is not about the accident, it's about the STS-107 astronauts and their mission.

There are some amazing stories about the Columbia mission, both happy and sad. Astonishingly, tiny worms survived the accident and continued to grow after their canister hit the ground. Even more amazing, off-the-shelf electronics boxes continued to record data for a couple of weeks after the accident, until their batteries ran down. Even if you've followed the space program extremely closely you'll discover in this book fascinating things about Columbia's STS-107 mission and her crew.

Many people ask why am I including information on the negative things in the lives of Columbia's astronauts—their human failings. I've been asked, "What good does it serve to dredge up those old feelings again?" or "Isn't it tacky to say bad things about a person who died as a hero?" To those people I've said it's good because it's the truth. The seven Columbia astronauts were incredibly well-trained people, but they were not perfect. They were humans. When they were alive, they always acknowledged their limitations and mistakes, and I will not dishonor them by pretending they were perfect heroes. This is their real story, not a sanitized version written by a publicity person. If you want to know the birthdates of the Columbia astronauts or their various awards, then read their NASA bios, which are included on this book's website and CD-ROM. If you want to find out about seven real and fascinating people and everything they accomplished on their flight, then read this book.

The seven Columbia astronauts were not the superheroes, saints, or patriarchs memorialized in eulogies, and they were not all the same. Some were religious, some were not. Some wanted to become astronauts from the time they were young kids, others only decided to try to become astronauts when they became adults. They had different interests and long-term goals. They were seven very talented but otherwise ordinary individuals who were trained to do an extraordinary task—to fly in space. This book is their story and the story of Columbia's last mission.

A few days before the end of the mission, I had the opportunity to talk to Columbia's astronauts in space. It wasn't just a plain, dry interview. I was talking to people I knew, and it was a pleasure to see the smiles on their faces when they heard many of my questions. Three days later when the accident occurred, I felt like I had lost seven friends.

Philip Chien
Merritt Island, Florida
August 2005

NOTES

All times are Eastern Time (standard or daylight) unless otherwise specified. NASA still hasn't gone metric, so English units have been used for most measurements, with the exception of cases where metric units are the overriding convention (e.g., 35-mm. film). Miles are statute miles (5280 feet) unless specified as nautical miles (6080 feet).

NASA is an extremely acronym-intensive organization, and in many cases acronyms are made out of other acronyms! Whenever possible I've spelled them out, even when only the acronym is normally used. When they're pronounced as words, it's noted the first time the acronym is used.

Chapter 11, "A NASA Primer," is recommended for first reading if you're unfamiliar with NASA terminology and organization, but hopefully still useful even if you're familiar with the space program.

The website http://www.sts107.info includes updated information, photos, audio and video clips, and technical documents. The companion CD-ROM [available for purchase separately], with over 1,000 photos and multimedia items, can be ordered from the author's website or with the form in Appendix B.

This book is dedicated to the STS-107 Columbia crew—

Rick Husband
Willie McCool
Dave Brown
Kalpana Chawla
Mike Anderson
Laurel Clark
Ilan Ramon

—and their families.

ex orba siencia
—from orbit, knowledge

"THE GREATEST HAZARD IN LIFE IS TO RISK NOTHING.
THE MAN OR WOMAN WHO DOES NOTHING HAS NOTHING, IS NOTHING."

—STS-107 Columbia astronaut Laurel Clark

CHAPTER 1

A M O D E S T
M I S S I O N

One of the tasks the space shuttle excels at is scientific research. It can operate as a precision pointing platform for telescopes or carry a pressurized laboratory filled with experiments. When gravity is removed, things happen differently, and that leads to new knowledge and insights. Scientists prefer the term "microgravity" instead of "zero gravity" because there are minute gravitational accelerations, enough to affect ultra-sensitive experiments. Microgravity research was a hot field that encompassed many scientific disciplines. From 1983 until 1998, NASA flew the European Spacelab pressurized laboratory aboard the shuttle 16 times. Some missions specialized in studying humans and animals. Others studied how materials grow and interact in space. Each flight lasted from 9 to 17 days, and the microgravity experiments were tailored to fit that length and the shuttle's capabilities.

The Spacelab program was winding down and NASA planned to transition to longer scientific experiments aboard the International Space Station (ISS). But there were delays to the ISS's construction, and many microgravity scientists were concerned about the gap between the last Spacelab flights and the full-time science activities on the space station, due to begin around 2004.

In 1998, Congress ordered NASA to add a dedicated microgravity shuttle mission. NASA complied with the congressional mandate by planning STS-107 and contracted with the commercial company Spacehab to provide their "Research Double Module" (RDM). The RDM was an improved version of the cargo carrier NASA used to carry supplies to Russia's Mir space station and three early ISS missions.

STS-107 was originally scheduled for launch in May 2000, but like Rodney Dangerfield got no respect, and had an unenviable 18 separate delays, finally launching on January 16, 2003.

Critics claimed that the only reason STS-107 even existed was Spacehab's lobbying so it could rent its research module to NASA. It's certainly true that Congress ordered NASA to fly STS-107 as a dedicated microgravity mission, but Congress instructing NASA to fly science missions was hardly unique. Many shuttle missions have flown primarily because of congressional interest. For example, the only reason the

ASTRO team got to refly their astronomical telescopes was because somebody told influential Senator Barbara Mikulski that three of the four telescopes came from her home district. Congress gave NASA the funds for ASTRO's reflight. It was incredibly valuable science—but the main reason it flew was an influential senator.

From the public and media's perspective, STS-107 was not very interesting because science in a laboratory is rarely glamorous. In fact, much of the outside attention was on the student experiments because they tend to be easy to explain and often have good visuals. Even many of NASA's supporters considered microgravity science to be "boring," and accused STS-107 of being a make-work mission because there was no central theme to the different experiments. None of STS-107's scores of experiments justified flying a shuttle mission by itself—but taken together they made for an incredible science buffet. If you need a central theme for STS-107, how about valuable science that can only be performed in space—a combination of microgravity experiments, Earth observations, and technology demonstrations. The only thing STS-107 was "guilty" of was not being as glamorous as other missions.

In 1995 the Clinton administration invited Israel to fly a scientific experiment and a passenger on the shuttle. The scientific instruments would have to be of a non-military nature, and its experimental data would be publicly shared with other scientists from around the world. The Israeli "payload specialist" would be given limited responsibilities, managing the experiment and otherwise staying out of the way of the other crewmembers. Similar offers had been made in the past to other countries, including France and Ukraine, but clearly the invitation to Israel presented unique political and diplomatic challenges. Finding a suitable mission would prove to be a tricky task for NASA, as most were dedicated to the construction or upkeep of the ISS. A Hubble Space Telescope servicing mission was out of the question because all seven astronauts on the Hubble Servicing missions are highly trained professional astronauts with many years of experience, and the packed timelines didn't have room for an add-on experiment with its own requirements. So the only choice was a so-called "stand-alone" shuttle flight. These had been more common in the past, but were disappearing fast as NASA concentrated its efforts on ISS.

Israeli Space Agency head Aby Har-Even says NASA told him, "You have to justify scientifically the experiment." Israel proposed the Mediterranean Israeli Dust Experiment (MEIDEX—pronounced "ME-dex"), in which the Israeli astronaut would look for large dust storms from orbit with a special camera. At the same time a specially equipped aircraft would fly over the same region of the Earth and observe the dust from underneath. NASA Earth scientist Dr. Jack Kaye noted that it was extremely interesting science, and he was grateful that Israel wanted to put the effort into it, especially the aircraft underflights. He also noted that NASA would not have had the resources to do such an intense two-week science program for a specialized investigation like MEIDEX. Scientists informally discussed the possibility, if the experiment worked well on the shuttle, of a follow-up version of MEIDEX designed for ISS.

In most cases, when international organizations fly shuttle experiments they build their own payloads. Part of the reason that any government sponsors scientific experiments is to develop domestic industries and jobs. But MEIDEX was the exception. While the Israeli scientists were interested in studying the Earth, they weren't as interested in building the hardware, so Israel paid US aerospace contractor Orbital Sciences Corporation (OSC) $1 million to build MEIDEX.

It's important to note that MEIDEX was only a secondary experiment; this meant it could use shuttle resources like crew time and pointing the shuttle in the correct direction, but had less priority than some of the other payloads. More important was the fact that the shuttle would fly when shuttle program managers needed it to fly, taking "big picture" NASA priorities into account. Scientists noted that the most desirable time for MEIDEX to fly would be the March to May timeframe, with October to December second in desirability. For the rest of the year (June to September and January to February), the odds of dust storms over the Mediterranean Sea for MEIDEX to view was far lower.

Including an Israeli was not the only "special" crew arrangement considered by NASA. The agency considered flying an all-female crew, supposedly for scientific research into how the female body in spaceflight responds differently from the male, but primarily for the added publicity, and as a way of encouraging young girls to study science and math. The only suitable mission for this purpose was STS-107, and scientists started to examine potential experiments concentrating on the female body. It was quickly pointed out, however, that only four members of the crew would be allowed to take part in invasive experiments like blood draws because flight rules prohibit the commander, pilot, and flight engineer from invasive experiments. (The logic is that there's always a possibility of an emergency that requires the shuttle to make a rapid return to Earth. It would be extremely undesirable to have an astronaut try to land the shuttle shortly after giving blood and still lightheaded.) So there was no legitimate reason to fly an all-female crew, even if medical research on the female body in space was justifiable for purely scientific reasons. Moreover, it turned out that many of the female astronauts were vehemently against the idea. They had spent most of their time growing up, going to college, and their entire professional careers having to prove that they were the equals of men. The concept of getting assigned to a mission solely because they were female had all the marks of a quota system. The whole concept went away quietly by the summer of 1999, opening up a spot for Israel's male payload specialist.

STS-107's flight requirements called for "an optimum science research flight," resulting in 85 different payloads. The exact number of experiments depended on how you counted them—for example, do a dozen different samples in a miniature space factory count as 12 experiments or 1? Unlike most science missions, where NASA asked scientists what experiments they would like to fly, there were plenty of payloads already approved which were waiting for a suitable mission like STS-107.

The experiments finally chosen had no single theme, but as mentioned above added up to a buffet of scientific inquiry. When I asked commander Rick Husband to tell me about the experiments, he said, "That's a pretty tall order, because there's a lot of them." Most of the experiments involved measuring the effects of microgravity. Biological experiments grew plants, monitored how animals adapted to microgravity, and manufactured protein crystal drugs. Materials processing experiments grew crystals with desirable qualities and studied how particles interact with each other. Combustion experiments set off tiny flames in a triple-sealed container for safety. On Earth flames are flame-shaped because hot air rises—in space a flame's more spherical and doesn't flicker. Other experiments were focused on technology development, and still others monitored the Earth's environment.

Safety was also paramount for experiments where potentially hazardous chemicals needed to be mixed. A glovebox, similar to an infant incubator, was used for safety when handling those chemicals. The astronaut placed the containers and tools into the box, closed the clear lid, and used a set of heavy rubber gloves to manipulate the contents.

In many cases the experiments had flown previously, and the scientists, having found something surprising, wanted to refly them to see if their earlier results might have been a fluke. For example, Dr. Fred Sack of Ohio State University, in a moss growth experiment flown in 1997, had been surprised to find that moss always grew in clockwise spirals (instead of at random) when gravity and light were removed. NASA agreed to fly the experiment again to see why it happened.

Of the reflights, many had flown on STS-95, best known as John Glenn's shuttle flight. On that mission, Glenn was wired to a special harness which monitored his body's activity while he slept. The STS-107 version was less complicated—each astronaut wore a watch-like sensor with a miniature computer. The computer kept track of the light level and the amount of activity, and after the mission the data would be used to extrapolate how well the astronauts slept and for how long.

Another reflight from STS-95 was the protein turnover experiment, which measures how the human body absorbs proteins in space. On that

John Glenn wears a sleep
harness on the STS-95 mission.

flight Glenn had kidded he was doing his best to avoid astronaut-doctor
Scott Parazynski, whom he had nicknamed "Count Parazynski." Glenn's
advice to the STS-107 crew was, "Just get ready to give a lot of blood. I
think I had 12 different samples."

Also planned for STS-107 was Triana, the brainchild of Vice President Al Gore. Gore has always been very conscious of the environment
and in a late-night insight, he suggested a spacecraft with a camera permanently facing the sunlit side of the Earth. The camera would be positioned at a gravitationally stable location between the Earth and the Sun
to return real-time high-resolution natural color images of the Earth,
which could be downloaded from the Internet. It was a nice idea, but
something Earth observation scientists weren't excited about. The exist-
ing GOES weather satellites accomplished much of what Gore proposed, and while natural color images do look pretty, false-color images,
with wavelengths invisible to the human eye, have far more scientific
content. Still, the vice-president is the chair of the National Space Council, and when he talks, you listen. NASA developed the Triana satellite,
named after the navigator on the *Santa Maria* who was the first member
of Columbus's expedition to spot the New World.

Originally the concept called for an extremely inexpensive satellite
by NASA standards, built by students as an educational project. But
eventually the decision was made to add more scientific value to the project and have the satellite built by an aerospace contractor—and the costs
skyrocketed. An oversight panel noted that while the project had a giant
cost overrun, the science was useful, so it still made sense to launch it as
planned in 2000. But Gore was running for president, and there were
complaints that if Triana was launched before the 2000 presidential election it could seem as if NASA was trying to curry favor with the new President. As a result, Congress instructed the agency not to launch Triana
before the elections, and it was put in indefinite storage.

NASA removed Triana from STS-107 and replaced it with a group of
secondary experiments flying on a space-available basis. The payload was

originally supposed to be called "Launch On Need Enabling Science, Technology, Applications and Research" (LONESTAR). "Lonestar State" is a nickname for Texas, and the folks at the Johnson Space Center in Houston loved the name, but somebody realized it was an election year and that Republican candidate George W. Bush was from Texas. They thought that replacing the Triana payload, conceived by the Democratic presidential candidate, with the LONESTAR payload, named after the home state of the Republican presidential candidate, wasn't a good idea. It was suggested that the team find a new name for their payload and they very quickly came up with FREESTAR (Fast Reaction Experiments Enabling Science, Technology, Applications and Research). The Israeli MEIDEX, which was handled by the same engineers and managers, became one of the six FREESTAR experiments.

While STS-107 was a stand-alone science mission with no connection to ISS, much of its science did complement ISS activities: For example, a two-week version of an experiment would fly on the shuttle while a longer version would fly on the ISS. In some cases, the shuttle version was more interactive because more astronaut time was available for "hands-on" attention to details, and based on that information, the scientists could develop a more automated version for ISS, where less crew time is available for each experiment.

There were also indirect connections between STS-107 and ISS. NASA needed a large oversize aircraft to transport the space station's components from their manufacturers to Florida. There was a payload transport canister available, but it was so large it would only fit in an Air Force C-5C Galaxy. That would limit the shipping schedule to when the Air Force could free up one of the two C-5Cs from its normal flights. During the Apollo program, NASA had used a set of "Guppy" aircraft—cargo aircraft modified with giant oversize fuselages—to carry giant rocket stages, but most of those aircraft were now retired. However, one Guppy was still flying in Europe, transporting components for the European Airbus aircraft from

The Guppy aircraft delivers a space station component to the Kennedy Space Center.

its manufacturing points to the factory where the pieces were put together. The European Space Agency (ESA pronounced "EE sa") was able to obtain that aircraft and made a trade with NASA—the aircraft in exchange for 450 kgs (990 lbs) on STS-107 for European payloads. STS-107 also carried one "Risk Mitigation Experiment"—a piece of hardware that's flown on a short shuttle mission before committing it for use as a critical application for a lengthy stay in space. Called the Vapor Compression Distillation (VCD) experiment, it was a prototype water-purification system. If VCD was successful, a future version could be used on long-duration missions to recycle wastewater, either on board a space station in orbit around the Earth, at a base on the Moon, or even on missions to Mars.

In one case, the many delays to STS-107 caused a payload to move to an earlier mission. STS-107 was originally supposed to include a cosmic-ray experiment as a prototype for a large antimatter telescope, but it was transferred to STS-108, which flew in December 2001, more than a year before STS-107. In its place on STS-107 NASA decided to fly the Space Experiment Module (SEM—pronounced "sem"), an educational payload which flies when there's excess space that can't be used by any higher priority payloads.

While it experienced a convoluted path to space, STS-107 was extremely fortunate in its team of designers, who did an incredible job of integrating the requirements for the 85 experiments.

The goal for the mission planners was to put as many experiments into the Spacehab module as would fit, use all the power available, use all the shuttle's communications capabilities, and use all the crew time. The planners devised a fairly close match to the mission's maximum capabilities. Each 12-hour shift had three or four astronauts, and their activities were carefully scheduled to use them as efficiently as possible. Unlike previous microgravity flights, where the pilots had little to do, everybody was kept busy throughout STS-107. The exception was two pre-planned 4-hour "off" periods for each astronaut, the equivalent of a short weekend break in which the astronauts could talk to their families and just look out the windows and enjoy the view.

The tight scheduling was a challenge. In some cases the astronauts just turned on automated experiments which operated on their own for the rest of the flight. In other cases the crew needed to participate on certain days but not on others. In some cases an experiment would run for about three or four days, but it didn't matter which days. Some used a lot of power and had to be paired up with other experiments that used less power to keep the overall power consumption below a maximum level. The three combustion experiments would run in series; after the first experiment was finished it was put away and the second experiment was installed and performed its tasks for several days. Then it was put away and the hardware was reconfigured for the third experiment

The space shuttle can fly in almost any orientation to perform its experiments. Unlike an ordinary aircraft, it can fly backwards, upside down, standing on its tail, or in any other direction—whatever is necessary to perform its tasks. Before the mission Husband told me, "There's going to be a fair amount of maneuvering in the timeline," and he wasn't kidding. STS-107 had 297 maneuvers to orient Columbia, almost a record for the shuttle. For the ultra-sensitive microgravity experiments Columbia stood "vertically," with the shuttle's tail pointed toward the Earth and nose pointed toward deep space. That orientation permits long periods without any thruster firings, for the lowest gravitational disturbances. For the Earth observation experiments like MEIDEX, the shuttle's cargo bay was pointed down toward the Earth, or toward the horizon to observe lighting and sprites in the atmosphere. The Low Power Transceiver (LPT) experiment required the shuttle to point down toward a ground station or up toward a communications satellite far above the shuttle. A solar measurement experiment needed to be pointed toward the Sun, but a microgravity experiment right next to it needed to be pointed away from the Sun as much as possible to avoid overheating. The StarNav payload needed to point toward bright stars. And so forth. Many of the payloads had to perform activities as they were passing over particular locations on the globe or at specific times during the mission. So precise timing as well as orientation scheduling were critical in planning the flight.

One of the more challenging tasks for the flight planners was what to radio back to the ground. All of the data was stored onboard on computer hard drives or digital data recorders. Some of the data could also be transmitted to the ground via a Ku-Band antenna, similar to a home satellite dish. The Ku-antenna could transmit video or computer data when it was in the correct orientation to view the Tracking Data and Relay Satellites far above the shuttle. Many of the experiments needed to transmit back video while they were running, others needed to transmit large amounts of data whenever it was possible. Public Affairs wanted video of the astronauts working inside the shuttle and time allocated for interviews where some or all of the crewmembers had to be available. All of these requirements had to be accommodated by the flight planners. On a typical day the Ku transmissions would be used for some video of the astronauts at work, a publicity event, and a bunch of scientific data. In the rare cases where the Ku wasn't being used for anything else and was available, it could be used to transmit views of the Earth from cameras located inside Columbia's cargo bay.

While it wasn't obvious to outsiders, STS-107 was an incredibly efficient and well thought out flight, maximizing everything that could be done on a 16-day shuttle mission. The efficiency of the mission team even extended to the color-coded signs for the crew to keep track of the length of time they were in space. Instead of a fancy high-tech digital display, the crew used cardboard. Red and black for Texas Tech grad Rick Husband's shift and navy blue for Navy officer Willie McCool's shift. Simple, but they got the job done.

Before the mission commander Rick Husband outlined exactly what he hoped to accomplish on STS-107 in three sentences: "We want to do the absolute best job that we can on every facet of the mission. We have worked very hard on our training, not only to fly the orbiter but also to conduct the experiments on orbit. We want to ensure the experiments we fly are conducted to the utmost of our ability, exactly the way the investigators on Earth have designed them to be conducted."

A rare photo shows the cardboard signs the astronauts used as their "calendar" in space on top of three storage bags. This particular photo was taken halfway into the mission--on the ninth day for the blue shift and eighth day for the red shift.

The STS-107 crew at Launch Pad 39A with the shuttle Columbia during their dress rehearsal in December 2002.

PART I

MISSION MAKERS

The most visible people on any space shuttle mission are certainly the astronauts—they're the ones who put their lives on the line and get to fly in space. And when there's an accident they're the ones who make the ultimate sacrifice, and become lionized as heroes.

After the Columbia accident, on February 1, 2003, there were scores of memorial services. At each ceremony, NASA officials, politicians, and religious figures stepped up to the microphone and talked about how wonderful the crew was, how they didn't die in vain.

The astronauts were described in gigantic proportions—as superheroes, saints, or the biblical patriarchs. They weren't. They were seven talented individuals who worked together as a team, and they had human flaws. Most of the speakers at the memorials had never met the astronauts and wouldn't have even known their names the day before the accident. Many didn't even know that the mission existed. In some cases the eulogies could have applied just as well to anonymous deaths after any accident or disaster. But some of the speakers did know the STS-107 crew and spoke from the heart, sharing anecdotes about them.

I knew the Columbia crew far better than the media who thought the mission was newsworthy only after the accident. I knew some of the astronauts better than others, some I only met a couple of times. They were seven individuals—with individual personalities, individual interests, and individual goals. They were not the superheroes that were referred to by people who never knew them. Rick, K.C., Mike, Dave, Laurel, Ilan, Willie—I'm sorry I didn't have the opportunity to get to know you folks any better, but at least I can share our time together with others.

Just as important as the astronauts who flew were the people who made the mission happen in a myriad of other ways. During the mission, commander Rick Husband acknowledged the efforts of the support teams on the ground—the people who worked in Mission Control to keep Columbia running in space, the ones who prepared Columbia for launch, and the scientists from inside and outside of the space industry who had experiments which flew on Columbia. Though only a few of these people are mentioned here, they represent a huge group of people, both inside and outside of NASA, dedicated to STS-107 and its goals.

CHAPTER 2

RICK HUSBAND:
THE BOY
FROM AMARILLO

The STS-107 commander was Rick Husband, an Air Force test pilot. Rick had a laid-back attitude and Texas drawl which showed his roots. He was a pure Texan—born and raised in Amarillo. Husband said he wanted to be an astronaut since he was 4 years old: "It was about that time when the Mercury program started up. I saw those things on the TV, and it just really excited me, it really grabbed my interest. And, seeing those rockets and learning about the astronauts, and seeing what they were doing, and then the models you could build—I remember a Gemini model that I put together. I thought everything about that was so fascinating. And in following that all the way through, from Mercury to Gemini to Apollo, watching the Moon landings and everything, it was just so incredibly adventurous and exciting to me that I just thought, 'There is no doubt in my mind that that's what I want to do when I grow up.' At the same time, I was very interested in airplanes and flying. And I'd be out in my backyard playing. And, any time I heard any kind of an airplane, you know, it's like, stop what you're doing and take a look and see, 'Where's that airplane? What kind is it? Where is it going? How high is it? How fast is it going?' And so, it's the kind of thing that has just been such a part of my life, what I wanted to do when I grew up."

Husband did admit, "There was a while in there when I wanted to be a Dallas Cowboy," but flying remained his passion. Husband

Four-year-old Rick Husband. Photo
courtesy of Evelyn Husband.

obtained his flying license at Amarillo's Tradewind Airport at the age of 17. He got together with Paul Lockhart, another teen who was interested in flying, and they had a conversation about what they wanted to do with their careers. Both decided they wanted to become Air Force pilots, and then hopefully astronauts. Both succeeded, Husband in 1994 and Lockhart in 1996. They flew on back-to-back flights, with Lockhart flying the shuttle mission before the STS-107 accident.

One of the things that anybody who knows Rick Husband will tell you is how devoutly religious he was. It was an extremely important part of him and his life, but unlike many he didn't try to push his views upon others. Husband had a life-long passion for singing and was in the choir at Grace Community Church in Clear Lake, Texas, where he lived. But more than anything else, Rick loved his wife Evelyn (he fell for her in college) and his two children—Laura and Matthew.

Husband earned his mechanical engineering degree from Texas Tech University in 1980, and a masters degree in mechanical engineering from California State University, Fresno, in 1990. He entered the Air Force, became a jet pilot, and was eventually selected to attend the prestigious Edwards AFB Test Pilot School.

Rick Husband and Evelyn Neely got married in February 1982, and a week later moved to Homestead AFB in Southern Florida. Evelyn recalls that Rick wanted to get up early one morning to watch the space shuttle as it traveled over their home. They saw Columbia on its third flight, as it soared across the pre-dawn skies. Almost 21 years later Husband would command Columbia on its final mission. Unlike many pilots, Rick didn't use much profane language—if any. When a NASA transcript says Husband said, "Oh shoot," or "Oh shucks," that's exactly what he said, not a cleaned up version.

After he graduated from test pilot school, Husband had the minimum skills NASA looks for in an astronaut candidate—a college degree in a technical field, a couple of thousand hours flying high-performance military jets, and test pilot school. His applications in 1987 and 1990 were turned down, and in 1992 he was invited to Houston for an interview with the astronaut selection board—and turned down again. It's rare for an astronaut to be selected on the first try, or even after an interview with the selection board. Many astronauts tried several times before they were finally selected. Husband made it on his fourth attempt, becoming one of the ten pilots in NASA's 1995 astronaut class.

Husband first flew in space in 1999, on STS-96, which was the first mission in which a shuttle docked with ISS. On that flight, he was the pilot, assisting commander Kent Rominger, but the term "pilot" is a bit misleading: Rominger did almost all of the flying, and Husband was in effect his copilot.

Of the ten pilots in the 1995 class, of course one had to be the first assigned to a flight, and one had to be the last. Husband, assigned to STS-96 in August 1998, was the last pilot in his class to get an assignment. But because of the changing order of the shuttle missions and delays to Space Station's construction, he got to fly over a year before the last pilot in his class flew. The first time I met Rick Husband, I teased him about getting the "honor" of the last flight assignment in his class. He said, "There's no doubt in my mind that they saved the best for last as far as the flight assignments. As it turns out I'll be flying before Jeff [Ashby], Michel [Tognini], and Pam [Melroy], and they all do a fantastic job. I'm really looking forward to flying on this one—so that really hasn't been anything which has weighed very heavily on my mind." Commander Kent Rominger quickly added, "From my perspective, in fact pretty much they did save the best for last. I think everybody on the crew would echo that sentiment."

The others in the STS-96 crew were NASA astronauts Tammy Jernigan, Ellen Ochoa, and Dan Barry, and Canadian astronaut Julie Payette, plus Russian cosmonaut Yuri Malenchenko. That flight was a maintenance mission to the ISS. At that point the space station only consisted of two pieces —the Russian Functional Cargo Block (FGB from its Russian acronym) "Zarya," which served as the control system for ISS during its early construction period, and the US Node "Unity," which was the centerpiece for all of the American components. They were mated on a shuttle mission in December 1998. It was a minimal place to visit at the time—no windows, no life-support system—just a shell that could maneuver as required.

The next component was the Russian Service Module, the living quarters and also critical for controlling ISS. The Service Module had to be in place before future assembly tasks could be done. STS-96 was supposed to fly after the Service Module was launched. The crew was responsible for activating the Service Module before the first permanent crew would live on board. But the Service Module was seriously behind schedule. NASA and Russia decided it was still important to fly the STS-96 mission to bring supplies up to the fledgling station and make repairs to components that had failed over time. That resulted in a last-minute change to the crew. Malenchenko, an experienced Russian cosmonaut who was also an expert on the Service Module's systems, was replaced with cosmonaut Valeri Tokarev just five months before launch. Tokarev was a rookie, but he was familiar with the Russian systems.

Interest in STS-96 was extremely high in Husband's Texas hometown, Amarillo. When Rominger introduced Husband at the preflight press conference, he joked about the level of interest: "Additionally part of his workload—there's a tremendous following from Amarillo that Rick's been having to put up with—actually he's been enjoying putting up with through this." Husband told me, "It's nice to be able to see the

The STS-96 crew poses with their Spacehab module. From left to right, Dan Barry, Tammy Jernigan, Valeri Tokarev (Russia), Julie Payette (Canada), Rick Husband, Ellen Ochoa, Kent Rominger.

hometown folks so excited about this and see a lot of attention generated there. Certainly from the point of all of the kids in school I think it's good for them to be able to see some coverage and give them a chance to be more involved in the mission." Rick Husband's mother, Jane Husband, and his wife Evelyn's parents, Dan and Jean Neely, live in Amarillo.

Many of Rick Husband's family and friends decided to travel to Florida for his first launch. Husband said, "We've invited quite a few—between our list and my mother and my wife's parents and my brother we've probably invited about a thousand folks. It will definitely be a large hometown crowd there."

The STS-96 launch was scheduled for May 20. But on May 8 a thunderstorm passed through Florida, pelting the shuttle's External Tank with hail. NASA made the decision to move Discovery back to the Vehicle Assembly Building, where platforms could be placed around the tank to give workers better access to the damaged areas. The move and extra time for the repair work resulted in a one-week delay. The launch was rescheduled for May 27 at 6:49 A.M.

Because of the launch time and the mission's timetable, the crew had to adjust their sleep schedules to be awake at night and asleep during the day. The crew flew to Florida for the launch in their T-38 training jets on Sunday May 23 at 11 P.M.

The city of Amarillo was incredibly interested in following Husband's first spaceflight—everything from what he ate in space, to fixing a leak, repairing a radio on ISS, choreographing the spacewalks, and undocking the shuttle from the space station.

After the mission, when he had time to reflect on his experiences, Husband said, "Launch morning was very interesting, getting suited up and walking out to the astrovan. There's a certain amount of anticipation you feel but also a lot of excitement. Really a great experience, hearing the experienced people on the crew talk about it. 'Hey, when you get out of the astrovan you're going to think the vehicle's alive'—and they're right, just the sounds that are going on out there. It's an amazing thing to think—an incredible amount of power and energy that has to be released to get that size of vehicle going as fast as it needs to achieve orbit. One really big impression is every other time I've been out to the pad they have a check-in station. But this time nobody was out there, everything was cleared out."

Because of the space station's location in space, the launch had to take place in the early morning, and the crew entered the shuttle before dawn. Husband said, "I could see the sunrise [from the pilot's seat], so I was telling everybody how pretty it was. The people on the middeck [without any windows] said that they appreciated the description of how things looked outside. I was very pleasantly surprised that I didn't feel nervous—all of the training we had was excellent preparation. I was just thinking, 'I'm ready to go,' and this is a really good feeling. We came out of the nine-minute hold, and at that point you definitely are on your toes. It's my responsibility to get the Auxiliary Power Units started and running. I definitely was checking and rechecking the switches. Everything went fine."

Husband's wife and children watched the launch from the Launch Control Center's roof, about 3.5 miles from the launch pad.

He continued, "The engines started up and you could feel a kind of a rumble. When the Solid Rocket Boosters (SRBs) lit off there's a definite shaking. It all went as expected. I saw a thin layer of clouds coming up the front window, I'm keeping an eye on the engines—and at the last second I'm going to watch us go zoom through that layer. I did, and they just went by in a flash, you could tell that we were accelerating. The rest of the ride was great, getting through SRB separation and having the ride smooth out. It was a very smooth and very quiet ride, almost a much more relaxed pace than what we're used to from training, where they're throwing simulated emergencies at us to keep us busy. As we went through 50 miles Rommel [commander Kent Rominger] said, 'I'd like to welcome the three new astronauts to the crew [Husband, Payette, and Tokarev officially became astronauts at that altitude]'—that was a great moment. We approached MECO [Main Engine Cut Off—pronounced 'mee coe'], the engines cut off, and everything was quiet. It was interesting hearing the External Tank separate and watching little pieces of ice go by—it was interesting seeing all of the different things, things I had never seen before on an airplane flight."

Husband said he adapted well to zero-G. "It wasn't too drastic. I felt a heavy feeling where the fluids are rushing toward your head and so you end up with a stuffy feeling in your sinuses. That was something you got more and more used to."

Because of the space station's orbit, Husband would only get the opportunity to see his hometown in daylight early in the mission—and only if the weather cooperated. He said, "Orbit 18 [a day and three hours after launch] we were supposed to go fairly close to Amarillo. Things were pretty cloudy, so I didn't get to see much of the Texas panhandle except under cloud cover. Sure didn't get to see Amarillo clearly during the daytime, but we sure got to see a lot of other beautiful places."

Two days after launch, on May 29, Commander Rominger docked the shuttle Discovery to the space station. Husband was sitting on the opposite side of the flight deck, monitoring the shuttle's systems. He said, "Seeing the station was a fantastic experience. It couldn't have gone any better—went perfectly. If the glass hadn't been there you could reach out and touch it."

Husband had trained as the backup for the spacewalk, but astronauts Dan Barry and Tammy Jernigan performed the spacewalk as planned. Husband helped the pair into their spacesuits and acted as the in-cabin choreographer for their activities outside. He said he didn't mind not doing the spacewalk "That's the way it should have been—no regrets. I felt very fortunate to go through the training. I wasn't disappointed at all. I knew I was trained as the backup, very happy to see Dan and Tammy go out."

Husband noted that the crew really enjoyed their microgravity experience. "One thing which was really fun, if two people were having a conversation and I needed to go upstairs, instead of having to say 'excuse me' and going between them all I had to do is go up and [float] over without interrupting. If there were two or three of us in some location talking about something, Tammy would say, 'Somebody has to be upside down,' so she'd flip. It was one of the very unique and fun things about the mission." Jernigan explained, "I felt that we had to spend as much time as possible [trying these things out,] since we wouldn't have that opportunity—or at least it wasn't as convenient on Earth. So I liked to eat upside down."

Husband said that sleeping in space was also a pleasure: "It was very comfortable, no pressure points or anything like that. That was one thing that's nice about zero-G—you just climb in your sleeping bag and float. I was able to go to sleep just fine every night."

The shuttle has an extremely low-tech alarm clock. It's a tradition for Mission Control to radio up music when it's time for the crew to get up. Each day, the music is dedicated to a particular crewmember or some activity on the mission. Flight day six—Monday May 31—was Husband's turn. Capcom Mario Runco chose the George Strait version of "Amarillo by Morning" as the wakeup music. After playing the tune, Runco said, "Good morning—I reckon. Amarillo's about 1800 miles north by northeast [from the shuttle's location at the time.]" Husband replied, "Howdy Houston. It's good to hear some music about my hometown. I'd like to say hello to everybody there and also to my wife and kids in Houston—Evelyn, and Laura and Matthew."

Later that day, Rominger and Tokarev talked to Russian news media. While responding to a question about the international nature of the crew, Colorado native Rominger joked, "The pilot [Husband] we kind of tease, because he's a true-born Texan, and they tend to be a little bit different from the rest of us Americans." Rominger quickly added, "Which is a very pleasant thing."

One of the most important tasks for the STS-96 crew was to transfer 3600 pounds of cargo from the shuttle to ISS. That cargo included everything from office supplies to clothing for the live-aboard crews once they arrived.

Ellen Ochoa was stationed inside the Spacehab cargo module (think space-rated U-haul) inside Discovery's cargo bay. She would go through a checklist and pick out the bags. Husband, Rominger, Barry, Jernigan, and Tokarev were the "movers." They flew the bags from Spacehab through a tunnel into the docking adapter, made a 90-degree turn and flew into ISS, either stopping at the American Node or going all the way through to the Russian FGB module. Julie Payette took the bags and stored them in their proper locations. Husband said, "I really enjoyed getting to pick up the bags and go all the way through the tunnel and up through the docking system into the Node and through the FGB. You had lots of long distances where you could travel."

The Amarillo area had the opportunity to view ISS with shuttle Discovery attached as it passed over the Texas predawn sky. I had written a story for the *Amarillo Globe-News* informing its readers when and where to look. On one occasion Husband looked down at his hometown as they looked up at him as he passed overhead. During an inflight press conference Husband told me, "Dan and I were on the flight deck just before we went to bed last night [Tuesday, June 1, predawn Amarillo time]. We were thinking that we were going to see Amarillo, but it turned out that as we went over it was still dark on the ground but we were in sunlight. Which explains how all of the folks on the ground were able to see the station as we passed over. I heard from my wife that my father-in-law and some of the relatives up there had seen us pass over. So that was real thrilling to

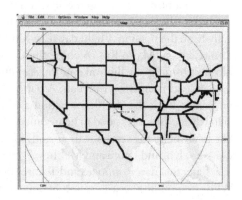

The path ISS and Discovery took over Amarillo, when the city's residents got to see their favorite son go overhead. Tracking map by Philip Chien.

hear that they had seen us. It's just a great experience being up here. I appreciate all of the support and all of the interest that the people in Amarillo have had over the course of this mission."

Before leaving ISS, the crew completed an important tradition. Husband said, "There's a ship's log [on the space station]. We made an entry and took a picture of the crew which we printed out and put in the log." The crew made sure to include their "crew motto" in the log—"If you're not having fun, you're not doing it right."

Husband was responsible for undocking the shuttle and the fly-around of ISS. Before launch he said, "I'm definitely looking forward to the undocking and fly-around. I think it's going to be beautiful seeing the station with the Earth in the background. We'll get to do the fly-around through some night and day periods, so we should get some fairly good photographs." It wasn't surprising that that was the portion of the mission Husband was especially looking forward to. The fly-around is the only opportunity the pilot gets to actually fly the shuttle. For the most part, the pilot is the jack-of-all-trades. The pilot has to have the same skills as the commander, just in case he has to take charge. He's also often the backup spacewalker. But the pilot's backup skills are ones he has to learn just in case: under normal circumstances, he doesn't use them. So the fly-around is a task that is specifically assigned to the pilot as a "thank you" for all of his hard work. It is also a bit of on-the-job training so that he can actually get experience that will be useful for future missions if he becomes a commander.

After Husband fired the thrusters to separate Discovery from ISS, he professionally radioed, "Houston, we have physical separation, executing the separation burn." After the mission Husband told me in more colorful detail about the fly-around, "That was a real thrill. We worked very well as a team—the separation went very smoothly. It was a really great experience from my point as a pilot actually getting to fly the orbiter. One point during the fly-around I was really enjoying myself. You're con-

A rare photo of the ISS ship's log. The STS-96 crew took a photo of themselves which they put into the space station's permanent log along with comments about their mission.

centrating very hard making sure everything goes very well but I just had to tell everybody, 'This is so much fun!' and then back to business. The rest of the crew thought that was pretty funny. From our standpoint all of us really enjoyed hearing somebody else on the crew expressing how much fun they were having—just made it that much fun for everybody else." Husband wanted his crewmates to share in his experience and have their own opportunity to fly the shuttle for a short time. He said, "During the fly-around I gave the opportunity to let each of the crewmembers make an input on the hand controller. I would tell them what we needed, and they'd make the input, and I think that's something they enjoyed as well. It was a really great experience. I was really enjoying myself."

The final separation from ISS brought mixed feelings. Husband said, "We were looking forward to the remainder of the mission, looking forward to some time to look out the window. But also sorry to say good-bye to the station because we spent a fair amount of time there and really enjoyed ourselves and it was very rewarding doing the work we did there and having it go so well. A sense of satisfaction that we left the station in as good shape as we found it, and a bit better."

Of the STS-96 crew, Rominger got to visit ISS again, on STS-100 in April 2001; Dan Barry got his turn on STS-105, in August 2001; and Ellen Ochoa returned on STS-110, in 2002. Tokarev spent more time on ISS than everybody else on STS-96 combined—on the long-duration Expedition 12 crew.

The STS-96 crew had one bonus activity before returning to Earth, deploying an educational satellite. Starshine was a hollow 19-inch (48-cm) sphere covered with 878 mirrors, each the size of a quarter. Students from around the world polished the mirrors. As Starshine tumbled in space, its mirrors would catch the Sun and create a flash of light visible from the ground. It was hoped that students would go outside to try to spot Starshine as it went overhead in the twilight sky and observe its position. If enough students participated and made enough accurate observations, then they could use the laws of physics to determine Starshine's orbit.

As the commander, Rominger was responsible for flying Discovery during the reentry and landing, with Husband's assistance. Discovery landed at the Kennedy Space Center (KSC) at night on June 6. Night landings are an unusual experience to view in person. Unless you have an infrared camera you can't see anything. The shuttle doesn't have any running lights so you don't see it until just a couple of seconds before landing, when you see a shift in the patterns of the giant search lamps that illuminate the runway. Then you suddenly see this black ghost-like object appear in the lights and land just a couple of seconds later. The main way spectators experience night landings is the twin sonic booms, two and a half minutes before the landing. (Night landings make the shuttle almost impossible to see, but they're just as easy to hear.)

Husband noted after the landing, "I was very thirsty. A little bit off balance; it wasn't a feeling of dizziness, but kind of an off-balance feeling. You just had to take it easy and work your way back into a one-G environment. I was very pleasantly surprised that the adjustment was very easy." Within a couple of hours, Husband was reunited with his wife Evelyn. (Their children didn't travel to Florida for the landing because of other activities they had scheduled.)

The day after landing Husband told me, "It was great, from launch to landing it was a new experience every minute of the day. The launch was fantastic, the views out the window were absolutely gorgeous, and getting to work with all these great people was probably the highlight of the whole thing. Seeing the station for the first time was very exciting when we were coming in for the rendezvous. Traveling from shuttle back to Spacehab into station and back and forth was a lot of fun. The work we did was both rewarding and enjoyable. We had a great time. This mission was one where we did just about everything you can do on a space shuttle—the rendezvous, the docking, an EVA [spacewalk], the work in the space station with repair and transferring supplies. Getting to do the undocking and fly-around, that was a tremendous thrill for me to get to fly the orbiter during the fly-around. And to deploy a satellite and get to spend some time looking out the window at the beautiful Earth. We had a great time, coming back in for entry was as big a thrill as any other part of the mission—just seeing the glow out the windows and feeling the Gs come back on and the beautiful landing that [Commander Rominger] made. It was incredible. It was a fantastic mission."

Husband was reunited with his two children, 8-year-old Laura and 2-year-old Matthew, when the crew returned to Houston. He said, "That was very nice, they're two very sweet kids. I was very, very happy to see them. They were very talkative and wanted to tell me about everything that had happened and the things that they'd done while I was gone."

A mystery showed up after the mission, when the crew was debriefed and compared notes. An Internet space rumor site alleged there was a problem with the air quality on the space station. Husband told me the truth: "It was kind of strange. While we were up there, if somebody had a headache or something like that at the time, we didn't think it was air quality, we just thought it was adjusting to spaceflight. It was only after the flight [and we compared notes] that we started wondering if maybe there was some problem with the air."

NASA and Russia had decided to rely on the shuttle's systems whenever possible before live-aboard ISS crews arrived. Husband's crew put a duct with a fan through the docking adapter that pushed fresh air into the space station. A return duct carried the stale air back to the shuttle, where it eventually went through the shuttle's air scrubbers that removed carbon dioxide and excess moisture. Similar techniques were used when the shuttle was docked to Russia's Mir space station.

Husband said, "There's even this theory that the air quality was okay, it was just a circulation problem. They've made some pretty good changes. The air return line from the shuttle was fairly close to the line which brought air in to the shuttle from the station. So they think it may be possible the air didn't get remixed with the orbiter air as well as it could have, so it may have been pulling a lot of the same air back into the station that had just been taken out."

Husband said, "If there was any problem with the air quality it was a very insidious thing, it wasn't something that you had this great realization after a while. Some folks did have headaches and things like that. Even with the investigations, they couldn't come to any conclusive decision whether or not there was an air-quality problem. But they made some changes to enhance the airflow regardless of whether we had a problem. In the FGB it was not something that was noticeable when you went in, in terms of air quality. [The STS-96 crew] raised [air quality] as a possibility. We didn't come back saying that we thought the air quality was all that bad. After the mission we thought there's a possibility there may have been an issue with the air quality."

Even before Discovery's landing there were already questions in Amarillo—"When's Rick coming home?" Amarilloans asked, "Can I meet him?"

A group of businesses and individuals underwrote the cost for a VIP homecoming trip, also extending invitations to Husband's crewmates. Tammy Jernigan had a commitment in Seattle and couldn't attend, and Valeri Tokarev had already returned to Russia. But Kent Rominger, Dan Barry, Ellen Ochoa, and Julie Payette had the time of their lives, as they were treated like royalty along with Husband.

The crew made a public presentation at the Amarillo Civic Center Grand Plaza, where Husband returned the flags and other souvenirs he had flown for various schools and organizations in Amarillo. Each of the astronauts was given a key to the city. Husband said, "No one at NASA doubts where I'm from." Rominger told the city that Husband had the nickname "BFA," or Boy From Amarillo. They also saw a performance of the musical "Texas" at Palo Duro Canyon State Park.

The trip ended with a tour of the Bell Helicopter Tiltrotor Facility. Not surprisingly, that was one of the biggest highlights from the astronauts' perspective. Bell Helicopter builds the V-22 Osprey, a combination aircraft/helicopter which can take off and land vertically from a small area, like a ship, and rotates its engines inflight to convert to a long-range aircraft. Afterwards Husband said, "The folks in Amarillo did a great job with the hospitality, it was just a fantastic trip. Everybody who was able to attend just really had a really super time."

After the STS-96 postflight activities wound down, Rick Husband was given a new desk job, which is the less glamorous part of an astronaut's life. He represented the astronauts in the selection of the Crew Rescue Vehicle for the Space Station and on safety boards. The key reason NASA has so many astronauts is that an astronaut's job is spent primarily on the ground in technical roles. Astronauts spend only a tiny part of their careers training to fly in space and a minuscule amount of time actually in space. But as experts on technical panels they can provide the perspective of somebody who's been there. Many have noted that astronauts contribute more to the space program when they're on the ground helping other missions and future projects than when they're assigned to a specific mission.

While Husband was busy with his desk job, he was certainly interested in getting another chance to fly in space. He said, "That is my plan, sure is. I definitely want to have the opportunity to fly on the shuttle several more times, and we'll just kind of see how things unfold." As the astronaut responsible for shuttle safety issues, Husband participated in safety meetings and was the weather coordinator in Florida for each shuttle launch.

On April 11, 2000, Husband was honored by his high school. He explained, "This is what they call the Sandie Hall of Fame. The school mascot is the Golden Sandstorm, which goes back to the Dust Bowl days, so we're called the Amarillo Sandies. They've had the hall of fame for quite a long time. I do remember when I was going to school there walking by and seeing the pictures on the wall." He added, "I sure didn't think my name would be on the list. I had no idea this would be in the cards for the future."

The Russian Service Module that was supposed to be docked to ISS for Husband's mission still wasn't there ten months later. Husband said, "For the early stages of the station game, that's kind of the way things have gone."

Because of the delays, NASA had to schedule yet another shuttle flight for maintenance and to raise ISS's orbit, which had become dangerously low. Husband's advice to the crew of STS-101, the mission sent to perform the maintenance tasks: "I've told them, if it's anything like our flight, the temperature is quite a bit warmer than on the shuttle. That was one of the biggest impressions I had. Just how nice the Unity (Node) looks, just how great it is to be able to have that much volume to work around in. I've been talking with the crew about their tasks and some of the things they'll have to expect, like rearranging the bags we left up there [as part of STS-96]. I think they're going to have a great mission, I know they'll really enjoy it. They've got a lot of work up there, and we're looking forward to getting those guys launched pretty quick."

In his role as the safety lead, Husband monitored the mission closely. He said, "I think we're in pretty good shape. I just got back from the flight readiness review in Florida. For all of the different systems that the folks talked about, I didn't have any significant concerns. The people seem to be doing a really good job staying on top of the issues with all of their respective systems. So I think we're in pretty good shape." He added, "I'm pretty happy working my [support] job right now, it's pretty interesting working as the safety guy. It's been a very good experience for me being able to learn about all the different people who work to get a mission off safely."

Besides his desk job, Husband continued his astronaut training. For example, on April 6 he was the commander for a launch in the shuttle simulator. Husband said, "It's a scramble for me right now, because I have a lot to learn about the [commander's responsibilities and tasks]. It's very enjoyable getting in there and getting up to speed with the systems and keeping some proficiency at handling the emergencies. I always enjoy getting into the simulator."

On October 27, 2000, he got the news that he was assigned to command STS-107. He told his wife, mother, and a few close friends. Husband's mother, Jane, was a friend of the editor of the *Amarillo Globe-News* and told the paper the good news. That day I got a mysterious e-mail. A reporter at the newspaper told me she heard a rumor that Rick Husband was going to command the next shuttle flight. I assured her that while the timing was right for Husband to get assigned to his second spaceflight, it wasn't likely for him to go directly to commander after only one flight as pilot because NASA had not done that since 1993. I added that I was quite certain his mother would always think of Husband as the commander, no matter what position he flew—because I had no doubt she was the newspaper's "source." Well, it turned out that Rick's mom was right and I was wrong, and Husband became the first astronaut in almost a decade to be promoted from pilot to commander after just one flight.

A couple of weeks later, I interviewed Husband about his new assignment. I told him I found out about it through my spy in Amarillo. He asked incredulously, "So you have a spy in Amarillo 'eh?" and I replied, "Well, you call her mom," and we both had a good laugh. That audio clip is on this book's website.

Actually Husband knew for some time he was being groomed to command a mission, and I probably should have suspected it when he told me he was training as a commander during the simulation back in April. Much of Husband's training before getting assigned to STS-107 was to help out commanders for other shuttle missions and to develop the additional skills he would need to command a shuttle mission himself. He assisted STS-106 commander Terry Wilcutt by attending meetings that Wilcutt didn't need to attend, and also got advice from his STS-96 commander Kent Rominger and experienced shuttle commanders

Charlie Precourt and Jim Wetherbee. In addition, he took classes and training sessions for the skills he would need as a commander.

About his rapid promotion, Husband modestly noted, "It's mostly just being in the right place at the right time. They kind of hit a brief period where they're a little bit short on commanders. I was fortunate enough to be positioned in the right place at the right time." That was true—the astronaut class before Husband included only four pilots. By the fall of 2000, many experienced commanders had retired from NASA or were moved to management positions. Two of Husband's fellow 1995 pilots had already left NASA after a single assignment. Many of Husband's classmates who got to fly sooner and had completed two shuttle flights as pilots were already getting assignments to command their own missions: Steve Lindsey, Dom Gorie, and Mike Bloomfield were assigned to space station missions, and Scott Altman was assigned to a flight to the Hubble Space Telescope.

Husband certainly could have been assigned to another mission as pilot before getting a command. However, STS-107 needed a commander. It was considered a less prestigious science mission in comparison with a flight to Hubble or the space station; it was certainly not the flight where an extremely experienced commander would be essential. STS-107 didn't feature any complicated on-orbit activities like a rendezvous with ISS, robot arm operations, spacewalks, or even a satellite deployment. If STS-107 didn't exist, Husband's second flight would probably have been as the pilot for a space station mission in 2001 or 2002, followed by a command on his third flight.

Of course, like every spaceflight, STS-107 included the most critical phases—launch, landing, and keeping the shuttle running properly while it was in space. But the lack of dynamic on-orbit activities would make it a suitable flight for a first-time commander. The key challenge for STS-107 would be to land a very heavy shuttle after a long mission. The longer a mission, the more the body adapts to being in space and the more challenging it is to land the shuttle while also adapting back to gravity. It's especially difficult if the shuttle has a heavy payload inside the cargo bay because there's less margin for error. Husband had the experience, qualifications, and capabilities, so it was a good fit for him.

Husband was ecstatic to get his own shuttle flight, even if it wasn't a high-visibility space station mission. He told me, "I think we've got a great mission even though we don't have a rendezvous or docking with the space station. We are doing a lot of really important science and research on our mission. Our contribution to the science community I think will be significant from the standpoint of all of the experiments we're doing. And we're doing such a broad-based variety of experiments that will be really doing a great mission for the science community in general. Getting some good concentrated science over 16 days."

Husband and his family spent the 2000 Christmas holidays in Amarillo, visiting his mother and in-laws. During their visit, they had 20 inches of snow. He said, "That was a super opportunity for our kids. Our daughter hadn't seen snow since she was two years old. Our son had not seen it at all. That was a real great time for us."

When I interviewed Husband in January 2001, the launch was scheduled for July—just seven months away. Husband said, "I've got a lot of confidence in the crew, there's some very, very talented folks. Dave [Brown] and Laurel [Clark], both being medical doctors, have a lot of background and experience [in life sciences]. We've been having some good interactions with the payload investigators. I really think we'll be in really good shape to execute the mission. The trend here [at NASA] is to reduce the time assigned to a mission. [The payload crew] is very knowledgeable about the payloads; I really don't have any concerns about being able to run all of these. I feel really blessed with all the talent we have on our crew. I'm just really looking forward to working with these folks and getting to know them better as we go through the training flow. They're a super bunch of folks." Little did he know at that point they would spend an additional two years together because of the many delays to come.

The crew was put together well before the 9/11 terrorist attacks, but even then there was talk about the additional security concerns because STS-107 had an Israeli crewmember. Husband told me, "Certainly, Ilan [Ramon] from our perspective is a full-fledged member of the crew. I've really enjoyed getting to know him—he's a super guy. That isn't something I'm too concerned about." As far as additional security for the mission was concerned, he said, "We've got some very capable folks here at NASA who are specialists in those areas. So I imagine I'll kind of let them sort out how they're going to handle it, and I'm sure they'll let us know what they plan on doing."

Husband said, "The big thing is trying to make sure I've got things down to make sure I can support the team. I don't want to be in a position where I'm letting my crew down. There are definitely some different aspects to being a commander, but I've got such a great crew. Willie [McCool]'s the pilot, he's a very energetic guy. He's got great ideas he goes after. I can suggest something and he'll take care of it, or he'll come up with a great suggestion and make it happen. Mike [Anderson]'s the payload commander. He takes care of the vast majority of the payload issues and keeps me informed there. Kalpana [Chawla], Ilan [Ramon], Laurel [Clark], and Dave [Brown] do a great job with their training and different aspects of flying the orbiter. I couldn't have hoped for a better crew. I've really appreciated their support for me, being a first-time commander." At that point Husband had been training with his crew for about three months, but of course, he had known some of them from the time he arrived at NASA.

Husband said, "I definitely am looking forward to [my first command]. I would say there's anxious anticipation looking forward to a mission after you've spent a long time going through a training flow and preparing for the mission. You come to the point where you definitely want to make sure you hold up your end of the bargain with respect to the rest of the crew. That's the thing I'm concentrating on now—doing as good a job for them as I know they will do for me."

One of the more unusual training activities for Husband was how to give injections and draw blood. While Dave Brown and Laurel Clark were both medical doctors, it would help to free up time for them to do more experiments if other members of the crew were trained to do the blood draws. When Husband was assigned to the mission, he told me, "We'll see how it goes. Right now I have no idea what it will be like. It's definitely something I want to be very well trained at and can do a good job. I'll make sure I'm as capable as I possibly can be before I actually go and do it on my crew."

NASA's life sciences staff gave Husband, McCool, Anderson, and Ramon the necessary training to give injections and draw blood. Before trying his skills on a fellow crewmember, Husband inserted a needle in a volunteer. So, what was it like the first time he drew blood from somebody? "I had the appropriate amount of concern for that person's well-being," he said. "You kind of approach it with, 'I sure hope I don't hurt this person.' The first few times are good to help you get used to it and get the hang of things and build some confidence and let you know that you can do this without hurting people." Husband practiced his skills on 8 or 10 volunteers before he got the opportunity to insert a needle into one of his own crew.

The lucky crewmember to have the honor of becoming the first to be stuck by Husband was Dave Brown. Husband said, "We're all starting to take turns taking each other's blood whenever we come through for a flight physical. I took Dave Brown's blood for his flight physical. Mike Anderson came in and drew mine. That's going along real well."

Husband said, "From the training I had gotten I wasn't nearly as nervous about it. I still kind of had that feeling in the back of my mind that I don't want to cause any pain to anybody. You just press on and do it the way you've been trained, it works out fine." Brown assured Husband that he did a good job and didn't cause any unnecessary pain. "[Dave] said it went fine, that's all he had to say about it. He was still smiling when I got through."

After several months of experience, Husband said, "I've got quite a bit more confidence now. I had a certain level of anxiety, but I've gotten to the point now where I'm not as reluctant or nervous about doing it. Gotten past getting used to it."

On the other hand, much of the training was far simpler than learning how to stick a needle into someone. Insiders joke that certain payloads are known as "commander science"—extremely simple tasks that any elementary school student can learn in a couple of minutes. In other words, just right for a test pilot jock. But there's a real reason for giving the pilots simple experiments. Most of the training time for the commander and pilot astronauts is to fly the space shuttle, especially in emergencies if something goes wrong. It makes sense to give the pilots simple science tasks to perform on orbit; tasks that don't require much training time and won't distract them from their more critical training. In addition, giving the simple tasks to the pilots that somebody has to perform frees up the scientist astronauts for more hands-on duties.

One of the "commander science" experiments Husband was responsible for was OSTEO-2—Osteoporosis Experiments in Orbit. OSTEO was a Canadian-sponsored miniature factory that grew tissues in small canisters. Husband said, "We've got different media that are set up inside the payload, set to simulate a bone culture. We send different fluids in. I'm essentially pushing buttons and turning knobs—it doesn't get much more complex than that!" Husband showed me how the payload works: "There's an activation to the payload—making sure the circuit breakers are closed, the switches are turned on. We do a status check and make sure everything's squared away. We've got a table which says OSTEO feed 1, feed 2 or something like that. It tells you which knob, which medium to meter into the cultures. It's not very long at all, so it doesn't take over 15 minutes per feed."

In August 2001 I asked Husband how he felt about the many delays—at that point it was already a year past the original launch date. He said, "Those things are pretty much beyond our control, so there's no sense really getting frustrated about it. Probably other than flying in space, one of

Rick Husband demonstrates the OSTEO experiment. Photo by Philip Chien.

the best things to do in the astronaut office is to be assigned to a mission and train for it, so we're happy for the additional time." He added, "There's no doubt we would have been ready [for the original launch date]."

Rick invited me to visit his crew while they were training in Florida, at the Spacehab facility in Cape Canaveral. Husband noticed me, looked up and said, "Hi, Phil," and promptly went back to his training with Ilan Ramon. They were working with their trainer, Lisa Anderson, and Flight Activity Officer Terri Schneider on the mission's timeline, with a computer program that simulated the Israeli-sponsored MEIDEX dust-monitoring experiment. Husband continued to concentrate on his training and only after it was finished did he come over to shake my hand and introduce me to Ilan and the rest of his crew.

So what was Husband anticipating on his second spaceflight? "There are so many things to look forward to. In series—I'm looking forward to the launch and getting on orbit and getting going—getting the ball rolling. Float for extended periods and experience that again. I'm looking forward to the great views we'll have out the window. I'm looking forward to executing the plan that everybody's worked so hard to put together over so many months. It's always a great thing when you work hard on something to see it come to fruition. It will be really nice to get up and put this plan in action and watch it unfold. And I'm certainly looking forward to landing."

Husband specifically cited the landing because it's the pinnacle of a test pilot's career. A shuttle landing is the commander's responsibility and something he will only get to do a handful of times in his career, if he's extremely lucky. It's been described as the most difficult job in flying, something where there's no second chance if things don't go perfectly. The pilot doesn't get to do the landing, all he gets to do is watch the commander, monitor the shuttle's health, and be ready if the commander has a problem. Only the commander gets to land the shuttle, and even if Husband decided to stay with NASA for a long time and got assigned to several more missions, landing the craft would be something he would have the opportunity to do only a handful of times at most.

While Husband was very aware of the risks involved in a spaceflight, he said he was very impressed with the level of detail and seriousness about safety. In June 2002, when tiny cracks in the shuttle's plumbing grounded his mission, Husband said, "If they found something with the hardware that needs to be fixed, then that's definitely the right thing to do. With my experience as the safety branch chief in the astronaut office prior to being assigned to this mission, I was very impressed at the level of detail and diligence that the people working on the hardware go to, to make sure we've got a good vehicle to fly. We're more than willing to let them do their work, and looking forward to flying a good vehicle when they're done."

The various slips to STS-107 put it after STS-111. That crew included Husband's high school acquaintance Paul Lockhart. Husband said, "Just more than anything I'm happy for Paul getting assigned to a crew and everything because he's such a super guy and I know he'll do a great job. We've known each other for quite a long time. We sat down one time in the summer of '78 talking about how we both wanted to become astronauts some day. So it's been a great thrill to watch it all unfold."

Because another astronaut was injured, Lockhart was able to fly in space twice in one year, on STS-111, in June 2002, and STS-113 in November. Lockhart slipped ahead of Husband to become the first Amarilloan to fly in space twice.

Husband was asked about how much he was looking forward to his second flight. He said, "I'm definitely looking forward to going. From what I experienced on my first flight, I'm very much looking forward to going again. Especially on this flight, getting to stay up for 16 days. This mission being of such a different nature, I'm looking forward to that. The same excitement, probably more. At this point, a little less nervous anticipation about the unknown as the first flight."

Proud of his Texas background, the Boy from Amarillo joked, "Everybody else wishes they were from Texas!" But Husband also appreciated the rich multicultural background in his crew. The many delays made it possible for the crew to get to know each other better, and become closer. Husband said, "We've got a fairly close group, it's been a real blessing being able to be together as long as we have. Even though we've had the launch slips, the benefit has been we've gotten to spend more time with each other, getting to know each other better."

Sharing training with Kalpana Chawla, an Indian-American vegetarian, and Ilan Ramon, an Israeli, made life interesting. Husband said, "It's made things more enjoyable in terms of variety. Whenever we have crew parties we make sure everybody's taken care of. It's been transparent."

For his second spaceflight Rick Husband decided to take a little bit of Amarillo into space with him again. The items included a banner made by the kids at the "Discovery Center," a T-shirt for the musical drama "Texas" that the STS-96 crew saw, a flag for the Jack B. Kelly company, which provides much of the Helium NASA uses, and a T-shirt for Boy's Ranch, a home for troubled youngsters.

Contrary to the image that NASA often tries to present of good company men who are almost like machines, many astronauts have an excellent sense of humor. At one point during the STS-107 mission, Husband crossed his arms while holding two gun-shaped tools, posing like James Bond.

After the accident, entry flight director LeRoy Cain said, "The thing that I would share with you is that [Rick] was—there probably wasn't a more perfect fit in the crew office for putting together and leading this particular crew. This was a marvelous group of individuals, very high-powered in terms of their capabilities. They were also very diverse and came from very diverse backgrounds. And Rick, with his kind of laid-back, very easygoing style, I think that lent itself very well to them being able to meld as a crew, which they did. This crew had a little bit more time in training, as you recall, because of some of the delays. And they really and truly, by the time they went and flew, they were a family in and of themselves. And it was just pure joy being around them and working with them."

Several months after the accident, STS-96 commander Kent Rominger said, "He was a phenomenal human being. He was just such a loving, caring, sincere human being. He really had life figured out— what his priorities in his life were. You had to love the guy. It was never about Rick, he was very unselfish. His priorities were his family and doing the right thing, and he had a very strong faith. He was just a really enjoyable guy to be around."

Rick Husband—license to fly.

CHAPTER 3

WILLIE McCOOL: RUNNING MAN

Navy pilot Willie McCool was selected as an astronaut in 1996 and assigned to STS-107 in October 2000.

I'll always remember Willie for his incredible optimism and outlook on life. Rookie astronauts have a motto—any spaceflight assignment is a good assignment, especially your first flight. Certainly some flights are more glamorous or challenging than others, and any pilot would like to get assigned to a flight where there are lots of piloting activities, preferably as quickly as possible. Since STS-107 was a microgravity flight with no spacewalks, no rendezvous, and not even robotic arm operations, it was considered a relatively "boring" mission. Astronaut Curt Brown was the last in the 1987 astronaut class to fly, and was assigned to a microgravity mission with just four maneuvers—"and the last maneuver was to point the engines for the deorbit burn," he ruefully recalled. He wondered what upper-level manager he had offended to get such a lowly assignment. But Brown went on to fly another five shuttle flights, including two of NASA's most prestigious high-visibility missions. Out of ten pilots in the 1996 class, Willie was the one who was assigned to fly on a microgravity flight, while the rest of his class were assigned to ISS or Hubble Space Telescope missions for their first flights.

No rookie astronaut is ever going to complain about getting assigned to a less glamorous mission, but was McCool the slightest bit jealous of his classmates who got to fly to ISS or Hubble? McCool told me, "Definitely not, if anything they should be jealous of me. First of all I get 16 days on orbit [6 more than a typical mission], I get to essentially be the commander during Rick [Husband]'s sleep time. So there's a degree of responsibility I have the pleasure of sharing with Rick that other pilots on other missions don't necessarily get to share. I get to do a lot of the science. I wish I could do [a space station flight]. But I'm equally happy being involved in the myriad of science experiments we have on board and looking forward to getting my hands in the glovebox and operating the MEIDEX cameras and maneuvering the vehicle around to support the Earth observation payloads that we have. Drawing blood—shoot, that's fine."

How did McCool feel the first time he stuck a needle into somebody? He said, "I wasn't comfortable from the standpoint of myself to the other person. I have no qualms about doing it for the betterment of science to my crewmates because it's for data. But the folks who volunteer to do it so I can [learn how to] stick a needle in them, I feel bad for them. I felt nervous, especially early on, for their sake." McCool had plenty of training, drawing blood from hundreds of volunteers. He added, "By the time we got to our crewmates, we were all experts at drawing blood."

McCool's optimism even extended to the mission's many delays, as he watched several astronauts in his class get assigned to flights after him, but then fly before he flew. McCool's attitude was that the continuous training helped. He said, "From a rookie's standpoint, the delays are probably good. I feel like going through the training flow essentially a second time a little less like a rookie and a little bit more like a veteran, and as a result I'm better prepared for the flight."

McCool was born in San Diego but moved several times while growing up: "Dad was a Marine and then a civilian," he said, and the family moved to wherever his father was assigned. Willie's first nine years were in San Diego, and he recalled fond childhood memories of the area, especially the San Diego Zoo. Then the McCools moved on to Minnesota, Illinois, and Guam before ending up in Lubbock, Texas, in August 1977. His mother taught at Texas Tech and didn't want to follow his father back to San Diego for another shipboard tour.

While growing up in Guam, McCool met Lani, the woman who would become the love of his life. McCool attended Guam South Elementary/Middle School and Lani was at another school, but they went together to Japan on the same school tour. She recalled it was love at first sight for the shy teenagers: "We had a crush on each other during that trip but never spoke until the following year, when we were in the same speech class in tenth grade at JFK high school." The next year McCool moved to Lubbock, Texas, but Willie and Lani stayed in touch, and after Lani's first marriage ended, they reunited. "It was as if we'd never parted and we continued right where we'd left off," according to Lani.

Willie attended Coronado High School in Lubbock for his Junior and Senior years, graduating in 1979. While most astronauts cite math and science teachers as their most important influences, McCool was an exception: "Believe it or not," he said, "this is coming from a guy who is very technically minded, my favorite course was English. Sharon Kingston was our English teacher. She was wonderful in her enthusiasm in bringing us mentally into the plays and stories that we read. Shakespeare's *Macbeth* is one which comes to mind. When the witches are going—'Macbeth, Macbeth!' She just made it fun."

McCool and STS-107 Commander Rick Husband shared a geographical connection—Lubbock is just 112 miles from Amarillo. In addition, Husband attended Texas Tech at the same time that McCool's mother was teaching there.

McCool attended the US Naval Academy, where he was a member of the men's cross-country and indoor/outdoor track teams. Coach Al Cantello was a major part of McCool's academy life, and the two remained friends after McCool graduated. Cantello already had two decades of coaching experience at the Naval Academy before McCool arrived, and continued to coach for more than two decades afterward. Cantello recalled, "[Willie] had one of these steely-eyed infectious personalities. He was upbeat, enthusiastic. When he was a freshman I asked him, 'What percentage of your life is running?' and he said 'I don't know—I'm here for academics, I want to do well in athletics, there's the military side—so I would say 30 percent.' I said, 'Willie, you'll never make it as a runner here.' When he was a senior he placed in a 10,000-meter run in the rain and got a medal. He came up to me and said, '95 percent,' and I said, 'What are you talking about?' said He answered, '95 percent of my life is running right now.' After four years, he finally realized what the commitment was to be a good runner here."

Cantello also noted that McCool gave up a precious free Saturday evening to tutor his son in math.

As a senior, McCool led the midshipmen to a victory over Georgetown and Syracuse on October 2, 1982, while setting a personal record of 24:27 in the 8K. When his classmates met for their 20th reunion, they held a memorial for McCool at the point where he was 15.5 minutes from the finish line, symbolizing the 15.5 minutes to go before Columbia's landing when the accident occurred. The cross-country team is planning to put a permanent marker at that location.

McCool running in a long-distance race.
Photo courtesy of US Naval Academy.

Cantello notes that McCool would have been number one in his Naval Academy class but for the fact that he was "put up on a conduct charge that was so bogus." According to Cantello, McCool drove home a midshipman who had a bit too much to drink, and was accused of fraternization, which cost him the number one slot. McCool's good-natured behavior to help out a woman who was drunk got him the demerit. So McCool graduated second in the Naval Academy's 1983 class of 1,083 students.

In 1985, McCool earned a master's degree in computer science from the University of Maryland. He had a condo in the Annapolis area and would cook spaghetti dinners for the Naval academy cross-country runners in later classes. McCool also earned a master's degree in aeronautical engineering from the US Naval Postgraduate School in 1992. He was selected for the Navy Test Pilot School at Patuxent River in Maryland, one of the key requirements for becoming an astronaut.

As a career Navy aviator from a career military family McCool spent most of his life moving from assignment to assignment. He said, "Bottom line is, I really haven't lived in one house for longer than three years until my family settled in Anacortes, Washington. That's where we truly call home. What convinced us [where to live] is the Navy telling us where we had to go and when we had to go." Lani said, "The reason we bought a house in Anacortes because we have the largest forest land west of the Mississippi. It's a very small town, but we have these wonderful trails, miles of trails. Willie loved running and loved nature."

McCool noted, "As my career progressed, things just worked out to my benefit in leading me into the astronaut program." McCool said that unlike many astronauts, who know they want to become astronauts from the time they were very young, he didn't make the decision until later in life: "I can't say that I had a lot of specific aspirations as a child. It wasn't until I was at the Navy Test Pilot School, when I saw my classmates start to make applications to the astronaut program. So I said, 'Hey I'm in line to do something which is a whole lot of fun, so why not try it myself?' It wasn't until I was probably 30 years old that I had my first aspirations to become an astronaut."

Once NASA selected McCool, he moved his family to Houston, but they decided to keep their house in Anacortes. Some astronauts come to NASA with the intention of staying forever, but McCool had a long-range plan for when he finished his astronaut career: He and his family always planned to move back to Anacortes.

McCool had a reputation for never being late for a meeting or appointment. Lani McCool said, "He was extremely conscientious about making other people wait for him." There was one exception—and just because it was such an exception, everybody remembered it. Lani explained: "We have a loft bed that's above our desk. A cockroach skittered against the wall. I was telling Willie he needed to take care of the cockroach and he was rushing, saying, 'I've got to go, I'm going to be late.' He was trying to give me a kiss goodbye, but I was insistent that he get the cockroach before he left. So he stayed, and he had to dismantle the desk to get to it."

Willie and Lani McCool were always interested in various charitable organizations. They donated money anonymously in the past, but realized that since Willie was now a public figure, they could also recognize their favorite charities in a special way. Four of the ten items Willie was permitted to fly with him in space were for charities—the UN World Food Program, For All Kids Foundation, Doctors without Borders, and Interplast. Lani said, "Flying something which represented their organizations was a way to show appreciation and actually have their emblems orbit the planet which they were trying to help improve. It was special going over Africa. That was one of the places we both wanted to go to after he retired. He wanted to give vaccines and get involved with humanitarian efforts." Why not? McCool had the training to give injections and draw blood.

Lani explained, "When I was in high school, my mother was in the Peace Corps and went to Africa. I've always followed a lot of the events that go on in Africa and where they need help." Willie flew a map showing areas with hunger around the world for the UN's World Food Program, based in Rome, Italy.

Lani said, "Rosie O'Donnell had started this "For All Kids Foundation" breakfast program for inner-city school kids who don't have money for lunches. I was talking to a friend and she mentioned it. Willie and I thought about adoption, but because of his career and time, we hadn't been able to." A US flag was flown for them. Tattered parts of it were recovered after the accident and returned to Lani McCool.

"Médecins Sans Frontierès"—"Doctors Without Borders" was honored with a T-shirt with their logo. Lani noted, "They're doctors who go to very, very remote areas." Interplast is a charitable organization that provides reconstructive plastic surgery to children and needy adults in developing countries. They gave Willie a book jacket to fly. Willie also asked Coach Cantello from the Naval academy for something special to fly. Lani said, "That's where Willie learned a lot of his discipline and passion." Cantello selected a blue and gold Navy pendant with 38 stars — one for each time Cantello's Navy runners beat the Army in cross-country.

McCool also carried items for his schools and organizations, and two natural items from his home state. The items from schools and organizations he honored included a blue kerchief for his Boy Scout Troop on Guam, a Coronado Mustangs Spirit Towel for Lubbock High school, and a patch for the Prowler community at Whidbey Island Naval Air Station in Washington. The McCools recognized Washington State with a pine cone to represent Anacortes Community Forest land, and a piece of pillow basalt to represent Olympia National Park. When Willie asked the Olympic National Park for something to fly, he received a brown box labeled, "Enclosed, a rock".

McCool and his wife Lani had three sons—Cameron, Christopher, and Sean. He said "They're pretty much used to me going on deployments and doing Navy type stuff, in that respect [flying in space] isn't a whole lot different."

McCool said his most enjoyable experiences were with his family, "back-country backpacking in the Olympic Mountains [in Washington] or the canyons in Utah and just enjoying life without outside distractions. And enjoying each other, and enjoying the environment. And we love to do that frequently, whenever we can. Unfortunately, I don't get enough of it here recently with all the training. But those memories prevail. And they're something that I look forward to doing in the future, when we get done with this mission." Lani said "He loved nature and loved the environment. Our greatest vacation was coming back to Washington and going camping. We would just hike as far away as we could and find solitude. That was Willie's vacation—not Hawaii, not a cruise."

As a Navy pilot used to catapult launches off aircraft carriers, McCool had an idea of what to expect for his first shuttle launch. He told me, "Just a little bit bigger airplane if you will, and maybe a little bit more dynamic than a launch off a catapult." That's a bit of an understatement. Unlike a catapult launch, which takes about a second, Columbia would continue to accelerate for eight and a half minutes. McCool said, "I'm looking forward to it. I'm comfortable with the cockpit environment, it's going to be interesting to see how the Gs impact my ability to stay focused if I need to take any actions in the cockpit."

STS-107 was planned for around-the-clock operations, with two shifts, so astronauts could do science continuously. McCool was the leader of the blue shift, which included Mike Anderson and Dave Brown. They would work midnight hours while the red shift slept. McCool had one of the most difficult jobs for a rookie astronaut—going to sleep just four hours after the most exciting event of his life [the shuttle launch]. McCool told me, "I think it's going to be very difficult. That's why we're focusing now in advance on doing everything very efficiently and on time. We hope to take whatever measures are necessary to get us into bed." After the accident, this statement was misinterpreted by *The New York Times*, which claimed McCool believed he'd have a hard time working on an around-the-clock shuttle flight.

Because McCool was on the "night shift," he was asleep most of the times when Columbia traveled over the United States, so he didn't have the opportunity to try to spot the places where he grew up from orbit. But he did get to see many incredible sights. At one point Willie and Laurel were videotaped as they talked about the amazing views while preparing for one of the blood draws. The tape survived the accident, and contained a beautiful description of the views in space. McCool told Clark, "I saw the sliver of the Moon and the light blue of the horizon started to

Willie McCool mimes using a camera to capture a photo of a spectacular sunrise in space.

come up, and the blue turned to deeper blue and got thicker and thicker—band, band, band—and the blue moved up and it was kind of white and then orange. I could distinctly see band-band-band, it wasn't a blend, it was a sharp distinction, you could see all the colors. The Moon was there, a little sliver. You could see the disk was silver [colored] in silhouette, so the rest of the Moon even though it was only a tenth of a Moon. Wow — oh my goodness, it was so thick too that whole thing… I snapped my feet out of the ergometer [exercise bicycle], jumped and grabbed the camera and ran to the window. You know how hard it is to take night shots. I turned the flash off and I'm freezing. I got four pictures of this and it just got thicker and thicker. And finally the Sun blossomed over the horizon, you get that orange ball coming up, coming up, coming up, and this blinding light so I had to come off. Oh my gosh, it was such a spectacular view." A video clip of McCool and Clark's conversation is on this book's website.

McCool said during the mission, "There's so much more than what I ever expected. It's beyond imagination until you get up here and see it and experience it and feel it. I would say the things that are new to me are the zero-G type changes to my body and the zero-G changes to the things that we do routinely day-to-day that you can only imagine, but you can't really quite imagine it quite right until you get up and do it. I can't describe the overwhelming sense of elation that I feel being up here. Especially the views. We describe them to everybody we talk with. They're just beyond all imagination. Photos just don't do it justice. The sunrises and sunsets, the moonrises and moonsets, going around the globe, seeing the Himalayan range, the Great Barrier Reef, it's phenomenal."

Lani said, "There's a beautiful picture of Willie. He did self-portraits. He pointed the camera in the mirror and right there is my wedding ring on his pinky. He wore my wedding ring on his pinky the whole time he was in space. Every time I saw him on TV, I'd see the glint on the finger. He wore his on a necklace and mine on his pinky."

Much of the humor in the astronaut corps centers around inter-service rivalries. Husband was Air Force, McCool was Navy, and capcom

Charlie "Scorch" Hobaugh in Mission Control is a member of the Marine Corps. At one point, Hobaugh asked Husband to reset the OCA (Orbiter Communications Adapter), pronouncing it as "orca" instead of spelling it out. The OCA is the box that controls the shuttle's radios. Hobaugh didn't get a response because Husband was in the back of the shuttle in the Spacehab module. Hobaugh called the shuttle's flight deck where McCool was located. After Hobaugh asked McCool to reset the OCA McCool replied "You're a whale of a guy, here it comes." and Hobaugh replied "Rick didn't get the joke, I guess you have to be a Navy guy." Husband overheard the conversation and got on the radio saying, "Get out of town Scorch, I told you I didn't hear you back there." Hobaugh retorted, "Good excuse," and Husband, playing the straight man, went "Huh?" A copy of the audio is on this book's website.

Willie's nametag from his flight suit was the only one found. Lani said, "His nametag was in almost perfect condition."

Lani said, "Willie was an excellent athlete, a poet, a photographer, an artist, and a wonderful teacher." During a private video phone call from space, McCool read Lani a poem he composed just for her :

> I've witnessed the beauty of Earth from space,
> far, far above,
> What a treasure it is to behold!
> But I would trade it to you for your embrace,
> my sweet love,
> because only you enrapture my soul.

Lani said, "Willie had a big picture, he was a philosopher. What are we here for? What are we supposed to do? He always had that big picture of being a human being and being able to contribute as much as we can to help others so there's harmony. He was an optimist. He gave me a ring with the word 'Hope' on it."

Willie on Columbia's flight deck. You can see Lani's wedding ring on his right pinky.

CHAPTER 4

DAVE BROWN:
PILOT, DOCTOR,
GYMNAST

Dave Brown became an astronaut in 1996. He was one of the rare astronauts who was both a medical doctor and a military jet pilot. But he'd also had probably the most unusual job ever held by anybody who eventually became an astronaut: He was a circus performer.

I asked him—at a news conference, in front of everybody—"A lot of kids dream about running away and joining the circus or becoming an astronaut. What was your family's reaction when you told them you were going to run away and join the circus, and what was their reaction when you told them you were going to become an astronaut?"

After a chuckle he replied, "When I was in college I was a gymnast. One day I was in my dorm and the phone rang, and the fellow on the other end said, 'Hi, you don't know me, but would you like to work in the circus?' I said 'Sure,' and I did for a summer, which was a pretty interesting job. When I called my parents and told them I was going to join the circus they were surprised, but when I told them I was going to become an astronaut they weren't surprised at all." It certainly makes sense for a circus to recruit gymnasts for summer jobs. Brown became an acrobat, 7-foot unicyclist, and stilt walker with Circus Kingdom in 1977. But he only stayed with the circus for the summer. He returned to college, went on to earn his medical degree, and joined the Navy.

Dave Brown (on the left) performs in the circus in 1977. Photo courtesy of Debby Kate Stahl Ramsey.

Brown was born in Arlington, Virginia, and educated in Virginia—from elementary school through medical school. His father, Judge Paul Brown, was elected as a traffic court judge, and was eventually appointed by the governor to serve on the state's circuit court, where he served for two decades.

As a child Dave Brown had his first experience with what would become a life-long passion—flying. He said, "I was interested in flying beginning at age seven, when a close family friend took me in his little airplane. And I remember looking at the wheel of the airplane as we rolled down the runway, because I wanted to remember the exact moment that I first went flying. And I do. I remember the exact moment when I saw that wheel lift off. So, that's really when the flying interest began for me. It was someone who took the time to take this kid in his airplane; and, boy, it sure set a bit in my head that's been there ever since." In contrast, Dave's brother Doug grew up loving boats, and became a CPA.

Even as a kid Dave enjoyed technical hobbies. He had a ham radio license, and spent a lot of time in the basement, using Morse code to communicate with ham radio operators around the world. He used his newspaper route money for a class trip to the British Isles, where he looked up one of his ham radio contacts. During another summer he was an intern at the Smithsonian Institution. His mother, Dorothy, recalled that Dave and a high school buddy got involved with a neighbor in a research project at the National Zoo, where they studied the behavior of the African antelope.

The Brown family loved outdoor activities. Although Judge Brown had been stricken with polio and had to wear a leg brace, he wasn't one to give up an active life. Dave's brother Doug Brown recalled, "It never stopped my dad from having a speedboat; he'd just climb over and get in and go." Dorothy and the boys loved to water-ski. Doug said, "Dave and I used to have these contests to see who could slalom and dip his right shoulder into the water without falling down. You had to lean over far

An early photo of the Brown family—Judge Paul, Dorothy, and brothers Dave and Doug. Photo courtesy of the Brown family.

enough so that you could get your shoulder to touch the water and come back up. Getting close is easy—getting to 'come back up' is hard. I'm sure we both did it at least once, but 99 percent of the time you'd wipe out." Doug said their father never considered his polio a handicap. "He never let it bother him, it was just part of life. I think that's just an inspiration in itself."

The Brown brothers attended McKinley Elementary School and Yorktown High School in Arlington. A suburb of Washington, D.C., Arlington had a population with a lot of people who worked for the federal government. Dave Brown told me, "A lot of my classmates were sons and daughters of army sergeants, senators, generals, admirals—a real smorgasbord of interesting people that were coming and going from the government." Brown said high school was the turning point of his life: "I always started things late. I guess I came out of my shell in high school. I decided on the spur of the moment I was going to go out for football and do gymnastics—normally contradictory sports. I think my interest in science blossomed in high school too."

Dave Brown recalled, "The teacher I remember best from my high school was my gymnastics coach, Jesse Meeks. When I showed up I said I wanted to play football and the same guy was also the gymnastics coach and I wanted to do gymnastics too." Brown noted that his coach didn't complain about him wanting to do such non-complementary activities, "He just said, 'Come on let's go.' There was none of this, 'We've never done this before, I don't know if you'll really be able to help the team as much.' He just said, 'Let's go, welcome aboard.'" Brown played football at Yorktown High when Katie Couric was a member of the cheerleading squad. She was in the class after Brown's and it was several years before she became famous as a news anchor. Brown joked he didn't need to become an astronaut to become famous because when he was in high school, Katie Couric cheered for him!

Brown went to the College of William and Mary, where he was an NCAA gymnast, and after earning his undergraduate biology degree went on to Eastern Virginia Medical School and a medical degree in 1982. Then he joined the Navy.

Brown was a flight surgeon attached to the aircraft carrier USS Carl Vinson, which was cruising in the Bering Sea in winter arctic conditions, for the first time since World War II. Brown had a passion for making videos. In his spare time he went around the carrier filming aircraft landing with his camcorder, doing interviews, and turning the footage into a training film for flight surgeons to prepare air wings and other personnel on cold-weather flight operations. Navy buddy Jeff "Goldy" Goldfinger noted, "Dave was awarded flight surgeon of the year because of that effort." McGraw-Hill used video clips of carrier operations for a video called "Flight Deck," earning Brown his first professional credit as a cameraman.

Goldfinger also recalled a less stellar moment from Brown's Navy career. On a later deployment, the Carl Vinson was in the Indian Ocean. The Navy F-14s would fly on patrol and escort any Soviet aircraft that came within 200 miles of the carrier. It was the height of the Cold War — Soviet and American forces were under strict orders to avoid anything that could escalate into an international incident. But pilots will be pilots, so they would "tease" their adversaries by locking their radars on each other's aircraft. You were allowed to lock your radar on another aircraft, but you couldn't put it into track mode because that would be considered a hostile action.

As a flight surgeon Brown was not required to actually fly in one of the carrier's aircraft, but he loved flying and asked if he could hitch a ride in the back seat of an F-14, the seat normally occupied by the Radar Intercept Officer (pronounced "ree oh"). Goldfinger said, "Everybody said this is probably okay, nothing's going to happen. Dave's in the back seat of the wingman's airplane. The two airplanes take off and all of a sudden we get intelligence information that the Soviets have launched a transport aircraft, and in about 20 minutes it's going to be within range of the carrier." The training mission became an operational flight, where the lead F-14 would use its radar to lock on to the Soviet aircraft and monitor its location.

Goldfinger continued: "Next thing you know, the lead aircraft's radar system takes a dive and goes stupid, so he becomes the wingman. Now Dave's airplane is in the lead, and he's responsible for joining up with the Soviet aircraft. But he's a surgeon; he doesn't know how to operate the radar. So the guy in front is trying to talk him through all of the button smashing. You can imagine the concern in the cockpit and back on the carrier, because they know there's a doctor in the back seat trying to perform an intercept on a Soviet aircraft. No matter how much the pilot tried to talk him through it, Dave never was able to get a good lock on the [Soviet aircraft]." As with other American-Soviet air encounters during the Cold War, it ended without incident, but by the time Brown's plane landed back on the carrier he had earned the brand-new nickname of "No Lock Doc", which stuck with him through the rest of that deployment.

The Navy gives a very small number of flight surgeons jet training so they can have a better appreciation of what goes on in the cockpit from a pilot's perspective. Brown said, "I got a brochure that showed a Navy physician standing on a flight deck next to an F-4 Phantom. I said, 'Boy, I've got to go learn about this.' I applied, and the first time they said, 'No, you're not going to do that.' So I thought, and I said, 'Well, I really *would* like to do this.' So I reapplied, and they said yes. I went as a medical guy off to Navy flight training and ended up flying the Navy A-6." Brown eventually accumulated over 2,700 hours in high-performance jets.

Of course not all of Brown's military career was so dramatic (or embarrassing) as the "No Lock Doc" incident. Dave described a less

Dave Brown in Navy flight training.

Dave in one of his personal aircraft, a Beech Bonanza. Both photos courtesy of the Brown Family.

embarrassing portion to a NASA interviewer: "When I think about one of the times when I was doing things that I just really enjoyed, it was when I was with the Navy in Nevada. I was working at a school there and was getting to fly two different high-performance jets. I lived in kind of a rural area. I'd ride my bicycle to work, 13 miles each way past all these ranches and cows and alfalfa fields. I actually rode my bicycle about 2,500 miles that year. And that was—for a guy that likes to fly airplanes and be outside and do interesting stuff and be around challenging people—that was pretty neat, that four years I spent in the Navy in Nevada."

Goldfinger recalled an amusing incident with Dave's bike. It had pedals that locked on to your feet, and on one occasion Dave couldn't get his feet out of the pedals in time as he slowed for a stop sign—and fell over when the bike came to a halt! When Goldfinger saw Brown riding a bicycle in space for an experiment, he sent Dave an e-mail cautioning him not to fall off. Brown replied, "As my crewmate K.C. would tell me I am continuously falling over up here [because microgravity is actually the shuttle constantly falling around the Earth]."

Brown was realistic enough to realize that his chances for becoming an astronaut were small. Initially he thought becoming an astronaut was too high a goal: "I remember growing up thinking that astronauts and their job was the coolest thing you could possibly do. But I absolutely couldn't identify with the people who were astronauts. I thought they were movie stars. And, I just thought I was kind of a normal kid. So, I couldn't see a path, how a normal kid could ever get to be one of these people that I just couldn't identify with. And so, while I would've said, 'Hey, this is like the coolest thing you could possibly do,' it really wasn't something that I ever thought that I would end up doing. And, it was really kind of much later in life after I'd been in medical school, I'd gone on to become a Navy pilot, that I really thought, 'Well, maybe I would have some skills and background that NASA might be interested in.'" By that point, Brown had four separate things that NASA looks for in its astronaut candidates: military experience, test pilot school, a medical degree, and jet pilot experience. He'd gone way beyond "maybe" and "might be interested in," and in 1996 he was selected as an astronaut. Brown never lost his love of flying airplanes. On the job, he flew NASA's T-38 training jets that the astronauts use for proficiency as well as for transportation to many of their work locations. Brown also loved to fly in his spare time, even if his personal planes weren't high performance. He owned two light aircraft—a Bonanza and a SuperCub. His home was located on Polly Ranch, a small airport near NASA where he could taxi his plane out of his backyard directly on to the runway. Dave would tell people, "I bought a hangar and a house came with it."

Brown had told friends, "When I'm in the T-38 and see Joe-Blow flying around in a Bonanza I think, 'That's just a little general aviation plane; I'm in my jet and I'm something else.' But when I'm in my Bonanza and see the T-38 I think, 'It's really cool to have your own airplane and be in general aviation and not be in that high-speed airplane—I can go anywhere in this little airplane.' It's great from both sides."

Brown's neighbor Al Saylor is a professional mortgage broker and skydiver. Saylor was flying in his Cessna 206 with another neighbor, Jeff Kling, and as Saylor tells it, "As we're climbing out of Houston going through about 4,500 feet, we heard a T-38 aircraft come off of Ellington Field with a familiar voice and a NASA call sign. Both Jeff and I immediately recognized Dave's voice and we started to grin." Kling was informed by air traffic control that there was a T-38 about six miles behind them. Saylor continued, "Dave recognized our call sign immediately and radioed 'Traffic in sight.' Just about that time we turned over our shoulders and saw the T-38 climbing at a very high rate of speed. And as Dave came right by us he radioed 'Hello boys' and did two 360-degree rolls as he went by. We were just thrilled that we got a private air show." Saylor notes that they were over the water and it was perfectly safe: "Dave was a safety-conscious guy. It wasn't thrilling because it was close—it was thrilling because it was our great friend and good buddy."

Brown also was quick to be neighborly. Kling recalled, "One time Dave came by just prior to tropical storm Allison. He knocked on the door and said, 'I need some help. Steve Robinson is out of town in Russia. We need to put up his furniture so if his house floods he won't lose everything.' So I said, 'Good enough, let's go be a work crew' and I thought to myself 'Here's Dave looking out for other people all the time.'" Many friends recalled that Brown was always more concerned about how they were doing or what was happening in their lives.

Kling recalled a neighborhood event: "We'd have these neighborhood fly-outs where we'd invite neighbors who didn't have airplanes and fill our empty seats and go out to dinner or something. We all flew down to Moody Gardens in Galveston to see the IMAX "Space Station" movie.

We flew all of our neighbors including Dave and [astronaut] Leroy Chiao." Chiao was one of the "stars" of the movie, which was partially filmed on his STS-92 mission in October 2000. The group was teasing him about it and encouraging him to wear dark glasses like a movie star so he wouldn't be recognized! Kling said, "Dave was saying 'I can't wait, this looks like so much fun,' and Leroy said, 'Actually this looks very real, this is as close as it gets. Without the zero-G, you can't tell the difference. They've done a real good job.'" Kling recalls, "It was kind of funny going down there with astronauts who had been in space and were about to go in space. They're just like the rest of us, thinking 'This is a real cool thing.'" Kling worked in Mission Control and was one of the flight controllers for Brown's mission, including the reentry. He recalled, "It was really cool doing all of the training and having him there." Kling remembered the debriefing from an extremely long simulation involving the entire team. Brown was using his camcorder to film the entire meeting. Everybody else attending the debriefing had technical inputs, Kling recalled, "Dave's asked if he has any comments and he says, 'I'm just really glad to be here and I'm ready to go fly.' Kling said, "That is so Dave—of all the things he could have said."

What all of Dave's friends and acquaintances will tell you is it was never about Dave, it was always about you. He was an extremely humble person, never one to brag about his fame or job. Kling said, "He was the most modest overachiever you'd ever meet." Neighbor Cindy Swindells recalls it was six months after she met Dave before she realized he was an astronaut. When he moved in and got acquainted with his neighbors he just told them he worked at NASA. But one day she saw him wearing his blue flight suit and it suddenly dawned on her he was more than just another typical NASA worker!

The souvenirs Brown flew on STS-107 included: a T-shirt for his elementary school, flag for his high school, and banner for his medical school. Additional items included a cloth patch for the Johnson Space Center Astronomical Society, a red-white-and-blue scarf from the base where he did his jet training, a paper resolution from the Guam legislature, a T-shirt for an elementary school in Colorado, and a flag for a school in Connecticut. Brown also offered to carry one monogrammed tea towel for a school in England. But the school sent two towels. Rather than disappoint the school by flying only one, Brown asked a friend to sew the two together to make them look like a single larger towel,.

Brown was the only single member of the crew, but was interested in settling down and getting married. At the time of the accident he was dating Janneke Gisolf, a graduate student in Amsterdam who was part of the STS-107 ARMS (Advanced Respiratory Monitoring System—pronounced "arms") experiment. His brother Doug said Dave was really serious about her: "When Dave came home for the holidays he put her picture on the dining room table where everybody could see it. They were discussing baby names." Dave was rather unusual for a single man—he got along well with his ex-girlfriends and invited many of them to Florida for his launch. They were at his pre-launch reception, but Dave couldn't be there because he was in quarantine to prevent him from catching a cold the before launch. He was only dating Janneke, but many folks wondered why so many of his girlfriends were there or at his memorials.

Without a family to support, Brown spent much of his income on electronics and other technical gizmos. He would refer them as BTUs–Basic Toy Units. When asked how much a particular item cost he might say, "Oh, about three BTUs." A BTU was the equivalent of $1,000. Friends remember he had a wireless Internet connection well before they became commonplace. Many of Dave's BTUs were video cameras and Macintosh computers. He was documenting all of his training and planned on making a movie about his mission to explain what made spaceflight so special and what drove astronauts to select that career.

Dave did, though, have a family of sorts–a Labrador named "Duggins," which he'd gotten as a puppy. Doug Brown explained, "Duggins flew with him in Dave's plane up here when Dave visited. Duggins was Dave's buddy." Duggins was originally co-owned by Dave and a Navy roommate, but when their assignments took them in different directions, Dave got Duggins. Because of Dave's many travels for his work he left Duggins with neighbor Cindy Swindells and her teenage son and daughter. Doug explained, "The neighbors were more than dog sitters. They did it so often Duggins became part of their house too."

Cindy Swindells said, "We got just as much fun out of watching Duggins as Dave did." As a thank you for taking care of Duggins, Dave would bring the Swindells presents from wherever he traveled. After returning from one trip Dave called to say "I've got a really good surprise—it's alive." Swindells noted, "When you tell that to kids they immediately think 'He's got us a puppy!' and everybody's excited and

Dave Brown and his best friend—Duggins. Photo courtesy of the Brown family.

waiting for him as he pulls up the driveway." She recalled, "Dave goes to the trunk and the kids are a little concerned because they knew Dave wouldn't put a puppy in a trunk and he pulls this box out of the trunk and it turns out its these two live lobsters. My kids were so mad. My daughter said, 'It's Sebastian' [from "The Little Mermaid"]. Dave's standing there realizing, 'Oh no, what have I done?'"

Cindy added, "I looked at my son and he's really ticked off that it wasn't a dog." She explained to her kids that they could either cook the lobsters or set them free in the creek. They compromised—they cooked the son's lobster and set the daughter's lobster free! That was the last time Dave Brown tried to bring back live presents for some time. Cindy said, "He always brought us back special gifts. When he went to Key West, we got key lime pie. When he went to Las Vegas he brought us poker chips. We got chocolates and live tulip bulbs from Holland. We have this huge collection of great gifts."

I met Dave Brown at a party in Florida, where he was working at KSC as an "Astronaut Support Personnel", one of the astronauts who helps other astronauts to prepare for their missions. He supported shuttle launches from May 1999 to September 2000. The first flight on which he helped the crew into the shuttle was STS-96, Rick Husband's first spaceflight. Brown also assisted the STS-93, STS-103, STS-101, and STS-106 crews. Dave was anxious to get his own flight assignment, and it showed. When I discovered that he was assigned to STS-107, he couldn't keep his delight hidden—the smile on his face said everything. But when I pushed him on the mission's details, he asked me to wait until the official announcement.

Saylor said Brown told him about an incident in Florida. Brown was in an elevator at the launch pad. This was no ordinary elevator in a building but a mesh cage, and it was stuck. Saylor said that Brown told him with a grin, "I'll tell you. I looked around the elevator and thought, 'Who's an asset and who's a liability?' " Brown was already thinking about who had the skills and emotional stability to help him get the rest of the people to safety.

Like McCool, Brown was on the "blue shift," which would go to sleep just a couple of hours after launch. He acknowledged how difficult this would be on what would almost certainly be the most thrilling day of his life: "Well, I think that probably my hardest job on this whole flight is going to be going to sleep the first night. The blue shift, which is three of us, we will launch actually in the 'afternoon' and 'evening' of our day. So, about four hours after we get to space, it's nighttime for us and we need to go to sleep. So, even though I will have had a long day, I think it's probably going to be pretty tough to go climb into bed and go to sleep. That's probably going to be the toughest thing I have to do the whole mission."

CHAPTER 5

K A L P A N A   C H A W L A :
T H E   G I R L
F R O M   I N D I A

Kalpana Chawla was a naturalized U.S. citizen, born in India. Her name is pronounced CULL-puh-na CHAV-la, which is rather difficult for most Americans, so she went by her nickname, "K.C." Her name in Hindi can be interpreted as "a dream, fantasy, something that is almost impossible to attain"—rather appropriate for a space woman who started off life in an environment where teachers told her it was impossible for a woman to become an engineer.

Chawla spoke English with a beautiful lilting accent. She was born in Karnal, fairly close to the Pakistan border. She told me, "The time I was in India, we didn't have riots in the areas where we lived, even though we lived very, very close to Pakistan, only 140 miles from the border. There are some pockets in India where these things happen, where the two sides are brewing. I do remember the 1972 war, the times during the two wars when I was there. Going in the ditches, which were not far from our house, any time the siren went off. Until it was clear the warplanes were gone and could not spot the city, because we were so close to the border. Those were, when you look at history, very short wars. The other good thing, if anything good can be said about war, was things were decided and then you were normal again."

Chawla didn't set out to become an astronaut; she wanted to design airplanes because she was interested in flying. She said, "I don't know why I always liked aerospace engineering. I was in tenth grade when I figured out what I wanted to do. My dad had a rubber tire manufacturing plant and he said, 'Oh, you should be a doctor or a teacher, those are much more respectable professions; why do you want to do this, mostly guys want to do it.' But I said this is what I really want to do. By the time I was in twelfth grade, I applied to engineering schools, and my mom volunteered to go with me for the interviews. The professors said, 'Why do you want to do aerospace engineering—electronics is more ladylike.' So again, the same arguments, 'This is what I really want to do.' While going to engineering schools, I read Kelly Johnson's book on the Skunkworks [the high-tech secret laboratory responsible for the U-2, SR-71, F-104, and F-117 aircraft], and that's what I really wanted to pursue—working on high-tech airplanes."

Indian universities didn't have the specialized courses required for aerospace studies, so K.C. examined overseas schools. In America, schools were not as dismissive about women in aerospace as the professors she had encountered in India. She received advanced degrees from the University of Texas and University of Colorado that eventually led to a job at NASA's Ames Research Center, near San Jose, California. Ames is one of NASA's aeronautical centers, concentrating more on aircraft and supercomputers than spacecraft. Chawla applied to become an astronaut and was selected in 1994. She said, "While doing my doctorate I figured, 'I'm going to try for the space program.' All of us know it's very fortunate for us to make it here. For me it's really amazing to have talked about it and made it—almost like winning the lottery or something." Among her classmates were astronauts Rick Husband and Mike Anderson, who would fly with her on her final spaceflight.

Chawla became a naturalized US citizen in 1990, but many in India still consider her a hero of India. They almost always ignore that she had to give up her Indian citizenship and become a US citizen before she could even apply to become an astronaut. Chawla said, "I think for a lot of people there's confusion," adding that before her first mission, people would call her mother and ask if they could talk to Kalpana please, because they thought she was still in New Delhi. She understands why people in India still feel a connection with her: "It's understandable. I was born there; I grew up there for 20 years. If some people feel possession as a result of that, it is not anything unexpected." But she was always quick to explain to people, "I am part of the US astronaut corps."

K.C.'s first flight was a 16-day microgravity mission, also on the shuttle Columbia. STS-87 was launched on November 19, 1997, at 2:46 P.M.,

NASA's 1995 astronaut class, with (L to R) Mike Anderson, Rick Husband, and Kalpana Chawla.

making her the first professional astronaut of Indian origin. Rakesh Sharma was the first Indian in space, but only flew as a token passenger on a Soviet spaceflight in 1984. Unlike Sharma, Chawla was a fully qualified career astronaut, with important responsibilities. Chawla's crewmates were commander Kevin Kregel, pilot Steve Lindsey, Winston Scott, Japanese astronaut Takao Doi, and Ukrainian payload specialist Leonid Kadynuk. Officially, Kadynuk was flying as a VIP passenger to operate some plant growth experiments, but the real reason was that the US offered the Ukraine a seat on the shuttle to improve international relations with this former Soviet republic. (There were rumors that the United States agreed to give the Ukraine a seat on the shuttle as part of a deal where the Ukraine gave up the nuclear weapons it had inherited when the Soviet Union broke up.)

On her first flight, Chawla said, she would be bringing along educational and personal souvenirs: "I'm taking items from my high school [Tagore School in Karnal], a T-shirt with the school logo and the college logo from the engineering college [Punjab Engineering College]." In addition, she carried pendants and small pieces of jewelry and an early photo of herself with her husband.

Before her first mission, she said, "I never thought, while pursuing my studies or doing anything else, about being a woman or a person from a small city or a different country. I pretty much had my dreams like anybody else, and I followed them. Fortunately, some people around me were always encouraging; they would say, 'If that's what you want to do, carry on.'"

Chawla was a Hindu, and her family was proud of the fact they were fourth-generation vegetarians. But K.C. decided to marry an American who ate meat. Jean-Pierre "J.P." Harrison said he tried to eat vegetarian when he was with her, but admitted it was difficult for him to give up meat. Jean-Pierre is a freelance flying instructor, and K.C. was his student. The two met shortly after K.C. came to the United States. He was incredibly proud and excited to watch his wife of 14 years launch into

Kalpana Chawla with her husband, Jean-Pierre Harrison, a couple of days before her first spaceflight.

space. He said she told him, "Stay out of trouble," when they parted on the night before launch.

Many of her family members traveled from India to the United States to see Chawla's first launch. Her mother, Banarsi Lal Chawla, said she had always told K.C. to do what she wanted to do, and told me, "The launch was amazing." Her brother Sanjay found it "very exciting." Sister Sunita said, "K.C. is the happiest person on this Earth." About seven hours after the launch, Columbia passed over Karnal, India, where Chawla grew up.

Chawla chose her vegetarian selections from NASA's large variety of menu items, with many rice and beans dishes. For her first meal in space, Chawla chose Rice Pilaf, Broccoli au Gratin, Tortillas, Strawberries, Dried Apricots, Trail Mix, Vanilla Instant Breakfast, and a Strawberry drink.

Her responsibilities on her first spaceflight included operating the shuttle's 50-foot-long robot arm to deploy and retrieve the free-flying Spartan satellite. She also orchestrated a six-hour spacewalk by two of the other crewmembers, as she explained before her flight: "I am inside keeping them on the timeline, and tell them where to go next." In addition, she was the primary astronaut for several important microgravity experiments which wouldn't be possible on Earth.

The key objective for the Spartan astronomical satellite was to observe the Sun at the same time as the NASA/European spacecraft SOHO (Solar and Heliospheric Observatory—pronounced "so hoe"). SOHO had a problem shortly before the shuttle launched, so NASA decided to delay Spartan's release for 24 hours while SOHO's engineers fixed the problems.

The 3,000-pound Spartan is one of NASA's least expensive satellites. Normally it has no way of communicating with the shuttle. It's a completely self-contained system operating on batteries for its two days of free flight. The shuttle drops Spartan off, then flies away and performs other tasks for two days before returning and using the robot arm to capture the satellite and put it back inside the cargo bay for the trip home. Spartan has a couple of versions with different instruments—some study the Sun, others study the stars, others are technology platforms. STS-87 marked the fourth flight of this particular version, and on all of the previous flights everything had gone properly.

At 4:04 P.M. on Friday, November 21, 1997, Chawla raised Spartan above Columbia's cargo bay with the robot arm and set it free. When Spartan is released by the shuttle's robot arm, it activates its thrusters and rotates itself. This indicates to the astronauts and flight controllers that the satellite is functioning properly. But this time it just sat in space without rotating. Experts on the ground quickly determined that Spartan had failed to receive a crucial computer command before it was set loose, either because of software glitches or crew error.

Most of an astronaut's training is actually preparation for when things go wrong. If there was no indication Spartan was working, the robot arm operator was supposed to re-grab the satellite with the robot arm. The robot arm grabbing the satellite would flip a switch that hopefully would activate Spartan.

Chawla reached out with the shuttle's arm, using a video camera on the arm to align it with the target on Spartan. If that had worked, then it would have been possible to reset the spacecraft and resume the normal mission. But she unintentionally hit Spartan with the arm, and it went into an uncontrolled roll. During the flight, K.C. told me, "It all seemed to go real fast. After I moved the arm back [and Spartan was spinning], I thought maybe Spartan was doing the maneuver it was supposed to do earlier. That was my immediate thought."

Commander Kevin Kregel flew Columbia around Spartan in an attempt to put it in a position where Chawla could try to re-grab the wayward satellite. But after an hour of trying and using precious propellant, Mission Control told Kregel to stop so he could save enough propellant for a later rendezvous attempt.

The crew quickly determined what had happened. They reviewed their procedures step-by-step, and figured out that K.C. had missed a step: Spartan was never turned on. And just as important—nobody else noticed. The way the software was written, there was no way of informing the crew that something had been missed. While training for STS-107 Chawla recalled, "We had figured out very early on that's probably what happened. We also knew that [with that] software, there's no insight [any indications that the commands had been properly entered into the computer]." Commander Kregel got his crew together and told them that while it was regrettable, a mistake had been made, the mistake was in the past, and there was nothing they could do to change that. The crew had to concentrate on the tasks remaining for the rest of the mission.

Kalpana Chawla operating the shuttle's robot arm on STS-87.

Mission Control devised a plan in which two of the astronauts, Winston Scott and Takao Doi of Japan, would manually capture the possibly tumbling Spartan during a spacewalk. The pair had been trained to capture Spartan manually in case of a problem with the robot arm.

For this "contingency" spacewalk Chawla had the same responsibilities as for the planned spacewalk, helping the spacewalkers in and out of their spacesuits and keeping them on schedule. After the Spartan rescue, the spacewalkers would try to finish as many originally planned tasks as possible.

With Chawla guiding them from inside Columbia's crew cabin, Scott and Doi grabbed the satellite with their spacesuit gloves on Monday, November 24. The two spacewalkers had problems aligning the satellite properly in its latches. Mission Control asked Chawla to step in and use the robot arm again. While the spacewalkers held Spartan steady she used the arm to grab the satellite. With the robot arm firmly grasping Spartan, the two spacewalkers could let go and get out of the way. Chawla gently pushed Spartan back in its latches for the trip home.

It was a bittersweet recovery for the crew. They were happy to rescue the $6 million satellite, but unfortunately it was coming home without any scientific data on its tape recorders. After considering relaunching Spartan for an abbreviated flight later in the mission, NASA decided against the plan on December 1 because there wasn't enough propellant for another rendezvous.

The Spartan spacecraft also had a couple of secondary experiments. The Video Guidance Sensor was designed so automated spacecraft could perform dockings in the future. While K.C. moved Spartan around the payload bay on a robot arm, one of the other astronauts aimed a laser toward Spartan to test the docking function.

During the on-orbit press conference, K.C. was asked several questions about Spartan. I asked her if she felt like Vasily Tsibliev, the former commander of the Mir space station, when a collision and fire took place and was blamed for what had happened. She laughed and said, "I don't know what that first name means, what you said there. And I didn't know everybody was pointing fingers at me until now. So there."

The STS-87 mission overlapped Thanksgiving, and NASA has a space turkey dinner. But it isn't anything exotic or special—just an ordinary $2.50 Dinty Moore no-refrigeration version. The problem was the six-person crew included Ukrainian and Japanese crewmembers who didn't celebrate Thanksgiving in their countries, and K.C. was an Indian-American vegetarian, so only half of the crew were Americans who would celebrate Thanksgiving with turkey. Six turkey meals were put on board, and obviously K.C. did not eat hers, but she did have Pumpkin Pie and Cranberry Sauce. During the mission I asked her if she enjoyed her beans and rice for Thanksgiving and she said, "It was very good. You get very used to the foods you eat, so you don't really know what you are missing out, I guess."

Later in the mission, Chawla was asked to speak to Indian Prime Minister I.K. Gujral. He repeatedly told Chawla how proud the entire country was of her, ignoring the fact she was a naturalized US citizen and India had nothing to do with her mission. Gujral talked about her mission cementing the bonds between the United States and India—again ignoring that she had to give up her Indian citizenship to even apply to become an astronaut. He made K.C. promise to take her entire crew and visit India after the mission.

K.C. did put in a request, through the Indian embassy, to take the whole crew on a visit the land of her birth, but the trip never took place. The Indian media was full of speculation as to why the STS-87 crew visit didn't happen, and they never discovered the real reason. K.C. told me in an interview in December 2001: "There were nuclear tests [May 11, 1998] and the relations [between the USA and India] were very cold, and I concluded it was better to just wait. I hope after this mission [STS-107] we can go as a crew."

For the rest of the 16-day STS-87 mission, K.C. used a laptop computer to operate the microgravity payloads from inside Columbia's crew cabin. Columbia landed at KSC on December 5.

The Spartan "Close Call Investigation Board" quickly verified that K.C. had missed a step in the procedures to turn on Spartan, and that none of her crewmates noticed the omission. It stated, "The Board's conclusion based on all the evidence available is the crew inadvertently omitted the Spartan Standby step." Everybody realized that Chawla had made a mistake—but it was a human mistake anybody could have made. But just as important was the software's design. Its programmers did not include any warnings to the crew that they were about to deploy the satellite without completing its activation sequence, nor did the circuitry that connected Spartan to the shuttle include any feedback mechanism. This was intentional in Spartan's low-cost design. Originally Spartan wasn't even controlled by a computer program—just a set of switches. Astro-

The STS-87 crew in space.
Kalpana Chawla is in the middle in
the back row.

nauts had made mistakes using that switchbox in the past, although not
with such embarrassing results, viewable by the outside world. When the
engineers decided to replace the switchbox with a computer program,
the program duplicated the functions of the switches but did not improve
the user interface. (Similar "user-unfriendly" setups are used to control
dozens of other shuttle payloads.)

The failure board recommended several changes including more
emphasis in crew training for when things go wrong, better cockpit
resource management in which the crew works together as a team, and
improvements to Spartan's computer software and instructions. A copy of
the Executive Summary of the report is on this book's website.

Chawla has always acknowledged her personal mistakes in the Spartan incident. Insiders note that it was her honesty and candor in acknowledging them that saved her career. She told me, "Something like that you
don't forget. Even though people who looked at [the software] said before
the next flight you should definitely make those changes [to the software
and procedures]. So even though they came up with a lot of recommendations you always feel eternally guilty. I think you can live ten lifetimes
after that and still feel the burden of something like that."

Some Indian organizations have tried to insist that it wasn't Chawla's
fault, and sloppy news reporters blamed her when unclear instructions
were really responsible for the Spartan failure. The same instructions were
used for Spartan on each of its seven previous flights, and while the instructions were not as clear as they could have been, that didn't excuse Chawla.

Before Spartan flew again, crewmate Steve Lindsey said, "I think the
[Spartan] investigation was done very well; they looked at every aspect,
and I think everybody learned a whole bunch from it. We've taken all of
the lessons learned from that investigation to heart and made some software changes, some hardware changes, and some training changes.
Everybody's learned a lot from it, and we're applying those lessons."

After STS-87, Chawla was given a technical assignment, like every
other astronaut. She worked with shuttle payloads, passing on recommendations to the engineers developing the payloads from an astronaut's perspective and giving them advice on how to make things better for the
astronauts to use in space. At various points Chawla, Clark, and Ramon all
worked with shuttle and space station payloads. One of K.C.'s payloads
was the ham radio station on board ISS, used by astronauts to contact
schools. Some of the team members noted she was extremely interested in
every aspect of the payload, even though she didn't consider herself a ham
radio operator and it wasn't one of her personal interests. And all the
while, many wondered if she would remain a support astronaut on the
ground indefinitely because of the Spartan incident on STS-87.

Spartan was reflown in 1998, on the mission which featured the
return to space for legendary astronaut John Glenn. Scientists talking

about the results from the mission were enthusiastic about Glenn's presence on their flight, even though he had absolutely nothing to do with Spartan. The scientists did not bother to mention the four astronauts who were responsible for controlling Spartan and making sure it returned its science—Curt Brown, Steve Lindsey, Scott Parazynski, and Steve Robinson.

Did K.C. believe that the Spartan incident would hurt her career? She said, "I think for me my attitude is not to think about something like that, just keep on trying to do your best. And whatever happens, happens." Bottom line—her bosses trusted her enough to give her another flight assignment. (Several of Chawla's classmates never got a second spaceflight assignment even though they didn't make any mistakes on their first flights that were visible to the public.) Managers assigned Chawla to STS-107 in July 2000.

Because of her death, many people want to ignore K.C.'s mistakes or pretend they never happened. Spartan only showed that K.C. was human, and wasn't infallible. After her death, many in Mission Control said their main regret was any resentment they held against K.C. because of Spartan, and how they treated her, and they were sorry they didn't treat her better.

Many of K.C.'s training team and colleagues talked about her intensity as she trained for STS-107, and how she paid attention to every little detail. Some wondered if her passion for learning everything she could about her payloads and other responsibilities was her own way of making sure that she would have a perfect performance on STS-107 to make up for what happened with Spartan.

Chawla talked about the anticipation before flying again: "I'm looking forward to this flight, of course, and part of the reason is having flown once before. One of my colleagues used to say, 'After you've gone to space once, you sort of get addicted and you want to have the same experience again.' And that's precisely what I feel, especially the part of look-

Kalpana Chawla studies procedures while training for STS-107.

ing out—looking at the Earth, looking at the stars at night, just looking at our planet roll by and the speed which it goes by and the awe it inspires. Just so many such good thoughts come to your mind when you see all that. Doing it again is like living a dream—a good dream—once again. Because of the delays, the nervousness has lessened quite a bit because now we're just waiting for launch day. So in some sense there's more room to enjoy the excitement that—oh, yes—this is coming."

K.C.'s passion was flying, especially acrobatics in tail-draggers, small planes where the middle landing gear is at the tail of the plane, and spent much of her free time flying acrobatics with her husband.

K.C. and fellow astronaut Sunita Williams flew Dave Brown's SuperCub from Maine to Texas for him. Friend Jeff Kling said, "It's not an incredibly fast airplane. It's a long time to ride in that little tail-dragger, but they did it because they loved flying. If you love to fly it's an adventure. K.C. absolutely loved flying."

KC and Dave take a moment to relax in a garden in the Netherlands. Photo courtesy of the Brown family.

CHAPTER 6

MIKE ANDERSON:
THE KID WHO
WANTED TO FLY

Mike Anderson was one of NASA's few African-American astronauts. Friends and family recall Anderson decided he wanted to become an astronaut at a very early age—and everything he did was aimed at achieving that goal.

Anderson was born in upstate Plattsburgh, New York, and grew up in Cheney, Washington, a primarily white suburb of Spokane. Keith Flamer, another black kid, met Anderson in junior high. Flamer recalled they'd be walking around talking about girls and music, and Anderson's favorite group was the Jackson 5.

Anderson told me, "I think it's one of the most beautiful parts of this country. It's absolutely gorgeous up there—great forests and mountains and just a very nice place to live. Like most kids growing up, I had very wide interests. I was interested in everything and I tried to take advantage of everything—everything from the sciences to music to writing to literature. But as I got older, my interests tended to become a little bit more narrow, and I found that science was something that really caught my attention; it was something I really could sink my teeth into. I was a big fan of physics and chemistry. I took everything my high school had to offer in those areas. Sports-wise I was a pretty small guy, so I stopped playing football by the seventh grade. The only sport I played in high school, really, was tennis." Fellow students recall that Anderson really studied hard and was a good student. Mike McKinley remembered, "We had calculus. If it hadn't been for him, I probably wouldn't have gotten through it. He was always willing to help me."

Anderson said, "I had a chance to go back to Cheney about three years ago. I was amazed that many of the teachers were still there. Some had retired, but they were still in the community, and I had the chance to visit with some of them. One teacher that sticks out in my mind was Mr. Vankuren, who was my math teacher for both advanced algebra and calculus. Just a tremendous guy, I really admired him. He had a tremendous work ethic when it came to the mathematics courses and he was really demanding of his students. And it paid off. I know, because when I went to the University of Washington, my math skills were really sharp, which helped make that whole adventure much more pleasant." When

Anderson visited Cheney, it was math teacher Lawson Vankuren who escorted him around to different schools and to a luncheon in his honor. Vankuren recalled, "It was really nice, several of his classmates really helped make him feel welcomed."

Anderson grew up in the 1970s, when many African-Americans styled their hair in large afros. Vankuren recalled, "Mike was wearing a great big afro in his yearbook photo." Anderson was one of only three black students in a class of 200. McKinley recalls, "I didn't know there were any [racial] problems. I never saw it." But Flamer had the same perspective as Anderson, growing up as a minority student in an overwhelmingly white school, and remembered things a little differently: "Mike was genuinely a very nice person, so he sought the best out of everybody. But he knew some of the things that were happening. Mike was sometimes surprised at the level of racism because it didn't happen to him as overtly. It was mostly the name-calling or joke-telling. Mike always had this ability to walk away from it, and take the higher ground: 'You know, there's nothing I can do about them, I'm just going to go about doing my business.' McKinley noted, "Mike would just walk away, but Keith wouldn't." Flamer acknowledged, "Mike was pretty well grounded—he didn't have to confront everything. It wasn't just from the kids, but also from the administration. When I was finishing my college applications, counselors would say, 'Learn a trade instead of pursuing college,' even though I had the grades and wanted to go. I think Mike overcame that because he knew exactly what he wanted to do since he was age five. To fly, number one, and two, to become an astronaut—that's all he wanted to do."

Anderson grew up on Fairchild AFB. He recalled, "My dad was in the Air Force. And being an Air Force brat and living on Air Force bases, I was always around airplanes. That was something else that really captured my imagination, just seeing airplanes taking off and landing every day, and flying over the house, and making all of this noise just was a fascinating thing to me as a kid. So, my interest in aviation and my interest

Mike Anderson's 1997 Cheney High
School yearbook photo. Photo courtesy of classmate Mike McKinley.

in science were, I guess, two of the things I really latched on to, and two things that I just couldn't shake as I grew older. So one day, just sitting down and thinking about it, I said to myself, 'How can I combine my two strongest interests—my interest in science and my interest in aviation?' At that time, we were going to the Moon and doing some really fantastic things with the space program, and to me that was just the best combination of the two. Here you have these men who are scientists, engineers, and they're also flying these wonderful airplanes and these great spaceships, and they're going places. And to me, that just seemed like the perfect mix and the perfect job. So very early on, I just thought being an astronaut would be a fantastic thing to do."

Anderson spent most of his time studying math and science and took every precaution to ensure that he didn't have any injuries that could affect his goal. McKinley said, "He'd wear goggles while mowing the lawn to protect his eyes—he didn't want *anything* to wreck his eyesight."

Anderson went on to the University of Washington, where he got his undergraduate degree in physics/astronomy in 1981. He said, "I picked physics because out of all the different scientific fields, I think physics is probably the broadest. It covers basically everything. It allows you to really take your interest and point it in any direction you'd like to point it in. So, I went to the University of Washington as a physics and astronomy major and just had a marvelous time. I found it very challenging, very rewarding."

Anderson always wanted to fly airplanes. He said, "If you're going to fly airplanes, the best place to be is the Air Force. So, I went through the ROTC program, and they provided me with a scholarship to help me pay for college. And after I graduated, I took a commission as a Second Lieutenant and came into the Air Force. Through my first four years, [I was] actually in the field of communications—communications and computers. But my real interest was flying airplanes. So, after four years of doing electronics, I put in my application for Flight School, got selected, and off I went. Afterwards, I was flying in the Air Force and

Upper-level physics student Mike Anderson (far right) poses in this 1980 photo with fellow students Michelle Brasseur Furman, '82; Cheri Moon, '81, and Charles Robertson, '61, '63, '82. Photo by University of Washington physics professor Mark McDermott.

enjoying that a great deal, but I realized I really needed to improve myself a little bit more academically. So I went back to school and picked up a master's degree in physics from Creighton University, in Nebraska."

Anderson became a pilot of large jets, including the KC-135, the military version of the Boeing 707 airliner. One of the most important planes he flew was the Strategic Air Command's airborne command post, code-named "Looking Glass," one of the ultimate deterrents to nuclear war. The Soviet Union or China could aim ICBMs at every command post in the USA. But it would be far harder to find and target an airborne command post, especially if it was in an aircraft that had the same dimensions, radar signature, and contrails as a civilian airliner. "Looking Glass" was filled with electronics and radios, capable of communicating with the White House, missile silos, and submarines with nuclear missiles. If it ever became necessary, the White House would have given "Looking Glass" the commands to wage World War III.

By the time Anderson was assigned to the Strategic Air Command, the Cold War was over. The world situation had calmed down, and the "Looking Glass" fleet no longer needed to fly 24 hours a day. But the crews continued to train and remained on constant alert, ready to take off at a moment's notice, and in any serious crisis situation, they were ordered to go airborne.

Anderson noted, "One day I said, 'Well, you know, I've been flying airplanes here in the Air Force for quite some time now, and I have a record there. And I studied science in school. And I'm really ready to put together a package and send it off to NASA and see what they think.' Fortunately, I got called down for an interview. One thing led to the next, and one day I got that call." The invitation came in 1994, and Anderson became a member of the 1995 astronaut class along with Rick Husband and Kalpana Chawla.

After finishing astronaut candidate school, Anderson was assigned to support preparations for shuttle launches in Florida. Since KSC is near Cape Canaveral and is often called simply "the Cape," astronauts assigned to Florida are nicknamed "Cape Crusaders." NASA announced in March 1997 that Anderson was assigned to his first spaceflight, the STS-89 mission to visit the Russian space station Mir.

Unlike the pilot astronauts at NASA, who are almost all former fighter pilots used to flying one- or two-person aircraft, Anderson had his experience in heavy aircraft with multi-person crews. Many have felt the multi-person cockpit crew on a larger plane is more like a shuttle cockpit crew than the typical one- or two-person fighter crew. That may be the case, but NASA still almost always selects test pilots with fighter experience for its pilot astronauts. Astronaut Terry Wilcutt, Anderson's commander on his first spaceflight, noted that Anderson was using the term "cockpit resource management" well before it became a buzzword in the shuttle program. Wilcutt said that Anderson, with his big-plane experience, understood the necessity for a crew to work together more as a team than as individuals.

I was introduced to Mike Anderson in November 1997 at a party before his first spaceflight. In a group where most people were drinking beers, Mike had a soft drink. As a devout Christian, he didn't drink alcohol, but he didn't try to preach his feelings or push his views upon others. One friend remarked, "Mike wouldn't tell you how you should act. He felt the best thing he could do was set a good example." Mike and I talked at that party about—of all things—how you could smuggle somebody into space. Would it be possible for a stowaway to hide on board the shuttle in some compartment ahead of time with enough food and water until the shuttle reached orbit? And how many people would it take to help the stowaway? Just idle speculation, but it made for an interesting evening.

About a month later, Terry Wilcutt invited me to sit in on his crew's final set of entry simulations for STS-89. NASA uses an extremely sophisticated flight simulator to train the astronauts for launch and landing. It's hydraulically operated and does an extremely good job simulating the most dynamic parts of a spaceflight. Wilcutt, pilot Joe Edwards, and mission specialists Bonnie Dunbar and Anderson were so intent on their tasks that they didn't notice when I poked my head in to see what they were doing. I went over to the instructor's station, where I was given an earphone to monitor what was happening.

Anderson's STS-89 mission was the next-to-last US space shuttle to visit Russia's Mir space station. The mission would bring US astronaut Andy Thomas up to Mir and return astronaut Dave Wolf to Earth. In addition, the crew was supposed to bring a couple of tons of supplies to the station. The other crewmembers were James "J.R." Reilly and Russian cosmonaut Salizhan Sharipov. Anderson, Edwards, Reilly, and Sharipov were the rookies on the mission, the Americans all members of the 1995 astronaut class.

Sharipov was assigned to the mission less than five months before the launch. He had some English skills, enough to get by when a translator wasn't available. He later recalled the many times that Anderson helped him to adjust to life in Houston: "Mike was very helpful to me. He taught me a lot. I came [to the USA] just seven months before STS-89. My English was not so good. Getting things was very difficult. Mike came to my home with things. He brought to my apartment food, vacuum cleaner, dishes, a cell phone and many things I took from Mike's home."

Sharipov noted that speaking a common language isn't always necessary. He said, "My wife doesn't speak English at all. Michael's wife doesn't speak Russian. They got along with each other without any problems—it was great to see they could understand each other. Sometimes language is not necessary to communicate."

As the flight engineer, Anderson's primary job on STS-89 was to assist Wilcutt and Edwards during the launch and landing. Anderson's seat on the flight deck gave him a good view of the instruments and displays.

STS-89 launched on January 22, 1998, at 9:48 P.M. It was a spectacular night launch, with the shuttle visible for several minutes as a bright star-like object until it was lost in the haze. Two days later, Wilcutt docked Endeavour to the Mir space station, with Anderson assisting as his navigator. Anderson recalled, "I think all flights that involve a rendezvous and a docking are very busy flights. As a first-time space flier, everything you do on a flight is just miraculous. You just can't believe it's actually happening. And for me, still to this day, when I think back to that flight, it's sort of like a dream. You just can't believe that actually happened. When we first saw the Mir space station, it looked like a star out there in the sky. But as we continued to do our burns and we continued to get closer to the space station, it started to get bigger and bigger. And it wasn't long before you had this huge, massive complex, this huge space station, just kind of taking up the entire window of the shuttle. As you've looked out there, you've kind of marveled at it. You were just in awe as to what was out there. It was just a tremendous experience to have a chance to do something like that. I think as long as I live, I'll never forget those moments. And, it was a truly miraculous time, and just a wonderful flight."

Joe Edwards recalled a time when he and Anderson were on the flight deck watching the Earth go by. In some ways it looked and felt very similar to what you experience flying in an ordinary aircraft, as Edwards later described it: "You could see the stark clouds and deep blue waters. We were about 225 miles high. As I floated up there and brushed up against him, Mike said, 'You know Joe, I really don't feel like we're that high,' and I told him, 'You know Mike, I was thinking the same thing, but I wasn't going to tell anyone.'"

Edwards recalled, "We were great friends and had lots of things in common. Both of us loved cars—classic cars, sports cars—and we spent

Mike operates the CEBAS (Closed Equilibrated Biological Aquatic System) experiment on his first spaceflight, STS-89.

more than our share of time visiting classic car shows. He ended up buying a Porsche."

The crew on board Mir when Anderson arrived included Russian cosmonauts Anatoly Solovyev and Pavel Vinogradov and American astronaut Dave Wolf. The key purpose of the mission was to exchange US astronauts. Dave Wolf was completing a four-month stay on Mir and Andy Thomas was just starting his long-duration spaceflight. Thomas commented, "Mike used to joke a lot about taking me up to Mir and leaving me there. He was an understated person—you didn't really have an appreciation of his skills because he didn't have a need to advertise them or to have a lot of bravado. It's only when you worked with him, and I flew with him quite a lot in the aircraft, that you started to get an appreciation for his background and his talents. He was looking forward very much to subsequent flights after [STS-107]—he used to talk about it, he was hoping very much that he could do a flight to the International Space Station. That was something he, he felt very strongly about and wanted to do."

Anderson was the only black person among the 104 people from 13 countries who had the opportunity to visit Mir during its 11 years in operation. He didn't think that was anything special, but he did note, "I just thought it was a neat opportunity to get a chance to go to Mir. It doesn't exist anymore, and nobody will ever have the chance to go there again. For the handful of astronauts and cosmonauts that had the chance to go there and really experience the first really successful space station is really momentous and will go down in history. So I'm just happy to have been one of the few people that have had the chance to do that."

Anderson did have some time to have fun during his first spaceflight—but as part of a scientific experiment. He squeezed out a blob of strawberry drink that floated in front of his face. He blew on it gently, demonstrating fluid physics principles and how his breath affected the liquid sphere. Anderson noted that the best part of the "experiment" was drinking the sphere of strawberry drink when he was finished.

Endeavour, with Mike Anderson onboard, approaches Mir.

The STS-89 space shuttle and Mir 24 crews gather for a group shot in January 1998. Clockwise from the right: Mike, Bonnie Dunbar, Anatoly Solovyev, Terry Wilcutt, Dave Wolf, Pavel Vinogradov, Salizhan Sharipv, Jim "J.R." Reilly, Joe Edwards, and Andy Thomas.

Coincidentally, as the U.S. long-duration astronauts were getting exchanged via a shuttle mission, Russia was preparing to exchange its long-duration crews via its Soyuz spacecraft a couple of days later. As Endeavour was preparing to undock from Mir, a Russian Soyuz was launched. The rapid and closely spaced exchanges resulted in three craft with a total of 13 people in space at the same time: Anderson's crew of 7 on the shuttle, 3 on board Mir, and 3 launched on the new Russian Soyuz. It didn't set a new record, but it tied the one set in 1995.

After his first spaceflight, Anderson was given a support position working on the space station's software, which he did for a couple of years, followed by a year supporting software testing in Florida for the various space station components being prepared for launch. Anderson was assigned to STS-107 in July 2000.

Anderson had an unusual background with both hard science (a degree in astronomy) and Air Force operations background (flying big jets). He told me, "It was great. On my first flight I was a flight engineer, so I could use my operational experience as an Air Force pilot to work on the flight deck, and that was great. On this flight [STS-107] it was a complete role reversal—a chance to use my science background. I really relish the opportunity to have a chance to do that. It's really been a challenge, a lot of fun,—to play scientist, so to speak." Anderson was assigned to the blue shift, along with rookies Dave Brown and Willie McCool.

As the payload commander for STS-107, Anderson was the astronaut responsible for working with the mission planners and scientists to get as much as possible out of the mission. By the time Anderson was assigned, some of the planners had already been working on the mission for three years. Anderson was quick to give them the credit they deserved for creating a flight plan that optimized the crew's time in space: "It's been quite a daunting task. But all of the hard work had been done by the people that had been working this flight for a number of years. It's a team of some really talented people that have really worked hard to iron out all of the rough spots and put together this mission as a nice package. By the time we were assigned to the flight, I think the hard stuff was done."

Anderson quickly realized that Ilan Ramon would have to do more than just token "payload specialist" activities, noting, "When we were assigned to the flight I started looking at the payloads and it became obvious Ilan would have to be fully integrated with the crew. Usually a payload specialist focuses on one payload, but we couldn't have that luxury on this flight. So Ilan is fully trained in all of the payloads. He's going to do everything that the other astronauts on flight are doing."

Was Anderson concerned about the challenges of a tightly packed mission with so many activities happening at once? He told me, "I think the challenge to this flight will be getting off to a good start. It's been choreographed very tightly, and there's really not a lot of room for problems. If we have a payload with a serious problem it can have a ripple effect or a domino effect throughout the rest of the flight. The key is going to be getting off to a good start on that first day. If we do have problems with payloads, we're really going to have to work hard to solve those problems quickly and not let them affect the rest of the flight. So we've really worked hard to ensure that the payloads are in as good condition as they can be and our procedures are as tight as we can possibly get them and our ground control team has thought of everything."

Anderson was busy concentrating on STS-107 and not thinking ahead about what he would like to do on his third spaceflight. He said, "It's hard to say—whatever the program wants me to do. Whether it's a long-duration spaceflight, or an assembly flight, or another research flight—I think they're all special. When I get back I'll talk to Charlie [Precourt, the chief astronaut] and see what he wants me to do." As far as a long-duration flight was concerned, Anderson said, "I think it depends on the individual; there are some people who think spending six months in space is the greatest thing. There are other people who look at the challenges that come with that and would rather not try it for various rea-

Mike Anderson holds up a Creighton University pendant during the STS-89 mission in January 1998. Photo courtesy Creighton University.

sons. There's a lot of travel, spending half of your time in Russia. And if you have a family, perhaps young kids, it may not be the thing you want to do in your career."

Anderson and his wife Sandy had two daughters, Kaycee and Sydney. The girls were fans of the Disney movie "The Lion King," and one of the wakeup songs they selected for their dad was "Hakuna Matata."

Anderson wanted to honor the long history of African-Americans in the military. He contacted the Buffalo Soldiers National Museum in Houston, which commemorates the black Calvary from the nineteenth and early twentieth century. Anderson said, "I approached the museum, and I'm going to be flying a regiment style banner for them. As you look at the history of African-Americans in the military it goes back well before the Buffalo Soldiers, but of course that's one proud aspect of the tradition of African-American service in the military—all the way to the African-American astronauts that you have in the military today. So I think it's kind of nice to bridge the old and the new."

Before STS-107, Anderson had an interesting confession—he never liked launches. He said, "I'm probably different than most astronauts. I really don't enjoy launches. I think a launch is a terrible way to get to space. But right now, it's the best way to get to space we have. Actually, the only way, and so I'll take that ride." He clarified, "When you launch in a rocket, you're not really flying that rocket. You're just sort of hanging on. I really shouldn't say that I don't like launches. I guess I should say, 'I understand the serious nature behind a rocket launch.' I mean, you're really taking an explosion and you're trying to control it. You're trying to harness that energy in a way that will propel you into space. And we're very successful in doing that. But, there are a million things that can go wrong. And, I think, when you really sit down and you study the space shuttle and you really get to know its systems, you realize that this is a very complex vehicle. And even though we've gone to great pains to make it as safe as we can, there's always the potential for something going wrong. So we try not to think about those things. We train and try to prepare for the things that may go wrong to do the best we can. But, there's always that unknown. And I guess it's that unknown that I don't like."

But even with the risks, Anderson was still 100 percent committed to flying on the shuttle. He emphasized, "The benefits for what we can do on orbit, the science that we do and the benefits we gain from exploring space are well worth the risk. So, I don't like launches. But it's worth the effort. It really is."

CHAPTER 7

L A U R E L   C L A R K :
S U B M A R I N E   D O C T O R ,
M O T H E R

I talked to Laurel Clark about her experiments and what she would be doing in space when I first met her in December 2001. But what sticks in my mind is a remark she made after we'd finished our first interview and were waiting in front of the Spacehab building. She noticed a pin I wore on my collar from the "Star Trek" television series and commented that her husband Jon was also a "Star Trek" fan. It's the little things like that you remember.

Clark was born Laurel Blair Salton in Ames, Iowa, where her father, Robert Salton, attended graduate school at Iowa State University. But she didn't remember Ames because the family left when she was just two years old and moved to her parents' hometown—the tiny hamlet of Delhi, New York, with a population of just 2000. She attended elementary school at the Delaware Academy until 1975. Friends recalled that even as an elementary school student Laurel wanted to be a doctor when she grew up. The family moved again, this time to Albuquerque, New Mexico, where her father attended graduate school at the University of New Mexico. Laurel attended Monroe Junior High School in Albuquerque for her seventh- through ninth-grade education.

Laurel was known as Laurie to her family and friends. She was the oldest of four children, with brothers Jon and Dan and sister Lynne. Her parents got divorced, and Laurel and her siblings lived with their mother, Marge, visiting their father during vacations. Laurel's family moved to Lake St. Louis, Missouri, for a year and ended up settling in Racine, Wisconsin. Clark attended William Horlick High School for her junior and senior years. Clark's mother and stepfather remained in Racine until 2002, when they moved to New Mexico, but Clark's brother, Dan Salton, still lives in the Racine area.

Laurie Salton graduated from high school in 1979. She told me with a laugh that she was a "boring straight-A student without many hobbies," but she was on the swim team and in the ski club. Laurel selected several of Horlick's advanced college prep classes. She noted two teachers who stood out in her memory: First was English teacher John Barootian, "An incredibly inspirational and demanding teacher. He was more demanding than any English teacher I had in college." Then there was her psychology teacher, Tom Lewis: "I found it fascinating because of him. I actually think that's part of the reason I moved from interest in zoology to medicine." Her high school teachers remember that even then she was interested in science, "Star Trek," and outer space.

Barootian says he taught a very vigorous course to prepare students for college. The course required reading several classics, a 3,000-word research paper, and a 1,000-word essay final on John Donne's "Death Be Not Proud."

Tom Lewis, her psychology teacher, said that he "remembered Laurie because she was just so good in all of her subjects—she was good in social studies, she was good in science, everything she did she achieved at. She was by far the best student I had in that class, and I had some really good kids in that class."

While Laurie Salton concentrated on her studies, she did try to become a well-rounded student. Lewis noted, "She made herself into a good swimmer—like everything she did. She didn't have a lot of natural talent, but she made herself into a pretty good swimmer." Those skills became important when she decided to learn how to SCUBA dive and went through the very demanding Navy diving courses.

Laurel joined the Navy ROTC to pay for her college education. She went to the University of Wisconsin in Madison and earned degrees in zoology and medicine before joining the Navy. She pledged the Gamma Phi Beta sorority in February 1981 as a sophomore and moved into the sorority house for her junior and senior years.

Sorority sister Martha Siepmann Wilson was Laurel's roommate for the second half of their senior year. Wilson said, "Laurel was always the sweetest person, warm and bubbly, big smiles, and very studious. She would be using flash cards all the time. She took a class on ornithology and she put all of the birds' technical names on the cards and would sit in the

Gamma Phi Beta sisters at the University of Wisconsin in 1981. Laurel's in the middle of the second row from the back, wearing a dark sweater and a big smile. Photo courtesy of Martha Siepmann Wilson.

bedroom and just flash through them. She'd always have time for the fun things; she was our social director for the senior year, so she was in charge of arranging the parties at the fraternities and getting us excited to go."

Wilson recalled that one of the parties Laurel planned had a theme based on the board game Monopoly, with each room in the sorority house designated as a street in the game and featuring particular refreshments or entertainment. (For example, there were martinis in the Boardwalk room.) Wilson recalled that while Clark was a very conscientious student, she also knew how to enjoy herself, even when it was for a good cause: "We had a lot of fun in college. We did a lot of charity things where we'd raise money for cancer, a place for underprivileged children, severely retarded children." And Laurel was popular, Wilson said: "She had lots of boyfriends and dates. She was very social—friendly with the girls, friendly with the guys. Everybody knew Laurel. You'd think that she would be a scientist-type personality but she was so friendly. I never really saw her depressed or sad."

Laurel Salton attended Navy dive school in Panama City, Florida, in 1989, where she met her future husband, Jonathan Clark. Clark was also a Navy medical doctor. The Navy dive school is not like a sports SCUBA diving class. It's an intense, grueling course where the instructors intentionally try to break down their students. Many drop out because of the stress. Jonathan Clark noted, "It's not as bad as Navy SEAL training, but it's pretty close. The instructors are yelling at you and you're doing pushups, you're getting punished. They're trying to stress you out and push you over the edge. If you can't handle it there in a controlled environment, when you get in the real world it can kill you."

Clark recalled his first impressions of Laurel: "I had admired her spunkiness. This was very physically demanding and mentally demanding, competitive. She really shined. I thought, 'Wow—what a real spirited person.'"

Laurel really earned the respect of the instructors in the dive school when remained calm in a literally life-threatening situation. Jonathan Clark recalled, "She was almost killed in dive school. She had become a legend because she kept her cool. Laurel was wearing a hardhat dive helmet, and the valve flooded when she was underwater. You're now trapped with a ball of water around your head. It's a very, very dangerous situation. She kept her cool and came to the surface and just held her breath until they could get the faceplate off. A lot of people said, 'Man, she was lucky she didn't get killed.' If it had happened deeper underwater, it would have been all over."

Laurel became a Navy doctor, specializing in submarine crews. She explained: "The Navy feels strongly that they like us to experience environments that our patients are working in to determine their physical fitness to do those duties, and to give us a lot better appreciation of what a head cold will do to somebody who's trying to dive. To be qualified as a submarine medical officer you're required to go down on submarines for a certain number of days and do various other things." Laurel said she needed to spend 30 days underwater on a sub to qualify, roughly half a dozen dives of four to six days per dive. She was assigned to Submarine Squadron Fourteen as the Medical Department Head in Holy Loch, Scotland, when she got engaged.

Laurel Salton and Jonathan Clark were married at Olympia Brown Unitarian Universalist Church in Racine in 1991, after he returned from the Gulf War. Laurel became pregnant in early 1994, and went through winter survival training when she was eight or nine weeks pregnant. That included climbing mountains and other stressful outdoor activities.

Laurel didn't set out to become an astronaut. She said, "I never really thought about being an astronaut or working in space myself. I was very interested in [Earth's] environment and ecosystems and animals. And that eventually shone through in my interests in zoology as an undergraduate. And then I decided to pursue medicine. I joined the Navy and was exposed to a lot of different operational environments, working on submarines and working in tight quarters on ships, and learning about radiation medicine. And it was really just sort of a natural progression when I learned about NASA and what astronauts do, and the type of things that they are expected to do, that I thought about the things I had done so far and became more interested in that as a career."

NASA's astronaut selection board interviewed Laurel in 1994, when she was six months pregnant. Jonathan Clark said Laurel told him, "That was a great deal, I didn't have to go through the medical tests, I just went there and had fun." NASA didn't select her, but on the next opportunity, in 1996, Laurel Clark was selected as an astronaut, along with Willie McCool and Dave Brown. Jonathan Clark was in the middle of a Navy tour of duty in Pensacola, so it was a year before he was able to transfer to Houston. He became a NASA flight surgeon, responsible for the health of the astronauts both on the ground and in space. (By coincidence, Jonathan Clark was a flight surgeon on STS-96, Rick Husband's first spaceflight.)

After astronauts go through the candidate school, they're given technical assignments—working to help prepare crews for current missions, making improvements to the shuttle or space station, or interacting with the Russians or other international partners on the space station, among other assignments. At one point, Laurel studied Russian, traveled to Russia to learn about the Soyuz spacecraft, and went through the Russian winter and water survival training courses.

Clark also worked on STS-107's payloads well before she was assigned to the mission. The payloads branch of the astronaut office is where astronauts, scientists, and engineers work together to develop scientific experiments for the space station and for shuttle missions. For STS-107, Laurel had recommend things like how to phrase a procedure better or what might be reasonable for an astronaut to accomplish within a given period. If she didn't have the answers herself, she could call on an experienced astronaut for that knowledge, gaining more knowledge for herself in the process. Describing this experience, she said, "I worked a lot of the different payloads and payload issues for this flight pretty intensively for a year before they even considered assigning [astronauts]."

One of Clark's other technical jobs was to represent the astronauts to the Human Research Policy and Procedures Committee (HRPPC). NASA has extremely rigorous standards for determining what's acceptable to perform on human or animal test subjects. (Many rookie astronauts have said—half seriously—for the opportunity to fly in space, they'd let themselves get stuck with needles all over and give whatever body parts the scientists wanted.) The oversight board examines the proposed experiment, verifies that it is worthwhile peer-reviewed science that can only be done in space, and establishes the limits. In one case, an astronaut performed a muscle biopsy on himself. He applied a local anesthetic and used a special tool to punch a hole in his calf and cut out a small chunk of muscle. Nothing as drastic as that was planned for Clark or the rest of the STS-107 test subjects. She said, "They give us every opportunity to understand what we're going to be asked to do. The investigators fully hope that we're going to participate in [all of the blood draws and injections] because that's how they get their scientific data. All of us take very seriously our role as scientific subjects for the scientists who designed these experiments and need the data." The scientists didn't need to worry. Not only were Laurel and her crewmates willing to give all of the blood, urine, and saliva samples the scientists wanted; in a couple of cases they gave additional samples when there were problems with earlier attempts.

One of the major responsibilities for every astronaut is public relations. Astronauts spend much of their time doing interviews, talking to civic groups, and meeting with school children. Some of the astronauts, including Laurel Clark, are better at this than others.

Laurel Clark inside a Russian Soyuz spacecraft trainer. Photo courtesy of Jonathan Clark.

The Rock Art Foundation, in San Antonio, Texas, specializes in Native American artwork on stones. They put in a request for an astronaut to speak at their annual rendezvous in October 1997, and Clark was selected to give the speech. She went there with her husband and son and was fascinated by their activities—so much so, in fact, that she and her family started going back to the annual meetings on their own every year.

Clark's sorority roommate Martha Wilson asked Laurel to come back to Wisconsin to talk to her fifth-grade science class. The class researched Clark and her career and suddenly realized that she was scheduled to visit them on her birthday. Wilson says she contacted the local television stations and newspapers to ask if they wanted to come to see an astronaut, but they weren't really interested because Clark hadn't flown in space yet and wasn't famous.

Wilson said Laurel told her students, "I love, love, love my job and I can't think of doing anything else." Afterwards several Gamma Phi Beta sisters in the area got together for lunch. "We celebrated her birthday at lunch, had a little birthday cupcake with a couple of candles in it," Wilson said. "It was great to see her again as a human being—not just as an astronaut."

Clark was one of about a dozen astronauts invited to attend a set of briefings where astronauts were informed about the science on STS-107 and what would be requested from them as test subjects if they were selected for the mission. (Her future crewmates Dave Brown, Kalpana Chawla, Mike Anderson, and Ilan Ramon were also in attendance.) In July 2000, she was assigned to STS-107

Spacehab manager Pete Paceley recalled running into Laurel in a supermarket one Friday night. He was amused to see an astronaut shopping for groceries, but besides the fact Laurel was an astronaut, she was also a mother. Paceley recalls that Laurel was kind enough to take the time to introduce herself to his kids.

Just before the STS-107 mission that launched his wife into space, Jonathan Clark was a crew surgeon for STS-111. That made his life

March 2000: Laurel reunites with sorority sisters after talking to Martha Wilson's fifth-grade class. Photo courtesy of Martha Wilson.

incredibly busy as an integral member of the STS-111 support team, an astronaut spouse supporting his wife as she prepared for her first space-flight, and a father all at the same time.

The Clarks' son—Iain—was nine years old when STS-107 finally took off. Several months before the mission, Laurel had said of him, "He mostly thinks [my training is] taking a lot of my time. He would just as soon have me home with him, especially now as we're getting busier and busier. He's also proud. When I went to talk to his first grade class this last year you could see it in his eyes that he was just very proud that it was his mom that was sitting there talking about what they were going to do. My son I think is a typical eight-year-old: He loves to spend time with his mother and his father. [Because of the delays] we've been able to spend some time with our families. I just got back from a week and a half spending time with him. While he wishes I wasn't going to be away, he is excited about what we're going to do and about going to see the launch."

As part of the personal items she was allowed to fly with her in space, Clark had an academic patch from William Horlick High School and her son's first grade class put all of their fingerprints on a sheet along with their photos. Items for her colleges included a Healthstar Medallion for the University of Wisconsin Medical School and a teddy bear mascot from the University of Wisconsin. The teddy bear was nicknamed Ursula Minor and wore a jacket with the College of Letters and Science logo and "Zoology" embroidered over the heart. Clark honored several Navy organizations she had a connection to, taking along a patch for the Naval flight surgeons at Pensacola, Florida, a US flag for the Navy dive school in Panama City, Florida, a medallion for the National Naval medical center in Bethesda, Maryland, a T-shirt for the Navy undersea medical institute in Groton, Connecticut, and a squadron patch for the Wake Island Avengers—VMA211. She also flew a patch for the Rock Art Foundation. A burned fragment of the Horlick High School patch and the University of Wisconsin medallion were found after the accident and returned to Clark's husband.

Friends and colleagues remember Laurel as an outgoing person who loved nature, would always ask you about your family and tell you about hers, had a different colored shirt with the crew logo for each day of the week, and even earrings with the crew logo. I recalled that she was patient with me when explaining the tasks she would be performing in space and her tasks.

Before the launch Clark said, "When we first get to orbit we have a lot to do. But the first free moment I have, I'm excited about just looking down on our planet. I've heard numerous [astronauts] talk about it. Even before I came here, when I found out that I had the opportunity to be an astronaut, I thought about how stunning the photos are that [astronauts] take. And knowing the difference between standing by the side of the Grand Canyon and looking at it and soaking it in as opposed to looking at a photograph, I can still only imagine how glorious and awesome that must be."

Space did not disappoint her. During the mission, rookies Laurel Clark and Willie McCool discussed the incredible views in space. As Willie put on his rubber gloves to prepare to stick a needle into Laurel for a blood draw, he described how amazing sunsets appear from space. Clark agreed and shared a similar experience with McCool. "There was one sunrise I was watching last night. I got over to the window, K.C. said, 'You've got to see this.' The Moon was just starting to shimmer on the horizon, I thought it was a reflection in water of the Moon, but I figured out it was moving. It shimmers as it passes through the horizon. And then it went up and the Sun got lighter and we watched the lightning storms. Same thing, the colors were incredible. We just saw the little disk of Sun and we ran for the [middeck] because we were supposed to go to bed. I'm like, 'I can't watch the whole sun come up.' Oh, it was so beautiful! And the layers of atmosphere. There's thin and thick, they don't just get thinner. There's some that are thin and then thick ones, more thin layers. It was absolutely incredible. Oh, and the Big Dipper, too, it was the sliver of Moon and the Big Dipper behind it." Willie said, "Let's get back to work," and Laurel sighed and held out her arm for the blood draw.

After the launch I talked to Laurel Clark's family. They told me Iain was excited to come to Florida, but wished one of them could go into space instead of his mother because he missed being away from her so much. Hopefully when he's grown up, Iain will realize his mother flew because she felt the science and benefits outweighed the risks to herself. Like every other astronaut, Clark was very aware of the risks inherent in spaceflight and that she might not see her son again. According to her husband Jon, one of Laurel's favorite quotes was, "A ship in harbor is safe—but that's not what ships are for."

Laurel with a photo of her son Iain during the STS-107 mission. Photo courtesy of Jonathan Clark.

CHAPTER 8

When the Clinton administration offered Israel the opportunity to fly a passenger on the shuttle, Israel had a tough choice to make—who? Should the first Israeli astronaut be a scientist, engineer, pilot, or even a poet or artist? Ultimately the Israeli Space Agency decided it should be a military pilot, just as the first Mercury astronauts were all military test pilots. The fact the Israeli Air Force was footing the bill for the astronaut's training probably had a lot to do with the decision, too. The requirements were a college degree in a technical field and some experience working with scientists.

Ilan (pronounced "i lan") Ramon was asked if he wanted to apply to become an astronaut and thought it was a joke. But when he learned the invitation wasn't a prank, he decided being an astronaut would be something interesting and different if he was lucky enough to be selected, and he filed an application.

Ultimately, Israel chose two candidates—Ilan Ramon and Itzhak Mayo—but didn't reveal their names. In the official Israeli government announcement, on April 29, 1997, the candidate who had been selected was identified only as "Col. A," a veteran F-16 pilot who had logged combat missions and was an electrical engineer. Israel has always been extremely secretive about the identities of its pilots, and their names were not revealed even after they had moved to Houston to begin training. (Their names were well known, however, to most folks at the Johnson Space Center.) Coincidentally, Laurel Clark's husband, Jon, was the flight surgeon who performed the physicals on Ramon and Mayo. It was about a year before Col. A's identity was revealed as Ilan Ramon.

Who was Ilan Ramon? He was an Israeli Air Force fighter pilot, a veteran of the Yom Kippur War in 1973 and the Lebanon War in 1982. Initially Ramon trained to fly the F-4 fighter. When the USA offered Israel F-16s, Ramon was one of the first Israeli pilots sent to the USA to learn how to fly the plane and study its systems. Ramon was sent to Hill AFB in Utah, attending classes with USAF pilots and learning from American instructors. Ramon's English was very good, but certainly not perfect.

In addition to Ramon's qualifications as a fighter pilot, he was also ideologically an extremely suitable candidate. Ramon's father and

grandfather were among Israel's founders, fighting in the War for Independence. Ramon's mother and grandmother were World War II survivors from the concentration camp at Auschwitz. Ramon said, "The fact
that I'm the son of a Holocaust survivor is more symbolic than usual. You
can think about it that I carry on the suffer [*sic*] the Holocaust generation
went through. I'm a proof, even after the horrifying times they went
through, that we're going forward."

While Ramon was very proud of his heritage, he was not an obser-
vant Jew. He was secular and did not keep Kosher or honor the Sabbath.
Many religious Jews will not work on the Sabbath unless it's necessary to
prevent the loss of a life. Obviously, working on the Sabbath is an important consideration for a space mission. Even before launch, much of the
training or travel can take place on weekends, and in space the timetable
is dictated by when the shuttle launches—not a religious calendar.
Ramon noted, "As you know, I'm secular and didn't get any special permission [to work on the Sabbath]."

From the beginning of the space age there were jokes about how a
Jewish astronaut should behave in space. In a spacecraft like the shuttle,
sunsets occur 16 times a day. Should a Jewish astronaut celebrate the Sabbath once every seven orbits? And what time zone should be used? A
prayer shawl is blue because the sky is blue—but what color should the
shawl be in outer space, where the sky is black? But the most important
question had to be—was it blasphemy to fly in space? At a memorial after
the accident, Rabbi Zvi Konikov joked about researching the religious
issues for a Jew flying in space, "Jerusalem, we've got a problem," spoof-
ing astronaut Jim Lovell's famous comment, "Houston, we've got a problem," during the Apollo 13 mission.

Actually many of the questions related to religious Jews and space
travel were addressed in the 1960s, even before the first Moon landing. A
group of rabbis specializing in the interpretation of the Torah got together
to discuss the issues. They recognized it would be some time before Israel

Ilan Ramon's official NASA portrait.

would launch their own astronauts, but Jewish astronauts from other countries such as the United States might want advice on how to practice their religion in space. The rabbis decided a Jew in space should use the same calendar and watch that is used on Earth, preferably using Jerusalem as their time zone (although there is disagreement here — some prefer the clock in Florida, where the shuttle launches, while others prefer Houston, the location of Mission Control). And while the sky in space is black, the Earth is blue, so a prayer shawl could be the same. Perhaps most important is the reason for flying in space — "as long as the motive is research and investigation and not to challenge God's authority in the universe," said Rabbi David Golinkin of the Schechter Institute of Jewish Studies in Jerusalem But at least one rabbi said that Jews should not travel in outer space because they couldn't honor the Sabbath properly.

There have been a handful of Jewish astronauts but they've always worked on the Sabbath as needed and none have kept Kosher diets. Many have brought mementos of their religion with them. Ramon, very proud of his country and its culture, wanted to commemorate Israel's heritage. While he was not religious, he recognized that as an Israeli he represented all Jews, and that while it might not be practical on an intense space mission to do Kiddush (a Sabbath blessing over a meal), the simple act of eating some Kosher food could be easily accommodated. (Only some of Ramon's meals were actually Kosher; the rest of his menu came from the standard shuttle food selections.)

Ramon also hoped he would be thought of as a representative for all of the Middle East. But he was an unlikely candidate to spread the concept of Israel/Arab unity, since he was an Israeli fighter pilot and the veteran of several wars with the Arab world. Ramon said, "There's no better place to emphasize the unity of people in the world than flying to space. My crewmates just talk about me and Israel. It goes the same for any country — Arab country, whatever. We are all the same people, we are all human beings. I believe most of us are good people. I'm not thinking of

One of Ilan Ramon's Kosher meals. This "Chicken Mediterranean" contains chicken, tomatoes, ckickpeas, potatoes, olives, onion, cornstarch, celery, tomato paste, lemon juice, corn oil, minced garlic, onion powder, sugar, salt, and spices. Photo by the author.

myself or my family as targets. This flight especially is going to take care of better life on Earth." Ramon noted, "We're going to space for research, for the benefit of the world and the people in the world. For better health, for better education, for getting children excited about space, about science."

Ramon said he appreciated that the mission included both Jewish and Arab scientists. Ramon never said anything negative about Palestinians in his public statements, and did say he hoped the Israeli-sponsored MEDIEX would benefit everybody — Jews and Arabs alike. About the Palestinians in Israel, Ramon said, "I'm sure some of them watch Israeli TV if not all of them. They're aware also of this flight. I hope they enjoy and they are as proud as the Israeli people."

There were suggestions that it would be extremely symbolic if a Jew and Palestinian citizen of Israel flew in space together. But a suggestion like that is absurd. First of all, NASA only offered one seat to Israel, and second, the Israeli Air Force paid for the mission, and it's doubtful if there are many Palestinians in the Israeli Air Force. If an Israeli was to fly in space, it was certain he would be a member of the Israeli Air Force, and certain that he would be a Jew.

Ramon decided to carry some symbolic items in space. He said, "I asked the Museum of Yad Vashem—which is the Holocaust museum of Israel—for an item to carry with me. They suggested a few, and I picked a drawing by Petr Ginz, a 14-year-old kid in the Theresienstadt ghetto during the Holocaust. He led a group of kids; they had their own paper. They would write from one to another. He was very talented, he was interested in science, he was a painter and writer and everything. Unfortunately, he didn't make it. He was murdered later on in Auschwitz. His paintings made it. By the way, his sister is still living in Israel. While I was handling this picture, I found out that his sister is the mother of one of my classmates. I'm not sure how they pack it [the picture]. It's small, a page." Because the original drawing was an irreplaceable piece of Holocaust history, the museum gave Ramon a copy to fly in space.

Ramon described the drawing: "The picture itself is of Petr as he imagined himself looking at Earth from the Moon. It was in the forties [1940s], so it's a very simple drawing, but it relates to space of course. It kind of symbolizes the spirit of this boy. I really feel I'm taking his imagination and fulfilling his wish of being there, of being his eyes."

One of the MEIDEX scientists was Joachim Joseph, a 71-year-old Holocaust survivor. Joseph was given a tiny Torah from a rabbi while both were imprisoned in a Nazi concentration camp in Germany in 1944. Joseph had just turned 13, and the rabbi secretly arranged a 4 A.M. bar mitzvah in the prisoners' barracks. "After the ceremony, he said, 'You take this, this scroll that you just read from, because I will not leave here alive. But you must promise me that if you get out, you'll tell the story,'" Joseph recalled. The rabbi was killed two months later. Joseph was freed

Petr Ginz's drawing of how he
imagined the Earth would look
from the moon.

Petr Ginz (1928-1944)
Moon Landscape, 1942-1944
Pencil on paper
Gift of Otto Ginz, Haifa
Collection of the Yad Vashem Art Museum,
Jerusalem

from the Bergen-Belsen camp in a prisoner exchange in 1945 and eventually migrated to Israel, where he became a scientist. Ramon found out about the Torah and asked if he could carry it in space.

Ramon also carried an ancient Jewish coin minted in Jerusalem in the first century A.D. for another Israeli scientist.

Ramon had a long record as an Israeli Air Force pilot, but he wouldn't discuss it, at least not publicly. Rumors put him in one of the most daring military operations in the history of the modern Middle East, "Operation Babylon." On June 7, 1981, a group of Israeli Air Force pilots flew several US-manufactured F-16s to Iraq. The F-16s bombed the Osirak nuclear reactor that Iraq claimed was a civilian power plant, but Israel believed was part of a nuclear weapons program. The USA had sold Israel the F-16s with the understanding that they would only be used for defensive purposes, and formally condemned the action, along with most countries. But privately President Reagan quipped, "Boys will be boys," and many countries, including much of the Arab world, were grateful that Iraq didn't get an operational nuclear reactor with the potential to develop nuclear weapons.

Ramon's biography says that in 1981 he was the Deputy Squadron Commander B, F-16 Squadron. He would have certainly been a strong candidate for the Osirak raid. But Ramon would not confirm his involvement. He said, "I'm 30 years in the Air Force and done a lot of operations. I can't refer to any of them. Use your imagination and you can put me there or put me out of there." According to the reports, Ramon flew the last plane in the squadron, the one that would be most exposed to Iraqi antiaircraft fire and defenses. The eight planes had to fly extremely low to avoid enemy radar, and at the limits of their range. The route took them over Saudi Arabia and Jordan, which in itself would have sparked an international incident. After STS-107's launch, high-level Israeli military officials told the Israeli media that Ramon was indeed one of the Osirak pilots.

By coincidence, one of the Israeli air traffic controllers during "Operation Babylon" was Yoav Yair. He later became a scientist at the Open University of Israel and one of the MEIDEX scientists studying thunderstorms from space. While preparing for the MEIDEX experiment, Yair and Ramon were talking about their past military experiences and realized they had participated in the same daring operation over 20 years before.

Even excluding the Osirak raid, Ramon's life as an Israeli Air Force fighter pilot was not without risks. He acknowledged that, in his career as a fighter pilot, he had two mishaps where he had to abandon his aircraft and parachute to safety.

When NASA originally invited Israel to fly a passenger on the shuttle, it was expected that the passenger would have very few responsibilities. The "payload specialist" designation for an astronaut was supposed to mean he would handle his own experiments and stay out of the way of the "real" crew, who had their jobs to accomplish. Token space travelers from France, Ukraine, and Russia, among other countries, have flown as international gestures of good will by the USA. In some cases, more qualified space travelers from those countries have also flown. Some of the foreign payload specialists had extremely poor English skills and were

A rare photo of the Osirak pilots. Ramon's on the upper left. Photo from the author's collection.

truly just add-on "cargo," without any significant functions as part of the crew. This could not be said, however, of Ilan Ramon.

Years into his preparation for STS-107, Ramon showed his fine sense of humor by saying this about his long stay with NASA: "The first launch date was November 30. I'm not going to talk about the year." He was avoiding mentioning the year because of the time it took him to finally get into space. He'd been chosen as an astronaut in 1997 and was with the initial thought that he would be on a flight that launched in 1999. But it took three years just to find an appropriate flight that would fill MEI-DEX's science requirements, so along with STS-107's many delays, Ramon ended up spending five years in training.

Ramon was pencilled in for STS-107 in 1999 for a planned launch in December 2000, well before the rest of the crew was selected. NASA made this decision when it was determined that STS-107 was the only suitable flight for the MEIDEX payload. It would be a year before the rest of the crew was added to the mission. There was also the issue of his responsibilities: When Ilan was assigned to STS-107, it was quickly realized that there was no way to justify flying a crewmember dedicated to just one payload on such a packed flight. It would be absurd for Ramon do nothing while the shuttle was out of the range of the MEIDEX areas, while the rest of the crew were busy working on the other experiments. So the decision was made that Ramon would have to operate many science payloads, like the rest of the crew. Fortunately, he was up to the task. Because of his long time training in Houston while waiting for the flight to materialize, Ramon received far more training than a typical payload specialist, even before he was assigned to STS-107. Once the actual mission was chosen, backup astronaut Itzak Mayo returned to Israel, as it would not be practical to give him all the intense training too

Rick Husband told me, "Ilan is fully integrated into every aspect of the mission. He is not an observer; he's a full member of the crew in every way. In my mind, Ilan is much more than just a payload specialist. He's working as many payloads as anybody else on the flight. He's gone through a significant amount of training."

Ramon noted that NASA is a very international organization, especially the astronaut corps. He told me, "When I first came to NASA about four years ago, I was amazed to walk in the corridors and see all of these international guys—I was sure I was not in the U.S. I was amazed more to hear American astronauts speaking Russian, and French, and Japanese. I really feel like I'm in an international environment. The most amazing thing for me was to hear an American astronaut, like for instance Charlie Precourt, speaking Russian. So I really feel that not only me but the whole astronaut corps today is an international environment and it's great." Precourt, the chief astronaut at the time, grew up in a bilingual English and French household. While attending the US Air Force Academy he spent time as an exchange student with the French Air Force academy. He's widely regarded as one of the most talented astronauts in terms of learning foreign languages.

Ramon said, "I was amazed, walking the corridors and seeing all these international guys. There are no borders; you can't recognize borders from space." One of the few borders that is visible from space is Israel's southern border with the Sinai. Intense irrigation in the Negev has resulted in lush farmland in Israel while the other side of the border is barren. From space it looks like a razor-sharp line, almost like a map's boundary.

I first met Ilan Ramon in December 2001, when I was invited by Rick Husband to observe the crew during a training session. When I interviewed Ilan in June 2002, he said he remembered meeting me six months earlier.

Ramon said, "I think we serve science for humankind and this doesn't have any borders as well." Since MEIDEX was a purely scientific experiment I asked him how he would he feel about NASA sharing the data with Israel's neighbors. Ramon replied, "Science is done for the humankind wherever they are. It's one of every scientist's obligations to share his findings. This goes for every experiment we are going to doing STS-107, including MEIDEX. We will share all of the information and all of the data. The Israeli scientists are working in collaboration with NASA scientists, so everything is going to be shared."

Ilan Ramon certainly wasn't perfect. He was extremely interested in the S*T*A*R*S student experiments, but for some reason could never seem to remember that the bees experiment came from Liechtenstein. At least one time he said it was from Luxembourg. But in space he tended all of his experiments passionately, including the Liechtenstein bees.

Much of the outside interest in the STS-107 mission was centered on the fact that Ilan Ramon was an Israeli, and the other members of the crew were all asked what they thought about him.

A photo of Israel taken on the STS-107 mission. The Israel/Sinai border is clearly visible, lush vegetation on the Israeli side—desert on the Sinai side.

Kalpana Chawla said, "Knowing Ilan is truly a privilege. He is easily one of the best people, if not the best person, I've met and worked with. I've learned tremendously from him. He puts all into the task and is very professional and excellent. He has this sense of humor where he makes fun of himself. At a personal level I don't think my views have changed about Israel. I just thought of people from Israel as a very smart lot."

"He is just so impressive in how he was part of the team. His accomplishments are so many. I think someone like him who has done so many things, yet when he introduces himself he says, 'I'm just a tourist,' and we all know that he is not just a tourist. He really is like all of us; he's got the same responsibilities we do on orbit. His attitude is always downplaying and totally at ease about this whole thing." She added, "What really stands out and impresses me is, he is easily one of the most accomplished people I've met. And yet he downplays his accomplishments. He is so at ease that he can say all this, and we know that he is as good as he is."

Willie said, "[Ilan]'s a wonderful, wonderful individual. Nationality really doesn't make a person. You can tell he speaks fluent English. It doesn't matter where he's from; it doesn't matter where any crewmate is from. You judge your crewmates by their character. He's just got a warm wonderful character, he's a team player. When somebody comes to a foreign country, you'd expect the hosts to embrace the newcomer. In many respects, it's been just the opposite. His family has opened up their home to all of the families on the crew frequently. Ilan and Rona would have us over on Fridays to celebrate the Israeli tradition of having Friday evening meals together. They would have all of us together at their home to experience the Jewish traditions, the Jewish food, the breaking of the bread, passing a little goblet of wine they share. In a sense, the Ramons embraced the STS-107 crew as much as we've embraced them."

Mike said, "Ilan's a great guy — he's very calm, very cool. He's hard to upset; he's got a very calm nature about him. Very confident, very easygoing. I'm hoping when we get to space and he's exposed to microgravity for the first time, it throws him a curve so I can see him show some expression of surprise. He's such a level-headed , well-prepared individual it's hard to imagine, but maybe spaceflight will be a surprise for him."

Laurel said, "I have long been in favor of [international] exchange programs. I do have a different feeling, I think, toward Israel, just because I've gotten to know Ilan and his family and the other research people involved in the Israeli payloads. Not so much the political issues, but knowing them as people, and very fine people. I had a similar experience when I went to work in Russia several times."

When Ramon was asked about whether or not he was concerned that his presence on the crew made the shuttle a greater terrorist target, he joked, "May I defer this question to K.C.?" On a more serious note Ramon said, "I think NASA security is doing everything needed since

September 11, and even earlier. I feel safe, I think everybody feels safe. We are here to conduct science for the benefit of humankind, it doesn't matter—no borders. It's one globe for us and we feel safe."

During the flight Ramon was so intense in his passion for getting as much scientific results from his experiments as possible that Mission Control had to remind him to stop for a second and enjoy his time in space. After one conversation, when Ramon offered to do some additional experiment runs while he was scheduled for some free time, capcom Charlie Hobaugh told him, "Based on what you told us we'll try to see what we can schedule for you. But do you realize there's a window onboard?" and Ramon replied "Yes sir, there are a few of them."

Ramon did see Israel from space. Israel is a fairly small country, so it's a challenge to spot—if you don't look at the right moment, you'll miss it. But at least it's on a seacoast and distinctive because of many prominent geographical features like the Dead Sea and Israel/Sinai border.

After the accident, several rolls of undamaged film were recovered and developed. Of the 92 photos released by NASA, 43 are photos of the Earth, and over half show Israel. There's no way to say for sure which astronaut took those photos, but most astronauts enjoy taking photos of their hometowns from space if at all possible. It's likely that Ramon took many of the Middle East photos. The crew also shot a fair amount of video of the Middle East. At one point Rick Husband narrated a video and pointed out the Sinai, Israel, and the Red Sea.

The United States offered Israel one space shuttle seat and after the 9/11 terrorist attacks it was extremely doubtful if seats would be offered to countries that might attract the attention of terrorists. Ilan Ramon had to realize he was almost certainly going to be Israel's only space traveler for a very long time. It was even less likely that Israel's economy could justify spending enough money to participate in the International Space Station. But Ramon was still hopeful. He said, "Of course I'm hoping it's a door opening and there will be another one. Either me or another [Israeli astronaut]. But I think it's more of a door opening for scientists from both countries to have mutual research and collaboration."

Ilan Ramon's flight was a feel-good moment for Israel. Ramon noted, "In Israel today there's a huge problem of economics and unemployment. This is the most important problem in Israel today. I think people are very happy to be distracted by my flight, maybe to forget a little bit of their problems and get out there with us." But unfortunately the Columbia accident changed that celebration into another reason for Israel to mourn.

CHAPTER 9

DODGING BULLETS:
THE ASTRONAUTS
WHO WEREN'T
SELECTED

The chief astronaut and head of flight crew operations select which astronauts are assigned to each mission. There are over 100 active astronauts at any moment, and in theory any of them can be assigned to any mission. In practice, each mission has its own technical requirements, so there's often a desire to assign astronauts with particular skills to various missions. As a rule, astronauts who are already assigned to missions or have just completed flights are not also assigned to another flight. And as in any organization, how well you get along with your boss comes into play too. The managers select a pool of suitable candidates who they think would be appropriate for each mission, and from that group downselect to the actual crew. In the case of STS-107, there was also an outside factor. Since it was the only suitable mission for Israel's MEIDEX experiment, that guaranteed that one of the seven seats would go to Ilan Ramon. So chief astronaut Charlie Precourt and head of flight crew operations Bob Cabana had to choose six from NASA's astronaut corps.

Israel asked its Air Force pilots who were interested in becoming their country's first astronaut to fill out an application. A group of officers selected Ilan Ramon and Itzhak Mayo, and their names were then forwarded to NASA for approval. In another case, the inclusion of another "outside" crew member was the subject of a discussion of logistics between NASA and the European Space Agency. ESA's ARMS payload, which involved studies of the heart and lungs, was complicated. To ensure the experiment's smooth operation and to maximize the science return, ESA suggested flying the experiment's chief scientist, Dr. Andre Kuipers of the Netherlands. With the delays in STS-107's schedule, however, NASA and ESA agreed that there would be plenty of time to train career NASA astronauts to take care of the experiment, and there was no need to fly a specific payload specialist just for ARMS. (Kuipers did get his opportunity to fly in space in 2004, on a nine-day mission to the International Space Station.)

Kuipers said, "Rationally you know in doing something, now and again it goes wrong. But it also happens when you get in a train or a plane or a boat. Planes crash, and it could be you in there. I don't really think about it. I knew this could happen to me, but I did not feel I had escaped death or something. If I would have been selected, the whole sequence of events would have been different. You cannot say I just escaped death, but I felt this is the risk and it can happen to me. But it never made me think, 'I should not do this.' I would never decide not to be an astronaut anymore. If they would have asked me to fly the shuttle the next day [after the accident], I would have done it, because the chances it would happen again would be very, very small."

In theory, any of the scores of mission specialist astronauts could have been assigned to STS-107's four mission specialist seats. But there is a desire to have at least some astronauts with medical or life science backgrounds on missions where there are many life science activities.

NASA held the first astronaut briefing on September 16, 1999, where two dozen astronauts were given a presentation on the mission's objectives. In March 2000, managers selected a dozen mission specialists as candidates for the four seats, and these potential crew members were invited to attend a series of "informed consent" briefings. These briefings are standard for any microgravity mission where the astronauts are the test subjects. In the briefings, the astronauts are told about the science experiments, justifications for the science, and what's going to be expected of them. In many cases, astronauts will be asked to give blood, urine, and saliva samples before launch, during the mission, and after landing. In other cases their diet and exercise may be regulated. In more extreme cases an astronaut may be asked to have a muscle biopsy in space or another invasive medical procedure. The purpose of the informed consent briefings is to assure the potential astronauts that the science is valid and worth the amount of pain they will have to endure to make it possible. The astronauts are encouraged to ask the scientists questions about the research and what will be done to their bodies. Oversight panels question the value of the science for all of the experiments involving human or animal subjects.

Dutch astronaut Andre Kuipers, a possible candidate for STS-107, flew on a Russian Soyuz to the International Space Station.

Astronauts are asked to sign waivers for the medical experiments but always have the right to back out at any moment. Laurel Clark explained, "It's still strictly voluntary — every single test, every blood draw. I can say at any point in time I don't want to, or I feel bad, or whatever." But since the astronauts do realize the science is worth it, they're willing to be poked and prodded as often as necessary for the experiment, and willing to go to the additional effort as required. On the other hand, when a NASA manager pushes an astronaut to fly on a mission, and that astronaut says he isn't interested, it can affect the astronaut's career.

One of the astronauts at the informed consent briefings was John Herrington, a Navy flight engineer. He said, "We sat down and listened to the medical briefs, and I looked around the room and realized, 'I'm not a flight doctor, and this is a life science mission, and there's a lot of people here who fill that [requirement] way better than I do.'" Managers came to the same conclusion as Herrington—there were other candidates more suitable for STS-107, so Herrington was passed over for STS-107 and assigned to STS-113 instead.

Another astronaut at the informed consent briefings was Mario Runco, who'd flown most recently on STS-77, in 1996. It was roughly the right timing for him to get another flight assignment. He said, "Mike Anderson and I became good friends, and even before the informed consent briefings, we would half-seriously kid each other about being assigned to the same flight and doing an EVA [spacewalk] together. We'd bump into each other in the hallways and say things like, 'You hear anything about our flight?' or 'When do we start our EVA training?'"

Runco felt his background in Earth science wasn't the best fit for STS-107's life science experiments. As he tells it, he reflected back on his three previous flights and what he had accomplished, and then came to a realization: "I knew then it was time to move on." Runco believes that if he had pushed, there was a good chance he would have been assigned to STS-107. Afterwards he said, "You thank God for the decisions you make if they

Possible STS-107 candidate John Herrington was assigned to STS-113 instead.

come out okay. I really think heavily on that, and think that I'm being looked after in more ways than you can count." Runco chose to retire from the astronaut corps but stay with NASA as an Earth resources scientist.

Four of Columbia's astronauts who took part in the informed consent briefings were selected for STS-107. Eight who took part were not selected. They're quite aware they "dodged a bullet," so to speak, but also aware that there's always some measure of randomness involved in how any astronaut gets assigned to any particular flight. And, like all astronauts, they are acutely aware of the very real risks involved with any spaceflight.

Herrington told me, "It could have been anybody [in the astronaut office]. You wonder about it. You love what you do, you recognize the risks involved." He added with a small chuckle, "But it will never happen to you." On a more serious note he said, "What's hardest is not for me, but for my family—my parents. It probably hit harder there than it did for me in terms of the recognition it could be me. It hit me very, very personally. But in terms of 'it could be me,' my mom probably more so."

Chief astronaut Charlie Precourt told astronauts Mike Anderson, Dave Brown, Kalpana Chawla, and Laurel Clark they were selected for the STS-107 science mission on July 25, 2000. Ilan Ramon had already been informally penciled in. Rick Husband and Willie McCool were assigned on October 27. The pilots and Chawla would not participate in the invasive medical tests like donating blood, but could perform the other science activities. At this point, STS-107 was scheduled to launch on July 19, 2001—just seven months away—but a variety of technical and management decisions would add an additional year and a half to their training schedule.

Four of the astronauts were almost guaranteed to get assigned to STS-107 based on their backgrounds and experience. As mentioned earlier, STS-107 was the only suitable flight for the Israeli MEIDEX experiment, earning Ilan Ramon his seat. Because of where Rick Husband ended up in his astronaut class, he was the most suitable choice to command the flight. Due to the heavy life sciences content, it was a pretty good bet that two of the four medical doctors from the 1996 astronaut class, like Dave Brown and Laurel Clark, would be assigned. The pilot's seat could have been filled by any of the 1996 pilots who didn't have a flight assignment at that moment. Many astronauts with the proper experience could have filled the final two seats for mission specialists.

A FINAL THOUGHT ON ASTRONAUTS

Columbia's seven astronauts were human beings. They were unlucky enough to get dealt a bad hand, but they knew there were risks before they decided they wanted to become astronauts, when they became astronauts, when they were selected for STS-107, and when they were strapped in on launch day. They realized that there were known quantifiable risks as well as unknown risks. But they considered those risks acceptable because they felt the benefits to themselves and to science were worth it.

Every astronaut is a talented individual, some more than others. The minimum requirements are a bachelor's degree in a technical field and decent (but not necessarily fantastic) physical shape. Hundreds of thousands of people meet those qualifications, and tens of thousands apply to NASA. Most who apply have far more than the minimum requirements, in many cases multiple advanced degrees and plenty of flight experience. About a hundred are selected for interviews, and from them around a score actually get selected each time NASA decides to add new astronauts. NASA administrator Sean O'Keefe joked that when he looked at some astronauts' biographies he wondered if somebody didn't accidentally combine the bios of three different people, because there are so many astronauts who have experience in multiple fields.

STS-107 had multitalented astronauts. But they weren't superheroes, saints, or patriarchs. They were just seven highly trained individuals who worked well together to accomplish a common goal. Many people have talked about how the STS-107 crew were better than others and they had fulfilled their destiny when they died. They weren't. The problem that destroyed Columbia could have happened on any shuttle mission.

The crew's total previous spaceflight experience was relatively low — only three flights (one each for Rick Husband, Kalpana Chawla, and Mike Anderson). Some members of their astronaut class had already completed that many flights by the time STS-107 finally launched. Anderson noted, "In recent times we're certainly fairly "young" [in terms of experience]. The key is to work together as a team. Those of us who have flown try to impart what wisdom we do have to the rookies and try to tell them what to expect from spaceflight and what to prepare for and what things they can do now to make the flight more successful for themselves." The most recent flight with as little total spaceflight experience was STS-40, in 1991, although there have been a handful of missions with more shuttle rookies. NASA certainly could have assigned astronauts with more flights, or added another experienced astronaut to STS-107. But there was no driving need. The 1996 astronaut class, which included Brown, Clark, and McCool, had a whopping 44 people. With roughly five flights per year and two rookies per mission, it would take about five years before all of them would get the chance to fly in space at least once. Assigning three 1996 rookies helped give additional astronauts flight experience.

Critics claim that NASA considered STS-107 to be an unimportant mission, citing the lack of experienced astronauts, among other factors. It's certainly true that STS-107 was less important than other flights, but that didn't make it unimportant. Certainly most of NASA's efforts were oriented toward the construction and maintenance of the space station, which required extremely complicated missions with multiple spacewalks, robot arm operations, and crew exchanges all going on at the same time. In some cases, planners will ask for a commander who already knows how to operate the robot arm, plus two astronauts who already have spacewalking experience. STS-107 didn't need an experienced spacewalker on a mission that had no scheduled spacewalks. There were also outside factors: for example, when one of the seats is allocated to someone from a foreign country, as was the case with Ilan Ramon. It's also important to note that while Dave Brown and Laurel Clark were rookies, their medical skills and operations experience in the Navy made them extremely suitable for STS-107's life science experiments.

Certainly the STS-107 crew got along with each other and liked each other, and that can't be said about every shuttle crew. While astronauts are professionals and try to get along with their crewmates, they're also humans, and there will inevitably be personality conflicts. And while you don't necessarily have to get along with somebody else, you do have to learn how to work with them to get your tasks accomplished. But the STS-107 astronauts really did bond with each other—partially because of their personalities, partially because of the very long time they spent together in training.

Within NASA, the STS-107 crew was well thought of, but they weren't considered anything extraordinary. As relatively new astronauts, they didn't have the close connections to others in NASA that those who'd been around for decades had—especially those who were involved directly or indirectly with the 1986 Challenger accident. There was an underlying sympathy for the crew—because their mission had been delayed again and again for so many different reasons, just as you'd feel sorry for anyone who's working toward a goal but always seems to experience setbacks.

There was also some pre-flight sympathy because STS-107 was perceived as a "less glamorous" mission—no spacewalks and no docking to a space station. It's possible some of the STS-107 astronauts could have been disappointed when they were initially assigned to a "plain" microgravity science flight. Or they may have been disappointed when many astronauts who'd originally been assigned to later flights got to fly before STS-107 finally got off the ground. But those disappointments are quickly put aside when you remember that you are assigned to a spaceflight and its science is important. Dave Brown told friends, "My seat's reserved"—even with all of the delays, he was assigned to a mission while many of his classmates didn't have flight assignments. Brown had told friends that he wanted to do a spacewalk at some point. None was planned for STS-107 but certainly Brown could have been assigned to do one on a later spaceflight. One astronaut commented, "It's *always* an honor to get assigned to a shuttle flight. That being said, some assignments are better than others." To which another astronaut replied, "When in doubt, remember the first statement." While STS-107 may not have been as glamorous to outsiders as other missions, it was a shuttle mission—one filled with plenty of science.

Everyone has always acknowledged that there are risks when flying on the shuttle. Before the mission Rick Husband said, "Spaceflight is a risky business." But so is flying in an aircraft or walking across the street. Certainly spaceflight is riskier, and will continue to be risky for the foreseeable future. And it's something that all astronauts are very aware of, from the time they fill out their first application. But each astronaut has also decided that the benefits are worth the risks.

Astronauts are asked dozens of times about the risks — by their family, friends, and especially by the news media. Those statements rarely make news unless there's a disaster. STS-113 commander Jim Wetherbee put it best: "I really wanted to give [my two daughters] a sense that if anything bad happened, I wanted them to remember two things. Number one, this is valuable to the people on Earth and it's a good profession and I wanted them to realize that I was doing it because of the benefits to humanity. Number two, I wanted to tell them if I didn't come back, what I wanted them to always think about in their future lives— if they had a tough decision to make—was what would I tell them if I were here, and they would come to the right answer. And that was my way of staying with them. Hopefully to make it easier if I didn't come back."

Seven ordinary people trained to do an extraordinary task—the STS-107 astronauts.

CHAPTER 10

IT ISN'T JUST THE ASTRONAUTS

While the seven members of the Columbia crew were the most visible portion of the mission, and the ones who lost their lives, there were thousands of managers, engineers, technicians, and support people involved with STS-107. Just a few of them are described in this chapter.

LAUNCH DIRECTOR MIKE LEINBACH

Michael D. Leinbach came to NASA in the fall of 1984 from an unusual educational background. He earned a bachelor's degree in architecture in 1976 and a master's degree in civil engineering in 1981 from the University of Virginia in Charlottesville. Coincidentally, Leinbach graduated from Yorktown High School in Arlington, Virginia, three years ahead of Columbia astronaut Dave Brown. It was a large high school and it's doubtful they would have known each other, but they could have passed each other in the halls.

When Leinbach first came to NASA in 1984, he was a structural engineer in the Design Engineering Directorate. He served as a lead design engineer for a variety of launch-pad systems, including the Orbiter Weather Protection and Emergency Egress Slide Wire systems. He was also lead design engineer for the Thermal Protection System Facility.

Leinbach became a NASA Test Director in 1988 and started to work specifically with Columbia on STS-28 a year later. That was Columbia's first flight after the Challenger accident, a top-secret military mission which launched a relay satellite for the National Reconnaissance Office. As a NASA Test Director, he was responsible for directing all daily operations at Launch Complex 39. Leinbach was promoted to Shuttle Test Director in 1991. In this position he was responsible for conducting the terminal countdown and launch of 17 Shuttle missions. At that point in his career he took a step sideways from the shuttle to work with payloads. He was the deputy director of the space station hardware integration office from 1998 to 2000, where he was responsible for the testing of all of the US space station hardware as the components were prepared before launch in Florida. One of the biggest challenges was the Multi-Element Integrated Test (MEIT—pronounced "meat"), where the space station components were hooked up with electrical cables and tested as a system. This was a critical phase of testing, since the next time the numerous pieces would be attached was in space. After his work on the space station, Leinbach returned to the shuttle side of operations, serving as the Assistant Launch Director for mission STS-101, in May 2000. He was promoted to shuttle Launch Director with STS-106 that September. As the Launch Director, he served as the "conductor" for the launch team, giving the final approval for liftoff and publicly wishing the shuttle crews good luck before each launch.

After the accident, Leinbach became the head of the Columbia reconstruction team, assembling the pieces of Columbia that were recovered in the field.

COLUMBIA FLOW DIRECTOR SCOTT THURSTON

For each space shuttle, there is a NASA engineer who serves as the "flow director." The engineer in this position is responsible for the day-to-day operations on the ground, from the time a shuttle lands after completing its previous mission until it's ready to launch again. The flow director also monitors the shuttle's performance while it's in space. Scott Thurston was in charge of Atlantis and Columbia.

Thurston grew up with the space program and went to the University of Central Florida, where he earned his degree in electrical engineering in 1989. He joined NASA in 1990 in the Shuttle Engineering Navigation and Tracking Systems Branch, advancing to senior lead engineer of Orbiter Navigational Aids. Thurston also served as project manager for all launch site equipment upgrades associated with the shuttle orbiter communications and tracking systems.

Thurston was also a management intern for the KSC director and a member of the Vehicle Integration Test Team (VITT—pronounced "vit"), the people who support the astronauts and astronaut activities at KSC. He earned a master's degree in engineering management from the Florida Institute of Technology in 2001.

MISSION MANAGEMENT TEAM CHAIR LINDA HAM

Linda Ham was the head of the Mission Management Team (MMT), a group of experienced engineers and managers who oversee day-to-day shuttle operations. Ham was born Linda J. Hautzinger in Salem, Wisconsin. She went to school in Kenosha, fairly close to where Laurel Clark lived in Racine, but they didn't meet until Laurel came to NASA in 1996.

Ham herself came to NASA in the early 1980s. She started as a propulsion engineer and was quickly promoted because of her performance. She became a flight controller, doing real-time operations inside Mission Control. In May 1991 she was promoted, becoming NASA's first female flight director. Each shift in Mission Control has a flight director, who oversees the real-time operations in Mission Control and on the shuttle. Rookie flight directors go through additional training and assist other flight directors before they're assigned to command a shift in Mission Control. After several missions, a flight director can become the "lead flight director" for a mission, in overall charge of the planning ahead of time. Ham became NASA's first female lead flight director with the STS-58 life sciences mission in 1993, with many experiments similar to the STS-107 experiments. Her second flight as the lead flight director was the STS-103 Hubble servicing mission, in December 1999. After that, Ham became an ascent/entry flight director, responsible for overseeing Mission Control during the super-critical launch and entries. In 2000 Ham was moved over to a management position within the shuttle program.

As a shuttle manager Ham was one of the members of the "Mission Management Team," a group of managers who oversee the shuttle's launch and flight operations. These managers watch over the work performed by Mission Control and off-line engineers who work issues during shuttle missions. The MMT also makes management decisions which affect shuttle missions, like whether to extend the mission to try to get additional science performed, whether or not a mission needs to be cut short if there's a non-critical problem with one of the shuttle's systems, and what paths to take when solving problems. A senior manager is assigned as the chairperson for the MMT. Wayne Hale, another experienced flight director, was scheduled to become the MMT chair in February 2003. Ham was the acting MMT chair until Hale was available to take that role permanently.

Some people questioned why Linda Ham was promoted so quickly within the shuttle management system, implying that she may have received promotions because of her gender more than her capabilities. Actually there are a very limited number of personnel in the Mission Operations Directorate, which includes the flight directors and their staffs. There are far fewer flight directors than astronauts, and when positions open up for new flight directors, there are several applications from the few qualified engineers. Candidates are questioned whether they're really planning on spending the next 10 to 20 years of their careers working at NASA, because of the intense amount of training time and dollars invested to certify new flight directors.

By the late 1990s, NASA had a shortage of shuttle managers. Many of NASA's more experienced flight directors were promoted to management positions, many had retired, and others—some with over 30 years experience—wanted to remain flight directors and weren't interested in management positions. Several qualified flight directors were promoted to engineering management positions, not just Ham.

Mid-level NASA managers view Columbia's debris after the accident. From left to right -- Wayne Hale, Linda Ham, Ron Dittemore, and Ralph Roe.

There's a myth that Linda Ham got her position because she's married to an astronaut. While it's true Ham's husband Ken is an astronaut, if there was any truth to the favoritism rumor, it would be exactly the opposite way around: She started to work at NASA in the early 1980s, he became an astronaut in 2000, by which time Linda Ham was already a lead flight director and an ascent/entry flight director. She had no influence on his selection or career as an astronaut. It's also absurd to even contemplate that a rookie astronaut could have any influence on the promotions of his wife, who already had over 20 years of experience at NASA.

Also unfairly, some of Ham's actions during the STS-107 mission were called into question after the accident. They're addressed in detail in Chapter 18, "Behind the Scenes." Many from within NASA and outsiders made Ham the scapegoat for what happened to Columbia, especially by those who didn't like the idea of a woman having so much responsibility and power.

SPACEHAB MISSION MANAGER PETE PACELEY

Pete Paceley was involved with the shuttle program for almost two decades. He originally worked for McDonnell Douglas and Teledyne Brown on the Spacelab program. (Spacelab was a European-built laboratory which flew inside the shuttles's cargo bay.) After the Spacelab program was shut down, in August 1998, Paceley joined the commercial firm Spacehab and was assigned to manage the Double Spacehab Research Module.

Besides sounding similar, Spacelab and Spacehab performed similar functions. Both were large pressurized laboratories where astronauts could work on science experiments. (Spacehab had newer hardware and was able to benefit from the Spacelab experience.) The original concept for Spacehab was a module that would occupy one quarter of the shuttle's cargo bay. It would expand the available pressurized space on the shuttle. Many more experiments could fly, while also retaining the capability of flying other payloads on the same shuttle mission. As a commercial enterprise, Spacehab had hoped to sell the space inside the pressurized module for a retail price of $1 million per microwave oven-size locker, but it had generated little interest. In theory it could sell up to 180 lockers on the six proposed missions. Many companies were interested in performing experiments in space, but the price tag for the ride was too high. NASA had a policy where commercial firms could work with aerospace companies and universities in a partnership arrangement where NASA would provide the ride on the shuttle for free. It was hoped that this would eventually lead to research that would have useful scientific benefits on Earth, insights from space experiments that would result in applications on Earth, or new hardware for use in space. NASA agreed to purchase most of the space in the four Spacehab missions that flew before the original contract for commercial payloads was finished (two flights were cancelled because of the lack of customers). But Spacehab never found a commercial company that was willing to pay $1 million to purchase a ride.

Later, Spacehab proposed that NASA should use its module as a cargo carrier to the Russian space station Mir. NASA had recently agreed to send several astronauts to Mir for four-month stays. As part of the agreement, Spacehab modules would carry cargo to Mir and return items to Earth for reuse. NASA agreed to rent the Spacehab module for six flights to Mir. One of those flights was STS-89, STS-107 astronaut Mike Anderson's first spaceflight. Most of those missions flew the Double Spacehab module—two modules put together to double the amount of cargo that could be carried. Each of the shuttle flights to Mir brought up both American and Russian supplies and returned finished experiments and unneeded items to Earth. Most significant from the Russian point of view, the Spacehab modules brought back failed hardware that could be examined to determine what had gone wrong. In addition, much of the failed Russian hardware that was returned to Earth was repaired, refurbished, and flown back up to Mir, reducing Russia's operating costs. The Double Spacehab was also used for three early flights to the space station, including STS-107 commander Rick Husband's first flight, STS-96.

Around the same time the Mir program was winding down, NASA was finishing the last of the two-week Spacelab missions. Spacehab felt there was a gap that needed to be filled and came up with its Research Double Module. Externally the Research Double Module looked like the Logistics Double Module used for the Mir and space station cargo flights. But internally it had more support capabilities, like power, communications, and cooling. The latter was especially important because three or four astronauts could be spending 12 hours inside, running some high heat load experiments. Lobbying by Spacehab and the microgravity community convinced Congress to provide the funds to NASA for at least one two-week microgravity research mission. STS-107 was the first flight of the Research Double Module, and Spacehab hoped it would lead to additional contracts with NASA for future research missions.

Paceley had worked on the Research Double Module for about five years by the time STS-107 finally flew—almost twice as long as originally anticipated. He admits that, throughout the many delays, he would occasionally think, "After so many [launch date announcements and delays], you just want to go fly this thing. After 10 of them, it gets old—you just want to go fly. It started being a joke."

FREESTAR MANAGER TOM DIXON

Tom Dixon was the FREESTAR manager at NASA's Goddard Spaceflight Center in Maryland. FREESTAR was part of NASA's "shuttle special payloads project," secondary payloads that fly when space is available. Dixon attended Loyola College. He started his space career with a contractor in 1986 and joined NASA as a civil service employee in January 1989.

One of the most fascinating portions of the special payloads division was the "Getaway Special" program—GAS. Any US school could pay NASA $10,000 to fly a scientific experiment in a 5-cubic foot GAS can; others were charged $27,000. Over a hundred student groups, commercial firms, and government agencies have flown GAS experiments. The more sophisticated version of GAS was Hitchhiker. A GAS experiment had to be self-contained and meet a variety of restrictions. In contrast, the Hitchhiker payloads had access to crew time, could request the shuttle be pointed in different directions, get power from the shuttle, transmit data during the mission, and request other shuttle "services." FREESTAR was a Hitchhiker payload with six experiments.

Hitchhiker made it possible for those without space experience to develop their instruments and fly them safely on the shuttle. Dixon recalled that it was a pleasure to work with scientists who wanted to fly simple experiments in space, and show them what they had to do to make their payloads work within the Hitchhiker program's rules. A wide variety of payloads from NASA, other government agencies, industrial firms, and schools flew within the Hitchhiker program, in what Dixon noted was a very family-like atmosphere.

Dixon was the manager of another Hitchhiker payload, Shuttle Ozone Limb Sounding Experiment (SOLSE—pronounced "sol ss"), which flew on STS-87, Kalpana Chawla's first mission. SOLSE looked sidewise through the atmosphere to examine the amount of ozone in the atmosphere at different altitudes. Most of NASA's Earth-observation payloads fly on freeflying satellites, but SOLSE was an exception, using the shuttle as a maneuverable platform to check out the sensors for possible use on future freeflying satellites. One of the key advantages for flying an experiment like SOLSE on the shuttle is that the experiment is carefully calibrated before the mission, and calibrated again after the experiment returns to Earth. With an instrument on a freeflying satellite, it's difficult to determine accurately how much the instrument has changed over time.

When Israel decided to conduct Earth observations, someone suggested Tom Dixon as an appropriate contact, since he had experience flying shuttle Earth-observation payloads. Dixon started working with the Israeli scientists in 1996. He told them it wasn't practical to use a camera from inside the crew cabin to take photos of dust because the shuttle windows have multiple layers with protective coatings. Dixon recommended another approach — a camera mounted within a canister inside the shuttle's cargo bay, and that was the origin of MEIDEX.

The Israeli scientists were primarily interested in getting scientific data, and not as interested in building the hardware. Dixon gave the Israeli Space Agency a list of aerospace contractors with Hitchhiker experience. The Israelis selected Orbital Sciences, which purchased a commercial Xybion camera and built the hardware necessary for the experiment to fit inside a "Hitchhiker Junior" canister. The heavy metal canister is about the size of a 55-gallon drum and includes a motorized lid to protect the camera when it isn't in use.

NASA's intense focus on the International Space Station's assembly and maintenance, even before the Columbia accident, had already led to the restructuring of the Hitchhiker program, and the Columbia accident was the final nail that shut down all Hitchhiker activities. Some of the hardware was given to other projects, and the rest was to be auctioned.

Dixon has moved on to the space environment testbed part of the "Living with a Star" program — spacecraft that study the Sun and how it affects life on Earth. He's still busy with his work but notes he isn't as passionate about it: "It's just not the same, not working on the shuttle anymore."

Ilan Ramon asked many of the members of the FREESTAR and MEIDEX teams to give him CDs to listen to in space so he could enjoy their favorite music and give them something that had flown with him in space. Dixon gave Ramon a CD by the band "Superchic[k]."

NASA MISSION SCIENTIST JOHN CHARLES

Dr. John Charles was the NASA mission scientist. That meant he was responsible for the NASA microgravity science experiments on the mission, but not the FREESTAR payload, or experiments from the European Space Agency or commercial firms. However, while he wasn't responsible for the other payloads, he was certainly interested, and helped coordinate various activities between the different organizations.

Charles was responsible for informing the media and the public at each of the status briefings about the science being performed aboard Columbia.

A long-time space fan, Charles was interested in the space program when he was young and followed the Apollo program. He participated in most of the shuttle missions where life science experiments were performed on the astronauts.

Charles was, however, too tall to fit into any spacecraft. Does he regret that his parents made him so tall he couldn't become an astronaut? Not really, because of all of the extra things astronauts have to put up with, he says. But he's happy with his contributions to the space program

SOFBALL SCIENTIST PAUL RONNEY

Dr. Paul Ronney is a professor at the University of Southern California. He has no direct connection with NASA, but as a combustion scientist occasionally uses NASA's resources for his research.

As part of its ongoing mission, NASA frequently asks the scientific community what experiments they would like to fly. Announcements are made at technical meetings and in professional journals, and feedback is solicited from the scientific community. Eventually a formal "Request For Proposals" announcement is made, where interested scientists are asked to submit proposals for how they would like to use the shuttle and what amount of support they would need from NASA to accomplish their science. The Microgravity Science Laboratory (MSL) included a variety of experiments, including a combustion chamber where carefully controlled experiments involving fires could be performed safely. Ronney proposed SOFBALL (Structure of Flame Balls at Low Lewis-Number—pronounced "sof ball") experiment to research how tiny flame balls act in microgravity.

NASA offered two of the seven seats on the flight to scientists outside of the astronaut corps. NASA astronauts would serve as the pilots, flight engineer, and two of the scientist-astronauts. The scientists responsible for MSL's experiments, both within and outside NASA, would select two representatives to fly as payload specialists. Ronney was selected along with Dr. Greg Linteris from the National Institute of Standards and Technology, and Dr. Roger Crouch, a NASA scientist who was not an astronaut. Eventually the science board decided to fly Linteris and Crouch, with Ronney as their backup.

SOFBALL flew on STS-83, in 1997. A technical problem cut the flight short, but fortunately NASA managers made the decision to fly the entire mission intact a couple of months later. Scientists got a bonus— some initial science to whet their appetites, and then the entire mission they had planned for. A total of 58 flame balls were generated on the two MSL flights. The SOFBALL scientists had expected flame balls would last a couple of minutes, but they were astonished to see the them still burning when the experiment automatically shut off. The obvious question was: How long could a flame ball burn in space? The SOFBALL team had always hoped for another mission, and NASA decided to refly the experiment for more data on STS-107. One of Ronney's goals was to send a flame ball around the world—through an entire 90-minute orbit.

Besides outer space, Ronney conducts combustion experiments on the ground in microgravity drop towers. The drop towers are extremely tall tubes with a vacuum inside. As the experiment falls, it experiences microgravity for a couple of seconds before it comes to an abrupt stop in a tub of sand at the bottom. This setup provides only a couple of seconds of microgravity, but is far less expensive and more readily available than a spaceflight. Ronney created the world's first flame balls in a drop-tower experiment in 1984. Later he discovered that a Russian physicist, Yakov Zeldovich, had predicted the theoretical existence of flame balls four decades earlier. One of the more unusual fuels that burns at the proper rate for the rapid drop tower experiments is the green foam used by florists and arts and crafts shops to display flowers.

Ronney said, "I think the crew liked the SOFBALL experiment in part because the flame balls have "personalities." Until you push the button and fire the spark, you don't know how many flame balls you'll get, and you don't know what they'll do once they start burning. Plus, humans are naturally drawn to fire, and flame balls are a bizarre type of fire that exists only in the low-gravity conditions of space."

FAMILY ESCORT STEVE LINDSEY

Before launch the astronauts select a couple of their colleagues as "family escorts." These astronauts take on the role of assisting the families in the hectic times leading to launch, while the crew is in space, and through the landing—until the crew is reunited with their families. In effect, the family escort becomes a foster parent for the family. The duties can include helping out with prelaunch receptions for family and friends, picking up the kids from school, taking them to tourist attractions when they come to Florida for the launch, and whatever else is necessary. If there's an accident, the family escort's title becomes "Casualty Assistance Call Officer" (CACO—pronounced "Kae coe"). Well before the mission, Rick Husband told me, "We've got lots of great folks who serve as family escorts here in the office." Normally there are two family escorts for each crew, but because of the additional security concerns, the STS-107 crew was given four escorts—Steve Lindsey, Scott Parazynski, Terry Virts, and Clay Anderson. Lindsey had flown on three shuttle flights, and Parazynski had flown on four. Virts and Anderson were rookies.

Lindsey and Husband both became astronauts in 1995, but Lindsey was lucky enough to fly more rapidly and more often. Lindsey first flew in space on STS-87, in November 1997, with Kalpana Chawla. Among other activities, Lindsey flew Sprint, a soccer-ball size remote-control satellite, from within the crew cabin during the second spacewalk.

Lindsey flew again in October 1998 on the STS-95 mission, which included a reflight of the Spartan from his first mission. There were a total of 85 different experiments on STS-95, but everything on the flight was totally overshadowed by the seventh crewmember—former Mercury astronaut and US Senator John Glenn, on his return to space at age 77. Unfortunately, with all of the hyperbole and enthusiasm for Glenn's presence, he got much of the praise the other astronauts actually deserved.

On his third flight, in July 2001, Lindsey commanded the STS-104 ISS mission. That flight featured the addition of the Joint Airlock to ISS, permitting spacewalks with US spacesuits while the shuttle wasn't present and additional flexibility for ISS assembly flights.

Lindsey set one of the fastest records for an astronaut arriving at NASA to command a spaceflight—just six years. And in those six years he flew on three missions. He was scheduled to fly on STS-119 in early 2004, another assembly flight to ISS, but after the STS-107 accident Lindsey and his crew were moved forward to fly on STS-121, the second shuttle mission after the Columbia accident.

Rick Husband had been a family escort for many of his fellow astronauts. Evelyn Husband said, "You take care of the crew member's family the way you'd want your family to be taken care of. And they really take that seriously. I think just from a practical standpoint for NASA it's a good choice because they want to keep the family happy so the crew can concentrate on what they're doing. It's very much appreciated." And she added, "Steve [Lindsey] was a huge help. Steve just made sure all of the t's were crossed and the i's were dotted as far as the logistics. We went over all of the [security] logistics for how we would get from place to place once we were in Florida. . . . On a daily basis one of the escorts would call and update us on the mission. They would brief us on different things that came up and set up our family conferences. We felt very connected to NASA during the mission."

Each astronaut also designates fellow astronauts to serve as their CACOs just in case something bad happens. Husband selected Lindsey. Mike Anderson chose astronaut classmate Carlos Noriega. Laurel Clark chose Jim "Vegas" Kelly, an astronaut classmate. (Laurel's husband Jon was a flight surgeon for Kelly's first spaceflight.) Dave Brown chose classmate and fellow Navy doctor Lee Morin. Willie McCool chose Navy friend Bob Curbeam. Kalpana Chawla chose Sunita Williams, another astronaut of Indian descent. Canadian astronaut Steve MacLean became the CACO for Ilan Ramon. After the accident, additional astronauts took up CACO responsibilities to assist the families.

Even to this day, CACOs still help out where they can. Evelyn Husband said Terry Virts and his family took her family to an Astros baseball game. She said, "Scott Parazynski still checks in regularly and does stuff with my kids, and so does Steve Lindsey. It's a job they have not abandoned. They've just tried to keep a connection with my children."

She added, "Scott Parazynski taught Matthew how to ride a bike. Matthew's birthday was in August, and Steve came over and played with all of Matthew's toys. Matthew got a rocket, and Steve showed him how to shoot the thing off. They do stuff like that, which is just terrific. Matthew got "Top Gun" for his game cube and wanted me to show him how to land. I said, 'Matthew, I have no clue how to land; next time Steve Lindsey comes over, he can show you.' So Steve played that with him one day when he came over. They can never replace Rick, and I'm not trying to, and we don't get to see [Steve] that often [in comparison with the months immediately after the accident], but it's nice that it's still there, it's not completely gone from their lives."

Lani McCool explained, "My first CACO was Bob Curbeam. I didn't like the fact that he was spending so much time away from his family. It just made me feel guilty. He was doing a lot of paperwork—which he did, and did a beautiful job. But I felt that I wanted to do things on my own. I've just always been independent, so I didn't really expect an intermediary for myself. I wanted information directly. I pretty much told NASA I didn't want a CACO, I just wanted to do things on my own, which was pretty much overwhelming, including the move [from Houston to Anacortes]. Stephanie [Wilson, an astronaut and one of Willie's classmates] just came up and volunteered."

STS-107 Family escorts Steve Lindsey and Scott Parazynski during the STS-95 shuttle mission.

CAPCOM CHARLIE HOBAUGH

Each shift in Mission Control has an astronaut who serves as the cap-com—the person in charge of direct communications to the crew. The STS-107 launch and entry capcom was astronaut Charlie "Scorch" Hobaugh. Duane Carey assisted as the "Weather capcom," coordinating with the observers in the field. Hobaugh flew with Steve Lindsey on the STS-104 ISS mission, which carried the Joint Airlock to ISS. Shortly before STS-107, Hobaugh was assigned to STS-118, Columbia's next planned flight.

In addition to the launch and landing, Hobaugh was the capcom for the "Orbit 1" shift in Mission Control, serving many shifts during the mission. On long shuttle missions NASA has four shifts in Mission Control, named "Orbit 1," "Orbit 2," and so forth. Each shift serves no more than six days in a row before getting time off for a weekend, although the "weekend" can occur on any day of the week.

Hobaugh enjoyed joking with Columbia's crew, and they clearly enjoyed their interaction with him.

CREW SECRETARY ROZ HOBGOOD

The STS-107 crew secretary was Rosalind 'Roz' Hobgood, a/k/a "The Great and Powerful Roz." The world found out about her when chief astronaut Kent Rominger mentioned her at the memorial ceremony on February 4, 2004, three days after the accident. She said, "I was kind of numb that day and wondered, 'Did he say my name?' I was sitting next to Jon Clark's brother—he got real teary-eyed and gave me a big hug."

Dave Brown's neighbor, Al Saylor, gave Roz her nickname in June 2002 as a variation on the "Great and powerful Wizard of Oz." Saylor said, "Oh, so you're the great and powerful Roz," and Dave Brown repeated, "Ah, the great and powerful Roz." Saylor didn't realize the nickname had stuck until the memorial ceremony.

Capcom Charlie Hobaugh before
Columbia's launch.

Hobgood was Husband's secretary since his STS-96 postflight activities. Rick was out of the office when Hobgood found that he was assigned to STS-107. She said, "I couldn't wait, so I called him on his cell phone to say congratulations."

There are seven secretaries who support the crews from the time they're assigned through their postflight activities. Rick Husband specifically asked Roz to become the crew secretary for STS-107. It was only her second time as a crew secretary; she had previously done the job for the STS-98 crew.

She recalled, "They were like part of my family: 80 percent of the time I'm not serious, I'd just go in there and mess with them. I'd tell them, 'You can tell them you really love your crew secretary—can you write my name on the back of the shuttle and fly it?,' just anything to disturb them. I'd see Ilan writing Hebrew and tell him he couldn't go because he was dyslexic (because he was writing from right to left) and everybody could see us laughing. Laurel would come in and give me these 60-second massages. Willie would come in with flowers; Mike always had a great smile on his face. K.C. was very articulate and very gifted in her way of approaching things and her mind. Dave was just 'happy Dave,' and he loved his dog and his planes and his photography. It was just a happy crew. It was a mutual love—it was cool."

The STS-107 crew posed with crew secretary Roz Hobgood for a special photo. The inscription reads, "To Roz, With never ending gratitude from *your* STS-107 crew!!" Photo courtesy of Roz Hobgood.

One of Hobgood's responsibilities was handling all of the invitations for the astronauts' families and guests. Each astronaut gets a limited number of seats, and they come in various levels of importance. The extended family and very close friends get a special reserved area closest to the launch pad. Camp friends, cousins, acquaintances, and others get passes for viewing sites further from the pad with fewer amenities. Each astronaut has to carefully select who's on his or her invitation list, and which category they fall in. (Some 2,000 invitations were sent for the entire crew.) It was frustrating when an astronaut invited "Friend plus guest" and the friend wants to bring additional people and Roz had to go to the extra effort to try to accommodate the extra guests, especially when passes were further restricted after 9/11. She recalled, "Dave Brown had a friend who was bringing a lot of people. Dave would come to me and say, 'Oh great and powerful Roz, what are we going to do?' You have a lot of people contacting you after the initial invitation goes out. It's coming closer and closer to launch and you have to survive on caffeine and candy."

She said, "The challenging part of my job is not the day-to-day job, it's the challenges—the heart-beating let's get everybody seated, make sure everybody's happy." Her responsibility was to ensure the guests contacted her instead of bothering the astronauts or their spouses. She said, "That part is the adrenaline part for me of working this and getting to know the people."

During the mission, the astronauts sent e-mails to their favorite secretary, and saw some video of her crew. She said, "I was so excited, I can't even tell you how it felt." She recalled, "I wanted to spend the time with them to hear about their flight. Not the stuff they tell the press—I wanted to hear what was the most magnificent sight they saw—I wanted to see the twinkle in their eyes. I wanted to be next to them and feel that energy."

After the accident, Hobgood questioned whether or not she wanted to keep her job, now that her crew was gone. She had lost almost as much as the crew's biological families. She took time off and eventually decided to return. One of her tasks was to log all of the cards, flowers, books, and other items—over 12,000—that were sent to NASA in memory of the crew.

Hobgood had been putting off having surgery until after the mission and through all of the delays. She said, "There was no way I was going to be laid up and not be their crew secretary." She finally had her surgery after the last funeral.

STUDENT TARIQ ADWAN

One of the least likely people to be involved with a space shuttle mission with an Israeli astronaut was Palestinian biology student Tariq Adwan. He attended Misericordia College, in Dallas, Pennsylvania, but considers Bethlehem his hometown. Not Bethlehem in Pennsylvania, but the original Bethlehem, in the Middle East. The space advocacy group The Planetary Society wanted to fly an "experiment for peace." One of the experiment's requirements was to have one Israeli and one Palestinian student work with the scientists. The professional scientists included Americans, Israelis, and Arabs.

Adwan's father is part of the Peace Research Institute in the Middle East (PRIME), an educational effort to encourage more cooperation and understanding among Israeli Jews and Palestinians. Tariq said, "I've participated in a lot of Palestinian/Israeli joint programs in Palestine, when I was still living in Bethlehem. One of the things I missed when I came to the United States was these programs, because I got a lot out of them. I got to know a lot of Israeli friends and hear about the situation from a different perspective." He found out about the Planetary Society's project and realized that he'd have the opportunity to work with Israelis again. Adwan said, "To have an opportunity to work in the United States with an Israeli partner on a scientific project was, for me, a fulfillment of this purpose— my desire for working and communicating with Israeli colleagues."

Adwan ended up studying in the United States because of a scholarship offered by a Jewish family in Scranton. He noted, "What a great opportunity, and I didn't want to miss it, so I took it." He said he was the only Palestinian at the school, although there's a small Arab-American community in the area.

Adwan's Jewish counterpart was Yuval Landau, a medical student attending Sackler School of Medicine at Tel Aviv University. Tariq and Yuval met in Florida for the STS-107 launch; they had exchanged e-mails for about three weeks but had not previously met in person. Adwan said, "I was very anxious, but very nervous at the same time. Throughout the joint programs there's a wide spectrum among the Israelis you can meet in terms of how open-minded they are, what are their feelings and how they perceive you. Once I met him and got to know him, it turns out he's an open-minded person and a very accepting person."

Yuval only eats Kosher food, and the pair could not find a Kosher restaurant in the area close to the Kennedy Space Center. They eventually found a rabbi who offered them Kosher food. Adwan notes that it's occasionally difficult for him to keep Halal, the Muslim food laws prescribed by the Qur'an. He can easily avoid pork products and alcohol, but it's more difficult to determine if ingredients in processed foods come from acceptable sources.

Adwan has mixed feelings about Ilan Ramon because of his multiple roles. Adwan said, "He was one of the astronauts so I view him as a scientist. As a scientist, he was our hands in space, turning on and turning off our experiment for us. We certainly look up to him and respect that and highly appreciate it." However, "As a militant, I don't appreciate him very much— but that's me, I'm a peace person—I don't believe in military actions."

During the mission Ilan Ramon said, "To the people of Israel, I wish we will have a peaceful land to live in very soon." Adwan said, "I think that's very significant, and those are very promising comments from a scientist like him. I hope that out of respect to him people actually start doing that, too. He sacrificed his life for science and peace. He set a great example for us to follow in those footsteps."

Adwan also said about his experience, "It definitely was a great opportunity for me to participate in a scientific-peace project. For me it was a wonderful fulfillment. It gave me an opportunity to promote peace on almost a global level. I hope this experiment really sets an example that Palestinian-Israeli collaboration can exist. If the right people are behind it, the sky's the limit. I hope this is just the beginning—not the end."

Tariq Adwan and Yuval Landau keep in touch—but only from a distance. Adwan said, "I wasn't really allowed to leave Bethlehem. It's very hard to leave Bethlehem and go to Tel Aviv, where Yuval lives. So we don't get a chance to see each other."

Would Adwan do it again if he had the chance? "Absolutely, I would not even think twice about it."

Israeli student Yuval Landau and Palestinian student Tariq Adwan. Photo courtesy of The Planetary Society.

January 16, 2003: Columbia's final launch. Photo by Philip Chien

PART II

THE LONG ROAD
TO LAUNCH

When the STS-107 research mission was first approved, it was anticipated that it would fly fairly quickly—all of the experiments had been previously approved, and some had already flown in space. It was expected to be a simple dedicated science mission, a bridge to the full-time science research that would begin shortly on the International Space Station.

But it turned out that STS-107 had to travel a long road to launch, with many delays. Out of necessity, it became the lowest priority mission in NASA's manifest. Once construction started on the space station and astronauts started living onboard, ISS missions had to be NASA's highest priority. Still, nobody could have predicted the many different twists and turns STS-107 took as it moved toward liftoff.

When Columbia finally did launch—on January 16, 2003—it was time for the astronauts and scientists to do what they had been training to do for several years—two weeks of intense research into the microgravity environment.

This part of the book begins with "A NASA Primer," a chapter which explains NASA terminology, and how things work in the organization and on the shuttle. The chapters that follow then describe the long and winding path up to and including the day Columbia finally left the launchpad.

CHAPTER 11 A NASA PRIMER

NASA uses terminology that seems extremely strange to outsiders and is often confusing to insiders. Why was Willie McCool the "pilot" even though he didn't fly the shuttle? Why did Ilan Ramon sit on the middeck for launch and landing? What's an MS2? Why was there such a long tunnel between the Spacehab laboratory and the crew cabin? Why did David Brown and Laurel Clark exchange seats for landing? And with all of the risks astronauts take—what's their take-home pay?

ASTRONAUT PAY

Astronaut pay has always been a funny issue. Many astronauts have said they'd gladly pay NASA for the opportunity to fly in space. Some take drastic pay cuts when they join NASA. Most are realistic and acknowledge that they need to pay their bills, just like everybody else. In many cases military astronauts decide to leave NASA primarily for financial reasons—each month they stay, they lose money.

Military astronauts are paid their normal military pay, a function of their rank and how long they've been on active duty. Civilian astronauts are paid decent but not fantastic salaries—anywhere from $50,000 to $110,000 a year, depending on their experience and the amount of time they've been with the government. For many incoming astronauts it's a step down in pay from their former jobs in private industry. It's important to note astronauts are paid the same amount as anyone who works for the federal government with the same experience and skills. A medical doctor astronaut is paid the same as a doctor working in a health clinic at NASA, or a doctor working for the National Institutes of Health.

Astronauts do get "bonus pay" while they're in space, but it isn't much—just $2 a day, and it isn't a special deal for astronauts: An astronaut in space is technically a government employee on official travel. He or she can file an expense report for travel-related expenses. But the government provides the transportation and living quarters (the space shuttle), clothing, food (freeze-dried or out of a bag), and entertainment (the view out the window, music CDs, and all the zero-G acrobatics you can manage). When that has all been subtracted, all that's left is the $2 per day. (One astronaut speculated that the $2 was presumably for dry cleaning.) The official travel orders describe an itinerary from home in Houston, Texas, to KSC in Florida, through launch, the mission, landing, and return to Houston. For STS-107, including the days before launch and a couple of days after landing, the total bonus pay is about $44. Some astronauts don't bother with the time it takes to fill out an expense form. Others get a kick out of spending the bonus on a special dinner with their spouse. Some people have suggested that astronauts should be paid by the mile, and since the shuttle travels at 17,500 m.p.h. that would add up quickly. But that idea hasn't been approved.

DEATH BENEFITS

There's a myth that astronauts can't get life insurance. As government workers, they are eligible for the Federal Employees' Group Life Insurance program. In addition, they can purchase extra insurance on their own, but the premiums are—not surprisingly—extremely high.

The death benefits for the Columbia crew are basically the same as any other government employee killed on the job. Kalpana Chawla was the only civilian on STS-107, and her husband received half of her salary, plus $24,350 per year. She was also eligible for Federal Employees Group Life Insurance. The other five Americans were all active military officers and got death benefits of $250,000—the same as any other military officer who dies whether in combat or from a disease. Their spouses get $935 a month plus $234 per child until the child reaches age 18. The military also provides $8,000 in cash for immediate needs and up to $6,900 for burial costs. Laurel Clark was a Commander in the Navy and a "captain selectee," and was scheduled to get her formal promotion and pay raise in the summer of 2003. She was given her promotion posthumously. There's also been an effort to obtain posthumous promotions for Willie McCool and Mike Anderson. Husband and Brown were already "maxed out" in their Air Force and Navy careers.

In addition, there's something called "the unwritten compact." The Johnson Space Center is a very close extended family. Beyond the formal benefits, every astronaut knows his or her family will be well taken care of by colleagues if something happens. It's a very tight-knit community.

TERMINOLOGY

The "STS" in STS-107 stands for "Space Transportation System." In the original concept, the name STS refers to the space shuttle with its Solid Rocket Boosters (SRBs) and External Tank (ET) as it's configured for launch. In practice, STS is just the designation for each flight. Flight numbers are normally assigned in the order they're supposed to fly, but

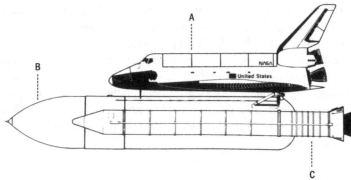

Ready for launch, the shuttle consists of three main components:
(A) the Shuttle Orbiter attached to (B) the External Tank and
(C) two Solid Rocket Boosters, with only one shown in this view.

the actual order often changes for a variety of reasons. STS-107 was the 113th shuttle mission, flying after several missions with higher STS numbers. The "orbiter" is just another term for the space shuttle vehicle.

Columbia was the first shuttle to fly in space, in 1981. It was designated OV-102 (Orbiter Vehicle). OV-101 was "Enterprise," a prototype used for early tests. The other operational shuttles were named "Challenger," "Discovery," "Atlantis," and "Endeavour."

"Payload" is a fancy term for cargo. The payload can include everything from food and office supplies to scientific experiments, both inside the crew cabin and in the cargo bay. Pressurized payloads are those inside the crew cabin or some other module such as Spacehab, while unpressurized payloads are the ones within the cargo bay.

When most people think about "shuttle delays," they're actually thinking about "launch attempts." A "launch attempt" is generally defined as when the ET is loaded with its propellants for launch. If the launch cannot take place that day for whatever reason, the launch is said to have been "scrubbed." Statistically, about half of the time there's a launch, and half of the time there's a scrub—usually because of unacceptable weather or technical problems. For STS-107, there were no launch delays: it launched on the first try, on January 16, 2003. But STS-107 *did* have many schedule slips. (A "schedule slip" is any delay to the scheduled launch date prior to the actual launch attempt. It can be caused by a technical problem, management decision to exchange the planned order for a group of missions, or other reasons.)

A "launch window" is the allowable period when a launch can take place each day. The window is determined by a variety of technical requirements. It can be extremely small if the shuttle is going to rendezvous with a satellite in space, or long if the mission doesn't have many constraints. The 2.5-hour STS-107 window was the maximum allowed for the shuttle. (The 2.5 hours is a constraint determined by the flight surgeons—as the maximum amount of time the crew should be allowed to lie on their backs with their knees pointed upwards inside their hot launch and entry suits.) The time the launch window opened for STS-107 was determined by the desire to have a daylight launch and daylight landing under all possible conditions— daylight for the planned end of mission, daylight if the mission had to be extended two days and land at the alternate landing site in California, and daylight if the mission was cut short for whatever reason. In addition, the Israeli MEIDEX experiment required good lighting conditions over the Mediterranean Sea. In contrast, a shuttle flying to the ISS has to launch close to the moment where the Earth rotates underneath the orbit of the ISS, so those missions typically have launch windows of only 5 to 10 minutes in duration.

ROCKET ENGINES AND PROPULSION

"Propellant" is an engineering term for both the fuel and oxidizer. For liquid propellant engines, the fuel is typically hydrogen (in the shuttle main engines) or monomethyl hydrazine (in the maneuvering thrusters); the oxidizer is liquid oxygen (in the main engines) or nitrogen tetroxide (in the thrusters). For solid propellant, the aluminum powder fuel and ammonium perchlorate oxidizer (similar to fertilizers) are mixed together and baked with rubber, which holds everything together in a semi-hard material that looks and feels like a gray pencil eraser.

The Solid Rocket Boosters (SRBs) are the two large rockets that provide most of the thrust (and vibration) at launch. After the SRBs complete their 2-minute 4-second burn, they're ejected. The boosters have parachutes and drop into the ocean, where they float vertically like an upside-down bottle. Specially equipped ships meet the boosters and pump air into them so they tip over into a horizontal position for transit. After the spent boosters are towed back to land, they're shipped back to the factory and refurbished for future missions.

The External Tank (ET) is the giant orange-colored cylinder on the side of the shuttle when it launches. Actually it's two tanks—one for the liquid hydrogen fuel and one for the liquid oxygen oxidizer. The ET is 153.8 feet long and has a diameter of 27.6 feet. It's similar in concept to the drop tanks used to extend the range of military aircraft. After the fuel in the drop tank is used, the tank is jettisoned. The ET is the ultimate drop tank, filled with 500,000 gallons of oxygen and hydrogen at launch. Those propellants flow through 17-inch diameter pipes into the back of the space shuttle to the three main engines. The orange color comes from the Spray On Foam Insulation (SOFI, pronounced "so fee"), similar to spray cans of foam insulation available in home improvement

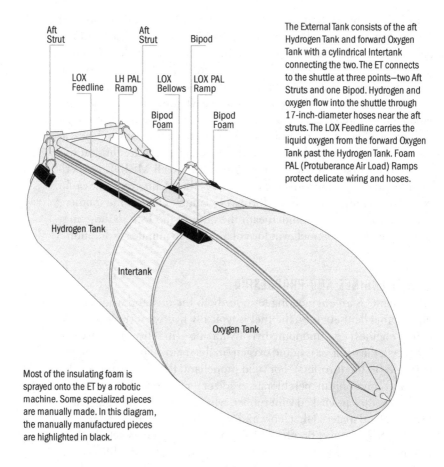

Aft Strut

Aft Strut

Bipod

LOX Feedline

LH PAL Ramp

LOX Bellows

LOX PAL Ramp

Bipod Foam

Bipod Foam

Hydrogen Tank

Intertank

Oxygen Tank

The External Tank consists of the aft Hydrogen Tank and forward Oxygen Tank with a cylindrical Intertank connecting the two. The ET connects to the shuttle at three points—two Aft Struts and one Bipod. Hydrogen and oxygen flow into the shuttle through 17-inch-diameter hoses near the aft struts. The LOX Feedline carries the liquid oxygen from the forward Oxygen Tank past the Hydrogen Tank. Foam PAL (Protuberance Air Load) Ramps protect delicate wiring and hoses.

Most of the insulating foam is sprayed onto the ET by a robotic machine. Some specialized pieces are manually made. In this diagram, the manually manufactured pieces are highlighted in black.

stores. The insulation is necessary because of the temperatures of the supercold liquid propellants: $-423°$ F for hydrogen and $-297°$ F for oxygen. Without the insulation, frost would form on the outside of the tank, like moisture on the outside of a cold freezer. The ice is dangerous for several reasons—vibrations would cause it to fall off during launch, potentially hitting a delicate portion of the shuttle. Ice would also affect the aerodynamic flow, and any ice that forms on the outside is excess weight that reduces a rocket's performance.

The ET has two attachment points on each side for the SRBs and three on the front for the shuttle. The lower attachment points are close to the shuttle's tail, the upper attachment point is close to the nose. Also at the bottom are the pipes that bring the propellants from the ET into the shuttle's plumbing. The ET forward attachment is a two-leg upside-down V-shaped support known as the bipod (a two-legged version of a tripod). The "bipod foams" are wedge-shaped pieces that insulate the bottoms of the metal legs to prevent ice from forming.

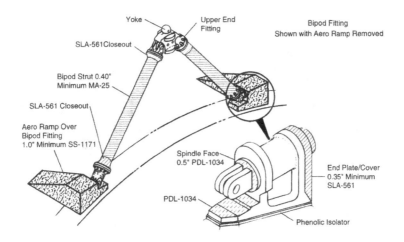

Yoke

Upper End Fitting

Bipod Fitting
Shown with Aero Ramp Removed

SLA-561Closeout

Bipod Strut 0.40"
Minimum MA-25

SLA-561 Closeout

Aero Ramp Over
Bipod Fitting
1.0" Minimum SS-1171

Spindle Face
0.5" PDL-1034

End Plate/Cover
0.35" Minimum
SLA-561

PDL-1034

Phenolic Isolator

The forward attachment fitting for the External Tank is a two-legged bipod. Each "foot" of the bipod is covered with a wedge of insulating foam.

The foam insulation is sprayed on the ET's large cylindrical surfaces by a robotic machine. Some specialized pieces of foam, like the bipod wedges, are manufactured by hand.

There have been a couple of changes to the ET over time. The lightweight tank started flying in 1983, and the superlightweight tank in 1998. The superlightweight tank uses a lighter Aluminum-Lithium alloy that shaves off several hundred pounds. Superlightweight tanks are used for ISS flights or extremely heavy launches. STS-107 was the exception, as it had a lighter cargo for launch, so it used the lightweight ET.

PROTECTING THE SHUTTLE DURING REENTRY

Thermal protection materials insulate the shuttle from the intense reentry heat. The areas that experience less heating are covered with white thermal insulating blankets. Areas that encounter more heat are covered with two versions of the black-colored tiles. Each tile is custom sized for its location. They're typically about 4 by 4 inches. The tiles start as sand—super pure sand that's converted into a "foam" with incredible heat-absorbing qualities and then coated with a stronger black coating. They're extremely light—about the same density as Styrofoam. You can put a tile into an oven and heat it until it is glowing white-hot and remove it with your bare fingers without getting burned because the tiles retain heat so well.

The highest intensity heat is on the tip of the nose and the leading edge (front) of the wings. They're covered with the best insulating material—Reinforced Carbon-Carbon (RCC). This gray material is also used

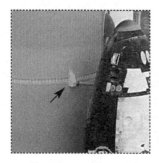

Columbia sits on launch pad 39A.
The arrow shows the wedge-shaped
foam on the foot of the left leg of the
bipod. Almost the entire bipod foam
ripped off 81 seconds after launch,
dooming Columbia.

for missile nose cones. It's extremely tough and can withstand a lot of
heat. But RCC is heavy, so it's only used in the areas that experience the
most heat. It would be more accurate to compare RCC with fiberglass
than tile. Each wing has 22 custom-shaped RCC panels, called simply
"RCCs." The RCCs are U-shaped and wrap around the front of the wing.
The T-seals, made from the same material as the RCC, interlock the panels. Each RCC has a pair of "carrier panels," access panels to attach and
remove the RCC from its mounting brackets on the front of the wing.
The carrier panels are aluminum with a couple of black shuttle tiles
mounted on top.

A typical installation of
a Reinforced Carbon-
Carbon panel on the
leading edge of a shut-
tle's wing.

TRACKING AND COMMUNICATIONS

People who track satellites refer to STS-107's orbit around the Earth as a "39 degree orbit." The degrees have nothing to do with temperature; it's the angle of the orbit in relation to the equator. A satellite's inclination determines how far north and south of the equator it travels. The angle, or inclination, when the satellite crosses the equator is the same as the maximum latitude the satellite reaches. If a rocket is launched due east, its inclination is equal to the launch site's latitude and benefits the most from the Earth's rotation. KSC is at 28.5 degrees north latitude, so a rocket launched due east ends up in a 28.5 degree orbit and can carry the maximum amount of cargo. The Baikonur launch site in Kazakhstan is located at 45 degrees north, but rockets launch slightly to the northeast to avoid traveling over China. That results in an inclination of 51.6 degrees. ISS is in a 51.6 degree orbit because of Russia's involvement. The shuttle launches to the northeast to reach a 51.6 degree orbit. The 39 degree orbit for STS-107 was chosen because it maximizes the amount of time the shuttle travels over the Mediterranean, the primary region of interest for the Israeli MEIDEX experiment.

The Tracking Data and Relay Satellite System (TDRS or TDRSS — pronounced "Tee dress") consists of a group of satellites in orbit which act like a switchboard in the sky. The shuttle and other satellites send their signals up to a TDRS instead of only when they fly over ground stations. The TDRS retransmits the data to a ground station in White Sands, New Mexico, and permits Mission Control to keep in contact with the shuttle for more than 85 percent of its orbit, as opposed to less than 15 percent for previous spacecraft that relied on ground stations. TDRS is also used by a variety of scientific satellites and some classified satellites. Besides the normal shuttle communications, one of the STS-107 experiments, the Low Power Transceiver (LPT), also performed its own experiments through TDRS.

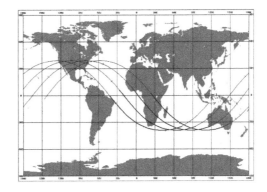

Columbia's orbit took it over the Earth up to 39 degrees of latitude away—North and South—from the equator.

The capcom is the astronaut in Mission Control who talks directly to the shuttle. The name is a contraction of "capsule communicator," from when the Mercury spacecraft was often called a capsule. It would be extremely confusing if each of the flight controllers talked directly to the astronauts, so the capcom acts as a central clearinghouse for the information that needs to be communicated with the crew. The STS-107 capcoms were rookie astronauts Stephanie Wilson and Ken Ham, and experienced astronauts Linda Godwin and Charlie "Scorch" Hobaugh.

The POCC (pronounced "pock") is the Payload Operations Control Center, located within the same building as the shuttle flight control team. Each of the Spacehab and middeck experiments has a team of scientists who monitor the science experiments as they're performed in space. They get some telemetry data, voltages, temperatures, etc., and in some cases the verbal descriptions by the astronauts or video of their experiments in operation. The POCC also includes engineers who monitor Spacehab's systems.

One of the most important positions in the POCC is the Crew Interface Coordinator (CIC). The CIC is an engineer who represents all of the scientists in the POCC to the astronauts, basically the equivalent of the capcom. But while the capcom is primarily responsible for the shuttle's systems, the CIC is responsible for the payloads. The three CICs for STS-107 were Lora Keiser, Brad Korb, and Beth Vann.

The FREESTAR experiments, including MEIDEX, had their own POCC at the Goddard Spaceflight Center in Maryland. When they wanted to talk to the shuttle, they communicated with the payloads officer in Mission Control, who passed on their information to the capcom, who in turn informed the astronauts.

TIME IN SPACE

One of the more common questions astronauts get asked by school children is "How do you keep track of time in space?" The simple answer is, "With a wristwatch." With "day" and "night" each lasting 45 minutes, you've got to come up with more practical ways of keeping track of time, and there is a space equivalent of a clock and calendar—"mission elapsed time" and "flight days."

Mission Elapsed Time or MET is the time zone and clock for the astronauts in space. The MET clock starts at launch. There's no way to predict whether or not there's going to be any delays before a launch takes place, so the flight plan is created with everything scheduled by MET. An event may be referred to as "1 day 10 hours 15 minutes MET." It's fairly easy to convert to and from Central Standard Time or any other time zone if you know when the launch took place. For example, STS-107 launched at 10:39 A.M. EST on January 16, so an MET of 1 day 10 hours 15 minutes is January 17 at 8:54 P.M. Eastern, 7:54 P.M. Central.

Another way to track time is by Flight Days. Flight Day 1 starts when the crew wakes up on launch day. (Flight days start counting with 1 and start a couple of hours before launch, MET starts with 0 at launch.)

There can occasionally be some confusion for when something happened during a mission if you don't keep track of what time reference you're using. As an example, the inflight press conference took place at a MET of 12 days 19 hours and 5 minutes, on Flight Day 14. In real-world terms, that was Wednesday, January 29, at 5:44 A.M. Eastern.

In space, astronauts keep track of time via MET and Flight Days and can literally lose track of the days on Earth because they're concentrating on the mission's timeline. It's extremely common for astronauts to talk about waking up at 17:30, meaning some number of days, 17 hours and 30 minutes after launch, or refer to "flight day 3" instead of "Saturday."

Entry Interface (EI) is the time when the shuttle reaches an altitude of 400,000 feet (75.7 miles), the top of the atmosphere, during its reentry. It's an arbitrary value because the atmosphere gradually becomes thicker as you descend. Many of the reentry events are measured in terms of the amount of time since EI. On STS-107, EI took place at 8:44:09 A.M. EST on February 1.

THE SHUTTLE'S LIVING SPACE

The shuttle's crew cabin has two levels, the flight deck and the middeck. There's also some space below the middeck for storage. The flight deck has the controls for flying the shuttle, with two seats for the pilots. The back portion, called the aft flight deck, has controls for operating many of the shuttle's systems and some payloads. There are two seats on the aft flight deck for additional astronauts for launch and landing. Those seats are folded up and stored while the shuttle is in orbit. Most of the windows are located on the shuttle's flight deck.

The flight deck is where the astronauts operated the FREESTAR experiments, looked out of windows at the Earth, and controlled the shuttle. It was also the location of Ilan Ramon's interview with Prime Minister Ariel Sharon and most of the press interviews. The middeck is where the crew lived with the galley, toilet, and some of the experiments. Most of the experiments were in Spacehab, a 2,200 cubic foot pressurized module in the shuttle's cargo bay. To maintain the proper center of gravity, the Spacehab module needed to be placed in the middle of Columbia's cargo bay, so a 15-foot tunnel connected it to the shuttle's crew cabin.

The middeck also included the hatch used to enter and leave the shuttle on the ground and a set of four sleeping bunks, each about the size of a very small hall closet. Three seats on the middeck were used by

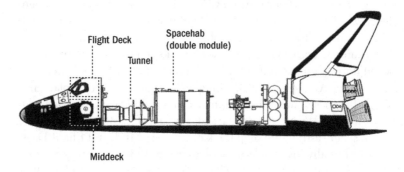

The pressurized areas inside the shuttle and Spacehab.

the crew during launch and landing; and folded and put away while the shuttle was in space.

Each shuttle has an airlock which permits two of the astronauts, wearing spacesuits, to exit into the vacuum of space. The airlock permits the rest of the shuttle to remain at normal atmospheric pressure while only the airlock chamber is depressurized.

Unlike the other shuttles, Columbia's airlock was inside the crew cabin, occupying a portion of the middeck. If a spacewalk was necessary, the Spacehab module would be sealed off, and Brown and Anderson would use the airlock to exit Columbia through a hatch on top of the tunnel from the crew cabin to Spacehab. The other five astronauts would remain within the crew cabin. The crew used the airlock and tunnel as their passage between the middeck and the Spacehab module.

There are two openings between the flight deck and the middeck. There's a ladder for use on Earth after landing, but in space the astronauts just float between the two levels.

The crew had access to the crew cabin (flight deck and middeck), tunnel, and Spacehab module. The cargo bay also included the FREESTAR pallet, and the Extended Duration Orbiter (EDO) pallet behind Spacehab. The EDO pallet had a set of spherical propellant tanks filled with supercold liquid hydrogen and oxygen to generate the power Columbia needed for its extended stay in space.

ASTRONAUT DESIGNATIONS AND DUTIES

Every shuttle mission has two pilots, known as the "Commander" and the "Pilot." On any other aircraft they would be called "Pilot" and "Copilot," but astronauts have egos and don't want to be called Co-anything. McCool was honest about his position, stating, "I have the privilege of actually being Rick's copilot, I sit in the right seat; it's a little bit of a

NASA misnomer when I'm introduced as the pilot." It's the commander who actually flies the landing manually; the pilot only gets the opportunity to fly for a short time before the landing. The pilot, often a rookie astronaut, is the jack-of-all trades, backing up the commander and helping with other activities. The pilot also has the "honor" of keeping the toilet working. After one or more missions flying as a pilot, an astronaut may be upgraded to commander, but that's by no means certain. Many pilots leave NASA without ever commanding a mission. In contrast, Rick Husband was upgraded from pilot to commander after only one mission, because there was a shortage of qualified commanders and he had the proper skills and abilities to step up to the greater responsibilities.

A shuttle crew normally includes three to five "mission specialists," who operate most of the experiments. Mission specialists are usually the astronauts who perform spacewalks, operate the shuttle's robot arm, and manage other critical shuttle functions. The flight engineer is designated MS2 and the other mission specialists are MS1, MS3, and if necessary MS4 and MS5. There's no particular seniority with the MS1, MS3, MS4, and MS5 assignments.

A shuttle crew can optionally include one or more "payload specialists." A payload specialist is a non-career astronaut who gets to fly on the shuttle. Normally it's a scientist or engineer dedicated to a particular payload. The title is also given to VIP passengers.

Payload Specialists are normally "one-shot" space travelers, although some have been lucky enough to fly twice. Many in NASA don't even want to call them astronauts because they typically don't work for NASA. But they do get to fly in space, so certainly by any reasonable definition they are astronauts. Payload specialists include scientists who have worked on an experiment for their entire careers, politicians flying on the ultimate taxpayer-sponsored junkets, and international passengers flying as a form of diplomacy.

In theory, payload specialists are only supposed to fly when there's a role that can't be filled by a career astronaut—for example, if the shuttle is carrying a scientific instrument that's so complex it isn't worth the time or effort to train a NASA astronaut to operate it. There's no reason the MEIDEX experiment had to be operated by an Israeli, other than the fact that the United States offered a shuttle seat to Israel. While Ilan Ramon was technically a payload specialist, his responsibilities were more like those of a mission specialist. Another NASA astronaut would have occupied Ilan Ramon's seat if the USA had not given Israel the opportunity to fly somebody in space.

By definition, the pilot astronauts are all test pilots with military backgrounds. The mission specialists include engineers, scientists, medical doctors, flight test engineers, and, most recently, teachers. The pilots and mission specialists are "career astronauts," government employees who remain with NASA. They're given technical assignments when not assigned to a specific mission and normally expect to make multiple spaceflights before they retire. An astronaut stays with NASA for an average of about ten years before moving on to a highly-paid career in industry or a management position. A few retire quickly after a single spaceflight, but others have stayed with NASA for much longer. The case of astronaut John Young is the rarest in terms of longevity: he worked for NASA for 42 years.

Seating on the shuttle is determined by each astronaut's responsibilities during launch and landing. The commander sits in the left front seat, the pilot is in the right front seat, and the flight engineer sits behind and between them. The STS-107 flight engineer (MS2) was Kalpana Chawla, and during launch and entry she read off the checklists and critical instruments to assist the pilots. The mission specialist in the fourth seat doesn't have any major responsibilities during launch or landing, but can assist with checklists, especially if there's an emergency. The other three seats are on the shuttle's middeck.

The astronauts on the flight deck have windows, while their crewmates downstairs can only stare at a wall of lockers. Generally, if a mission specialist has already seen a launch or landing from the flight deck, he'll defer to colleagues who haven't had the opportunity, especially the rookies. Mike Anderson, the flight engineer on his first flight where he got a great view of the launch and landing, sat on the middeck for Columbia's launch and entry.

Two of the other mission specialists swap seats for the launch and entry. Dave Brown had a lot of experience working at KSC, and was also a jet pilot, so it made sense for him to sit on the flight deck just in case there was an emergency during Columbia's launch. Dave and Laurel swapped seats for the reentry so Dave got a good view for launch and Laurel had the good view for entry. Payload specialists always sit on the middeck.

The commander, pilot, and flight engineer are often called the "orbiter crew," since they're responsible for maintaining the shuttle's systems. The rest of the astronauts are called the "payload crew" or "science crew" because they're responsible for most of the experiments. These are only broad designations; in many cases the orbiter crew will perform experiments, and the payload crew backs up the orbiter crew for some functions. K.C. was Rick Husband's backup and Dave Brown was Willie McCool's backup for orbiter activities. NASA's flight rules exempt the orbiter crew from invasive experiments like blood draws. The logic is there can be a major problem at any time resulting in an emergency reentry. If that happens, the pilots and flight engineer have to be at their peak performance, and not lightheaded after a blood draw. The rest of the astronauts just sit through the reentry, so there isn't as much of a concern.

Extravehicular Activity (EVA) is a spacewalk. IVA, or Intra-vehicular Activity, is anything inside. Normally the term IVA is only used when discussing EVAs, otherwise it's assumed. Most current shuttle missions include EVAs as part of ISS's construction or maintaining the Hubble Space Telescope. STS-107 was the exception—a mission without planned spacewalks. However, every mission has two astronauts trained to perform spacewalks in case of an emergency. Anderson and Brown were trained to close the payload bay doors manually if there was a problem with the motors, and retracting the Ku antenna that sticks out over the side of the cargo bay. Laurel Clark was assigned as the IVA, responsible for getting them in and out of their spacesuits.

OTHER NASA TERMS

LiOH (pronounced "lie-hoe") is the chemical abbreviation for Lithium Hydroxide. LiOH absorbs the carbon dioxide the astronauts exhale. It's extremely simple, foolproof, and has been used since the beginning of the space program. All that's needed is a fan to push the cabin air through the canisters filled with LiOH crystals. As each canister is saturated, it's replaced with a fresh canister. The seven astronauts used 3.5 canisters per day.

Microgravity is what non-technical people call "weightlessness." Many people refer to astronauts and things floating around in space as "zero gravity" or believe that a spacecraft in orbit is beyond the pull of Earth's gravity, but that's a fallacy. There's almost as much gravity on the space shuttle as there is on the Earth's surface. An object in orbit is falling continuously around the Earth at the exact rate needed to counteract gravity. It's similar to the dropping sensation in a rapidly descending elevator or going over the top of a roller coaster, but constantly falling at just the right rate to cancel out gravity. Scientists prefer the term "microgravity" because it isn't actually zero. Very small gravitational forces can be measured, even though they're a millionth of the normal amount of gravity.

The Shuttle Training Aircraft (STA) is a modified Gulfstream II business jet. The right side of the cockpit is an ordinary business jet cockpit but the left side looks like the shuttle's cockpit. The plane has been modified to dive at the same angle as a shuttle coming in for landing. When the shuttle pilot flies the STA, a sophisticated computer translates the astronaut's commands into how the shuttle would react during an actual landing approach. The STA is just one of the many simulators used to train the shuttle pilots. NASA has a fleet of supersonic two-seat T-38 jets the astronauts use for flight proficiency and transportation.

Finally, there's the sometimes confusing terminology to describe NASA itself, and its facilities and operations. The location of the Kennedy Space Center (KSC) can be very confusing. The physical location is easy—it's the point that sticks out from the middle of Florida's Atlantic coast. But KSC is not located at any particular city. While most news services use the dateline "Cape Canaveral," that city's actually slightly to the southeast. The nearest incorporated city is the sleepy town of Titusville, which touches the north side of the space center. The Cape Canaveral Air Force Station (CCAFS), with launch pads for other rockets, is to the east of KSC. KSC and CCAFS share many of the same resources, and many NASA facilities are actually located on the Air Force side. Insiders use the term "the Cape" to refer to both KSC and CCAFS. The town of Cocoa Beach, which has long been associated with the space program, is further to the south of Cape Canaveral. Melbourne, the location for the control center for many of the European experiments, is located 35 miles to the south of KSC.

Politics determined that Mission Control ended up in Houston Texas at the Johnson Space Center (JSC). JSC is also the home of the astronauts and where they do most of their training. However, the actual hardware's located in Florida at KSC. KSC is responsible for maintaining the shuttle between missions and preparing it for launch. JSC is responsible for operating the shuttle from the time it launches until it lands. Other NASA centers are responsible for other functions. There's some overlapping functions between the different NASA centers and some responsibilities are shared between the centers

"NASA TV," a dedicated television feed intended primarily for the press. Besides its original purpose for providing information and press conferences to the media it's also part of NASA's education outreach. It is offered by some satellite television providers and cable systems and on a streaming feed on the Internet. During shuttle missions, there's almost continuous coverage of the mission with the air-to-ground audio and a public affairs commentator. Unfortunately many of the commentators seem to be in love with their own voices and talk over the astronauts, repeating *ad nauseam* from their prepared scripts and not mentioning what's actually happening—even when it deviates from their scripts.

Neverthless NASA TV has proved to be a valuable resource for space fans who want to follow what's happening as it happens instead of just condensed status reports or the limited amount of coverage available through the mass media.

CHAPTER 12

THE EIGHTEEN
DELAYS

STS-107 went through an amazing series of delays—18 of them. Depending on how you count certain contractual holdups, you can come up with a different total, but using any method, it was an extremely large number. Only a few were caused by technical problems preparing Columbia (the actual spacecraft) for STS-107. Many were caused by problems that affected the entire shuttle fleet, and most of the delays specific to Columbia were caused by a ripple effect resulting from delays to Columbia's earlier missions.

Congress gave NASA funds for a two-week research flight. In March 1998, then NASA administrator Dan Goldin announced that STS-107 was scheduled for May 2000. NASA agreed to pay the commercial company Spacehab $27.5 million for the use of their Research Double Module. That price would also include various management functions that Spacehab could perform less expensively than NASA.

Nobody could have predicted that there would be several months of delays waiting for a payload on one of Columbia's earlier missions, additional months to inspect Columbia's wiring, and additional postponements because of the higher priority for other missions. Ultimately the 18 delays added two and a half years to the original schedule.

STS-107 had lower scheduling priority because it was a stand-alone science mission, without an urgent deadline that had to be met or the requirement to coordinate with other flights. In contrast, if a space station assembly flight slipped a couple of months, it would delay subsequent linked space station flights—anywhere from a couple of weeks to a couple of months. If it was a space station crew-exchange mission, where a new long-duration crew would fly up on the shuttle to replace a crew on orbit finishing their long-duration stay, then the issue became even more complicated.

The other non-space-station mission in NASA's manifest was the STS-109 flight to service the Hubble space telescope. The other shuttles were dedicated to the ISS, while the STS-109 Hubble mission and STS-107 were both assigned to fly on Columbia. STS-107 was less "delay-sensitive" than other flights, and in effect became NASA's less important child, getting less priority than its siblings in the shuttle manifest.

With each slip to STS-107's launch date, a couple of months at a time, NASA paid additional money to the science teams to keep them intact, and to Spacehab to keep its team together. It's similar to reserving a vacation condo for two weeks, then deciding you want to change your travel plans but without giving the owner enough warning to rent the condo to somebody else. NASA found itself paying Spacehab $60 million in "late fees"—more than the original $27.5 million contract price! But overall the amount was small in comparison with the shuttle's budget, and far less expensive than the delays to other missions.

For a shuttle science mission the mission specialist astronauts have to be selected about a year before launch to give them enough time for their training. The pilots are assigned a couple of months later. The mission and payload specialists were assigned on July 25, 2000—Mike Anderson, Kalpana Chawla, Dave Brown, Laurel Clark, and Ilan Ramon. At this point the launch was scheduled for June 2001—just 11 months away.

The press release announcing the crew wasn't released until September 28, 2000. The release started, "A cadre of 20 astronauts and one Russian cosmonaut has been assigned to four Space Shuttle missions targeted for launches in 2001," including Ramon with the American astronauts. The release did not mention that Ramon was from Israel. But the attempt to direct attention away from the sensitive news backfired. On the day NASA issued its less-than-honest release, the Associated Press reported that an Israeli astronaut had been assigned to an upcoming mission.

On October 27, 2000, commander Rick Husband and pilot Willie McCool were added to complete the crew. Launch had slipped a month—to July 19, 2001—just seven months away.

As the crew assignments were completed, Columbia was in California going through the space equivalent of a thousand-mile checkup. The Orbiter Maintenance and Down Period (OMDP) was supposed to last nine months, but was extended to 17 months, primarily to do an intensive inspection of Columbia's wiring.

NASA paid Boeing $145 million for the OMDP and received an overhauled shuttle. The most significant improvement was the installation of the glass cockpit. Other improvements included removing 1,000 pounds of no longer needed items, improved protection from space debris, a change in the infrastructure to add a docking system for flights to ISS, and—ironically, as it turned out—enhanced heat protection for the RCC panels on the wing's leading edge.

The RCC panels are not designed for strength and could easily be penetrated by a micrometeoroid or piece of space debris. A thin layer of Nextel-440 fabric behind the RCC panels would reduce the risk. With that in place, NASA calculated that the middle RCC panels, which receive the most heat, could have a quarter-inch hole and still reenter the Earth's atmosphere successfully. The outer RCC panels, exposed to less

Columbia flies piggyback on a specially modified Boeing 747 during its return to Florida in March 2001.

heat during reentry, could have a hole as large as 1 inch in diameter—far larger than any likely piece of space debris. As it turns out the hole in Columbia's RCC, which doomed the shuttle, was hundreds of times the size that the Nextel layer was designed to protect against

Columbia returned to Florida on March 6, 2001, eight months later than planned. That resulted in an additional three-month slip for the STS-107 launch date. The original plan was to fly the STS-107 mission on Columbia first and then the STS-109 Hubble Servicing mission. But there was an outside factor. Each month the Hubble flight was delayed cost taxpayers an additional $10 million—roughly five times as much as delaying STS-107 again, so managers gave Hubble higher priority. In March 2001 the decision was made to swap the order of the two missions. Space shuttle director Ron Dittemore said, "Columbia came back from Palmdale later than we expected. We decided that Hubble Space Telescope servicing mission had priority over STS-107, so we put it ahead in the queue." That resulted in another four-month delay.

In December 2001 Rick noted that the mission's delays started almost immediately after he and Willie were assigned, and the various slips had added up to almost exactly one year. Husband told me, "The week after Willie and I were assigned they moved the launch into August (because of the OMDP delays), it stayed there a while and went to October. Then they put Hubble in front of us, which put us in April, and then we went to May, and then to June." But there were more delays to come.

Privately Laurel felt that the delays helped; they permitted far better refinement for the procedures for the experiments that weren't as good as they could have been if the launch had occurred in 2001 as planned.

Additional delays to Columbia's processing, pushed back both STS-109 and STS-107. Columbia finally flew the STS-109 Hubble Servicing mission in March 2002, with STS-107 scheduled to follow in July 2002, but during the Hubble flight one of Columbia's two Freon cooling loops had a partial clog.

The flight rules call for a mission to get cut short if a Freon loop fails, but since it was only a partial clog, managers decided to monitor the situation and keep the flight going. If the clog didn't get any worse, then the mission could continue, but if the situation deteriorated, the mission would be cut short. Fortunately, that didn't happen, and the mission was completed as planned.

After STS-109, NASA estimated about four months to refurbish Columbia and prepare it for STS-107, including the additional time to service the Freon loop. The Extended Duration Orbiter pallet was installed in Columbia's cargo bay on April 20, followed by the Spacehab Module and FREESTAR bridge on May 24. Most of the experiments were already installed and ready to fly. Time-critical items would be installed a couple of days before the launch.

The shuttle processing team said they could have Columbia ready by July 26, 2002. But managers felt a target date of July 19 was more desirable because of outside factors such as the schedule for other rockets which needed the same equipment, like radars, and the engineers agreed to aim for the earlier date. Columbia manager Scott Thurston said, "It's very tight. The guys stepped up and said they want to go for that. We're working almost every weekend. We were looking at slipping a week to give us another six days of breathing room, but due to other constraints, that couldn't be done."

The schedule to make the July 19 launch date for STS-107 required work seven days a week with two shifts each day, as well as some third-shift work. July 4th was still on the calendar as a holiday, but if anything fell more than a day behind, workers would have to give up their Independence Day to keep the launch on schedule. Thurston added, "Right now [mid June] things are very good. We look really good for making roll-out. We aren't working any major issues."

By June, after all the delays, everyone was anxious to go, and STS-107 was the next mission on the schedule. Columbia was in the Orbiter

The Spacehab Module and FREESTAR are installed in Columbia's cargo bay on May 24, 2002.

Processing Facility (OPF), the hangar where the shuttles are maintained between missions, and ready for transfer to the Vehicle Assembly Building (VAB) to meet its already assembled SRBs and ET.

On June 12, Rick told me, "We're glad we've got a pretty good solid date now. We've taken good advantage of the [additional] training time we've had available." He shouldn't have spoken so soon.

The same day a problem was discovered inside the shuttle Atlantis. The shuttle's internal plumbing for the liquid hydrogen and oxygen propellants includes pipes connected with flexible sections covered with accordion-like bellows. The flexible sections allow the pipes to move without breaking during the launch vibrations. Each pipe-to-bellows interface has a pair of "flowliners" to keep the propellant flowing smoothly past the bellows. There are baffle-like holes on each flowliner. Eagle-eyed quality inspector David Strait noticed a fine hairline crack on the edges of the holes in Atlantis's flowliners.

The shuttle Discovery was in the middle of its OMDP. When it was inspected in light of the findings from Atlantis, tiny cracks were found in its flowliners, too. Apparently NASA had been flying with these cracks for some time—they were so small that they had gone unnoticed for many years.

Columbia's engines were installed and it was ready for rollover from the OPF to the VAB, so the only way to inspect Columbia's flowliners would be to remove its engines. Managers met to decide whether to proceed with STS-107 as is or to remove its engines to inspect the flowliners and accept the additional delay.

Columbia's rollover was scheduled for June 24, but that morning technicians were informed that it was on hold due to the flowliners. Managers wanted to examine the situation more thoroughly before deciding whether to continue preparing for STS-107 or make repairs. Management decided to be safe and err on the side of caution, rather than fly with an unknown situation. A NASA press release noted that the delay would be just "a couple of weeks."

Rick Husband said after news of the latest delay: "The best thing to be able to do is fly a mission. The next best thing after that is to be assigned to a mission. We're thrilled to death to be not only assigned but to be the next crew up. We have had a fair number of slips through the course of our training, but we've made good use of those in preparing for the flight."

While the crew loved training together, it did put various restrictions on their lives. NASA's rules prohibit astronauts training for a spaceflight from hazardous hobbies like sports parachuting and acrobatic flying. Once K.C. was assigned to STS-107, she couldn't do any acrobatic flying in her spare time until the mission was completed. With the many delays, her hobby had been put on hold for over two years.

The many delays to the flight also caused a significant change to Laurel's long-range personal plans. If any of her male crewmates wanted to have another child, it wouldn't be a big deal, but if Clark became pregnant, she would have to be removed from STS-107. Jon Clark said, "She really wanted to have another kid." Laurel's decision to put off having another child until after STS-107 must have made the many delays even more frustrating.

Columbia's engines were removed, its flowliners were inspected, and, sure enough, they had the same tiny cracks. Everybody quickly realized "a couple of weeks" was incredibly over-optimistic—no matter what decision was made for corrective actions.

The team analyzed what would happen if the decision was made to fly as is—would the cracks grow larger, stay the same size, or—worst-case scenario—actually break off a piece of metal, resulting in debris getting sucked into the engines? Engineers examined three repair techniques—weld the cracks closed, drill a small hole at the end of each crack to prevent them from growing any larger, or create a "super slot"—drilling out an entire chunk of metal between two holes to make them one large slot.

There was an additional complication. Columbia's flowliners were manufactured from 321 CRES (Corrosion Resistant Steel). The other shuttles had flowliners made out of Inconel, a Nickel alloy. Columbia's holes were larger and spaced closer to each other than the holes on other flowliners. Different repair techniques might be required, and engineers would need to do a separate analysis for Columbia.

On July 26, managers made the decision to fly the two space station missions, STS-112 on Atlantis and STS-113 on Endeavour, as close to their originally planned dates as possible. This meant STS-107 would slip behind them. The new planning date was December 1. Dittemore acknowledged, "STS-107 has been around for a long time, and it's been slipped for a number of months—years. That crew and all of the scientists and engineers that have been preparing for many years have continued to wait. We're going to ask them to be patient and wait a little longer."

A closeup view of Columbia's flowliners in the 17-inch wide propellant lines. Photo from the author's collection.

Quality inspector Tim Appleby examines Columbia's flowliners.

After completing the repairs to the other shuttles, engineers analyzed how to repair Columbia's flowliners and decided that welding was the best choice. Thurston said, "We've got the flowliners all fixed and put back together. The welder mentioned it seemed easier [on Columbia] than with the Inconel [on the other shuttles]. We were done with them pretty quick."

NASA has three Orbiter Processing Facility hangars where most of the maintenance is performed between flights. With four shuttles and three OPFs, it's desirable to keep one shuttle "vertical" at all times, like a juggler always keeping one ball in the air. A shuttle is "vertical" (i.e., not in an OPF) when it's in the VAB, at the launch pad, or in space. But when all four shuttles are on the ground in between missions or in the middle of long-term maintenance operations, the lowest priority shuttle has to be put into temporary storage. Discovery's long-term maintenance tied up one of the OPFs. Atlantis, Endeavour, and Columbia would have to share the other two OPFs. There's an area in the VAB where a shuttle can be stored. It's basically a carport, with no capabilities to do any significant mainte-nance. As the lowest priority shuttle, Columbia spent three separate peri-ods in storage in the VAB when an OPF hangar wasn't available. When a hangar was available, technicians took advantage of the time to perform minor maintenance operations that weren't absolutely necessary, or items that didn't need to be accomplished until after STS-107.

Thurston told me about one of the more unusual unplanned main-tenance tasks: Columbia was the first shuttle to fly in space, and improve-ments were made to the subsequent shuttles based on experience learned while building Columbia. As a consequence, the "younger sib-lings" got the benefit of some significant improvements in design. One of those improvements was safety wired nuts. Bolts on the later shuttles have tiny holes drilled crosswise. After the nuts are tightened, a wire is threaded through the holes and tied off, and even if vibrations loosen the nuts, the safety wire prevents them from coming off. During Discovery's long-term maintenance, engineers discovered that a nut on a dynatube (metal pipe) on its thrusters was not as tight as it should have been. On Discovery, Atlantis, and Endeavour this wasn't a major concern because their nuts had safety wire as a backup, but Columbia's bolts didn't have holes for safety wires. There were no indications that the torque was incorrect on Columbia's nuts, but there was also no way to verify the torque settings. (Engineers did not want to test the torque nuts on Columbia because they would have to depressurize the system, which would involve hazardous operations and potentially another delay to the mission.) Instead, the engineers came up with an innovative solution—a crow's foot. The clamp looks like crow's toes clasping a branch. It clamped on to the dynatube on both sides of the nut, so even if the torque on the nut was too low and it started to turn, the crow's foot would pre-vent it from turning any further—sort of a belt and suspenders approach.

Crow's feet clamps (like the one illustrated above) were added to nuts on Columbia's thrusters, as shown in this edited photograph. The crow's feet ensured that the nuts would not come loose even if their torque settings were incorrect.

The team that developed the crow's foot was recognized for their work after the STS-107 launch.

There was one more delay to STS-107, but in comparison to the other delays it was minor. On August 23, NASA managers made the decision to move STS-107's launch date from early December 2002 to January 16, 2003. That move was necessary because of the decision to use the SRBs already assembled for STS-107 for STS-113. The additional time for those exchanges, plus the desire not work over the end-of-year holidays, resulted in selection of the later date.

Columbia's payload bay doors were closed on October 31; the next time they would be opened would be after the shuttle reached orbit.

Columbia was supposed to spend about four months in its hangar after STS-109, leaving in June for a planned July launch. Because of the flowliners and other delays, Columbia didn't exit its hangar for launch until Tuesday November 18—twice as long as planned. It's important to note that once Columbia's flowliners were welded and its engines reinstalled, everything proceeded toward launch with no further delays.

Flow manager Scott Thurston acknowledged, "Some of the crew have been assigned to this mission for several years." He wasn't kidding— by this point the pilots had spent three times as much time in training as planned when they were assigned to the mission. The others twice as long.

Columbia was moved a couple of hundred yards from the OPF to the VAB. Two days later, it was mated to its ET and twin SRBs. And on December 9, 2002, Columbia finally made it to its seaside launch pad. The giant crawler-transporter carried the 2.9 million pound shuttle plus SRBs and ET from the VAB to launch pad 39A. The 3.4-mile journey is at a snail's pace—about half a mile an hour. Because of post-9/11 security, the media were not given access to take photos.

Columbia's rollout to its seaside launch pad on December 9, 2002, as seen from inside the Launch Control Center.

But as Columbia's rollout was taking place, yet another problem on another shuttle became a potential threat to the schedule. While examining the 17-inch liquid oxygen feedline on Discovery, engineers found a crack in a Ball Strut Tie Rod Assembly (BSTRA—pronounced "bee stra") ball closest to the engines. The BSTRA acts like a joint, similar to a hip to leg ball and socket to give the plumbing hose some flexibility. If the BSTRA ball failed, the plumbing could flex back and forth too much and fail. The BSTRAs were in the same area as the flowliners, and there were similar safety concerns. Each of the shuttles has 18 BSTRAs, and many of them are in extremely inaccessible locations. In a worst-case scenario, NASA would have to disassemble the aft end of each of the four shuttles to get access to the plumbing for inspection, and if repairs or a new design was required, the shuttle fleet could be grounded for years.

The BSTRAs are made from a Cobalt-Chrome-Tungsten alloy and range from 1.25 to 2.24 inches in diameter. Engineers put a BSTRA through 140 simulated flights to collect enough evidence before coming to the conclusion that the cracks were self-limiting and would not grow any larger. The BSTRAs would not delay STS-107, but it took a lot of testing before engineers were convinced it was safe.

At long last the hesitations and disappointments and postponements had come to an end. Once STS-113 landed in early December STS-107 became the highest priority mission. During its many delays from 2000 to 2003 STS-107 had been scheduled for launch in every single month except March. Now the flight really was going to happen.

The astronauts flew to Florida for their dress rehearsal with the launch team in mid-December. Rick Husband said, "Everybody's waiting for the launch—the crew, the family, and we're ready to go. I'm sure everybody's excited to go—at least myself and my family. My family had to sacrifice a lot, even just to go backwards in July, more so because of these delays. We're happy to be able to go next month." This time his confidence didn't jinx the schedule.

CHAPTER 13   TRAINING FOR THE MISSION

An astronaut's training starts from when he or she first arrives at NASA as an "ASCAN"—an astronaut candidate. Astronauts are given briefings about the shuttle and its systems. They go through survival training just in case they have to bail out of the shuttle or their T-38 training jets. Much of an ASCAN's training involves classroom lessons on orbital mechanics, the shuttle's computers, basic photography, how to handle interviews and public talks, and the other necessary skills.

One of the least glamorous portions of training is learning how to use the shuttle's toilet. It looks like a toilet on an airliner, but it has a set of handles to hold the occupant in place, similar to a roller coaster ride's safety bar. Since the waste won't drop away from your body without gravity, a strong airflow is used to pull the waste away. You need to position yourself very precisely over the opening for a tight fit.

Another even more embarrassing skill is learning how to use a diaper again. There are times when it isn't possible for an astronaut to go to the bathroom for several hours, most notably during launch and landing and spacewalks. For those activities, astronauts wear adult-size diapers. Part of astronaut candidate training is to overcome the toilet training instincts learned as a toddler to avoid needing a diaper.

The mission specialists are assigned before the pilots because they have more to learn. Columbia had over 80 experiments, and while no astronaut was responsible for all of the experiments, each had many different responsibilities. Most of the training takes place at the Johnson Space Center. The astronauts also travel to other locations, including Cape Canaveral, Florida, for Spacehab training, and NASA's Glenn Research Center, in Cleveland, Ohio, for the combustion experiments, among others. In many cases the astronauts fly their T-38 training jets, but in other cases they're just like any other business traveler, rushing to catch a plane at Houston's Hobby Airport.

In the early stages of training for a mission, each crewmember works on his own with instructors, learning about the shuttle's systems, each of the experiments, and the other tasks he will be responsible for in space. The crewmembers are given background information on their experiments and operating instructions. They're also given detailed information

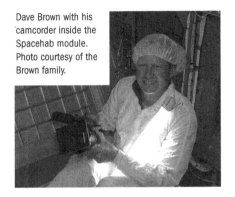

Dave Brown with his camcorder inside the Spacehab module. Photo courtesy of the Brown family.

Willie in the cockpit of a T-38 training jet.

on how the experiment works, just in case there's a problem and they're called upon to reconfigure the experiment or make repairs in space.

Throughout the training, Dave Brown used his camcorder to record his experiences. He even brought the camcorder into the training simulators. Dave planned to make a professional-quality interactive movie documenting the mission and the crew's passion for flying in space.

The astronauts get training for specialty skills, like giving injections and drawing blood. The new "vampires" trained on mannequins and volunteers until they had enough confidence in their skills and were comfortable with sticking a needle into a fellow crewmember.

So how did the astronauts feel about having pilots draw their blood? Mike Anderson said, "When they told me I had to give blood I looked on the crew to see who was going to be sticking me and realized we had two doctors I felt pretty comfortable about that. Since then we've trained Rick, Willie, and Ilan to draw blood. And they've all drawn my blood and they do a pretty good job so I'm pretty comfortable with it right now. Rick does a great job. I think pilots have very good hands so they do things very smoothly, and that's what it takes." Dave Brown noted that the pilots were "Very good—they haven't missed yet on me."

Astronauts take flight physicals on a regular basis, and that's a good opportunity for the newly trained "bloodsuckers" to practice on their teammates. Husband told me, "I did one on Mike for his physical, that went fine. Ilan and I both went in and took turns—I stuck one arm and Ilan stuck the other.

Rick, Willie, Mike, and Ilan's medical skills were limited to giving injections and drawing blood. Two astronauts on each crew are assigned as "crew medical officers." If there are no medical doctors on the crew, the assigned crewmembers are given paramedic training. Doctors Dave Brown and Laurel Clark were STS-107's crew medical officers. If someone gets sick or injured in space, the medical officers could provide any-

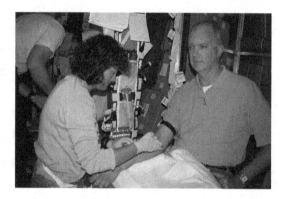

Laurel Clark does a
blood draw on volunteer
John Charles in training.
Photo by Phab4 manager
Angie Lee.

thing from first aid to minor surgical procedures, operating under the remote supervision of the crew surgeon in Mission Control.

Each of the payload crewmembers (Anderson, Clark, Brown, and Ramon) had intense physicals before the mission and were scheduled for several physicals after the mission for studies to show how their bodies functioned before the flight and how their bodies readapted back to gravity. Laurel Clark told me that the scientists wanted to keep the astronauts' arms in good condition, so the only blood draws on the crew ahead of time were for flight physicals and the baseline studies for the science experiments.

In some cases the training to operate an experiment is extremely simple. Scientists realize the simpler they make their experiments, the more likely they are to succeed. Crew time is an extremely important commodity, and if the scientists can minimize the amount of time the crew has to interact with their experiment, that's best for everyone. In many cases all the astronauts had to do was flip a switch to turn an experiment on early in the mission and the experiment would operate on its own for the rest of the flight. In other cases an astronaut might be called upon to inspect the experiment at various intervals, clean filters each day, or make adjustments. The astronauts are laboratory technicians, performing the science for the scientists.

As part of their training, the crew flew to the scientists' laboratories. For example, a two-day training session in Europe for the Biopack experiments included three hours of classroom briefings and instructions and 12 hours of hands-on training with the hardware.

For the Biopack experiments, a crewmember would use a screwdriver to turn a set of valves. The single turn would permit chemicals to mix and "activate" dozens of samples. After a specified period, the astronaut would turn the valve again and another chemical, like a fixative, would get mixed in and stop the experiment. In other cases, the astronaut would load an experiment into a centrifuge or incubator, and after the experiment was completed place it in a freezer.

Willie McCool learns how to operate the Biopack experiments.

Laurel trains on the ARMS experiment.

Other experiments were more complicated and required more interactions by the crew. For the European ARMS experiment, the astronauts would put on a special sensors that monitored their heartbeat, respiration, and other body functions while they rode an ergometer [an instrumented bicycle without wheels]. In space it would require time to set up the experiment, to put on the gear, to exercise as instructed, then time afterwards to take off the gear and put everything away.

For many of the experiments, the astronauts were required to adjust parameters while watching the experiment run. The scientists responsible for those experiments gave the astronauts briefings explaining the scientific goals, how they expected the experiment to operate in space, and what to do if the experiment met those expectations or something unexpected happened. In some cases the scientists could even interact with the crew during the mission, sending feedback for later experiment runs. Dr. Paul Ronney, the principal investigator for the SOFBALL experiment, said, "Mike Anderson, Dave Brown, Kalpana Chawla, and Ilan Ramon [operated] SOFBALL. Even though K.C. was the only one of the four with a PhD scientific background, all of them seemed really excited about my experiment, which I had been working on since 1984. They asked me lots of questions about the science behind SOFBALL. They didn't just learn how to do the experiment, they worked very hard on the development of the crew procedures to minimize the chance of mistakes and extract every possible bit of data. I was especially pleased to see how Ilan, an Israeli military pilot with no scientific background, attacked the training with a vengeance." But the astronauts aren't Einsteins; while they learned how to operate the experiment, they left understanding the theoretical science and explaining why things acted the way they did to the scientists on the ground.

There was also 'just in case' training for activities which hopefully would not be needed. No spacewalk was planned, but Dave Brown and Mike Anderson were given a minimal amount of spacewalk training, in

Kalpana Chawla learns how to operate the Combustion Module. Photo courtesy of Dr. Angel Abbud-Madrid.

Dave Brown wears a spacesuit inside a vacuum chamber.

case an emergency developed during the mission. Much spacewalk training is spent underwater in a giant swimming pool with a mockup of the shuttle's cargo bay. Weights are added to the spacesuits to make them "neutrally buoyant"—they won't sink or rise. Brown and Anderson also practiced using their spacesuits in a vacuum chamber.

A couple of astronauts were given Inflight Maintenance (IFM) training. When things do go wrong Mission Control comes up with incredibly creative techniques to solve problems. During STS-107, the IFM training was put to good use with a variety of repairs to the Spacehab cooling system, and Biopack, Mist, and StarNav payloads.

The crew spent three weeks at NASA's Goddard Spaceflight Center in Maryland learning how to operate FREESTAR, including Israel's MEIDEX. One of Dave Brown's first tasks in space was to activate FREESTAR. He said, "I've got about a one-page set of procedures where I make sure that the shuttle comm [communications] configuration's correct, and make sure the payloads are properly powered. And then I [have] got four switch throws. There's four 'on' switches, and I go right down the line and one, two, three, four, and turn them on. So, it's kind of an important task, but it's pretty straightforward as a first job."

The other FREESTAR activities were more complicated. For LPT, the pilots needed to aim Columbia toward a satellite high above the shuttle or toward a ground station. For SOLCON, the shuttle was pointed toward the Sun. For SOLSE-2, the operator would send the commands to open the door which protected the instrument and aim Columbia. MEIDEX required similar adjustments, but with more attitudes: It was aimed sideways while looking for lightning storms and sprites, and down toward the Earth for the dust observations. From the crew's perspective that was one of the most interesting portions of the mission to look forward to—they would look out of Columbia's windows and search for large dust clouds. If they found any, then they would command the MEIDEX camera to pivot back and forth to center the dust storms on their video display.

Willie McCool in a photography class learning how to use a camcorder.

Mike Anderson inside the Spacehab training module.

The most intense training is the simulations, or sims, where every-thing's duplicated as realistically as possible. NASA has several special-ized shuttle simulators—the "fixed-base simulator," which includes the crew cabin, the "motion-based simulator," which is a hydraulically con-trolled flight simulator, and a variety of other specialized simulators for the various shuttle functions.

The motion-based simulator could be thought of as the ultimate amusement park ride. It has a small room almost identical to the shuttle's cockpit. Each of the windows has a video display of the simulated view the astronauts see during the launch and landing. The simulator sits on top of a set of computer-controlled hydraulic jacks that rotate the simula-tor and change its orientation. Inside, with the video effects, it really feels as if you're flying, even though you are not going anywhere. Less sophisti-cated simulators are used to train airline pilots, and there are "virtual roller coaster rides" at amusement parks that use the same principle.

No amount of training will expose the crew or Mission Control team to every single possible emergency they could encounter. The prin-ciple behind the simulations is to give everyone a broad range of skills and experience working together so when something unexpected does happen the astronauts and mission control can work together smoothly as a team. If a problem comes up, then a specialist may know how to han-dle it. But even if something unanticipated happens, the crew has the teamwork skills they need to work the problem together.

Pilots Rick Husband and Willie McCool and flight engineer Kalpana Chawla went through countless sessions in the shuttle's flight simulators for both launches and landings. Dave Brown joined them for the launch sims, and Laurel Clark joined them for the entry sims. Dave or Laurel would sit in the fourth seat on the flight deck and could help during an emergency as an extra set of eyes and to monitor checklists. There was no need for Mike or Ilan to participate in any of the launch or entry sims because they wouldn't be on the flight deck during launch or landing. Instructors tossed hundreds of simulated problems at the astronauts and the Mission Control team in the hope of catching them in a mistake, but always with the intent of teaching them how to deal with real-life problems in space.

I sat in on a set of landing simulations for Mike Anderson's first flight, STS-89. Commander Terry Wilcutt invited me to watch the simulations from training team leader Gail Barnett's workstation, where I was able to monitor what was going on inside the simulator's cockpit, what was happening in Mission Control, and the activities of the training team in the control room.

The seven runs I observed marked the final time STS-89's crew would get in the simulator and train how to land the shuttle while various simulated emergencies occurred. Besides a normal end-of-mission landing, this training would also be useful if an emergency landing was required.

The simulation supervisor (sim sup) is the person responsible for orchestrating the action. The sim sup can create false leads and other problems that appear to be something else to try to fool the astronauts or flight controllers.

The Mission Control team participates in the simulations using their actual consoles and software. The only difference is instead of interacting with the shuttle in orbit they're connected to the simulators in Building 5, less than half a mile away. Astronauts have joked that the main difference between the simulator and a real shuttle is the simulator's switches needed for emergencies are worn while on the actual shuttle they look brand new because they've never been used.

I watched from the control room as Wilcutt's crew land the simulator—safely—all seven times. Each landing simulation starts when the shuttle's at an altitude of 200,000 feet, about 14 minutes before touchdown, and shortly after the point when Columbia broke up. (The shuttle *does* take only 14 minutes to descend from five times the cruising altitude of a jetliner.) After each simulated landing run, there's a debriefing where the performance of the crew and flight controllers is discussed and comments are made about everything that occurred.

After several months of training together, the STS-89 astronauts and Mission Control were a well-honed team, operating on instinct. But they made a mistake, even just a couple of weeks before launch. The astronauts were asked to enter an illegal "state vector" (the numbers that determine the shuttle's exact location and velocity at a given moment in time), and nobody caught the mistake. The simulation team was extremely excited that they had successfully slipped an error by Mission Control and the astronauts without anyone realizing it. But even without a properly entered "state vector" the astronauts were still able to land the shuttle. During the debriefing, the mistake was discussed and changes were recommended to the procedures.

For the final run, the simulation team gives everyone a break. It's always nice to end the simulations on a positive note, and give the crew a more realistic landing simulation, since in most actual shuttle landings, everything's *not* going wrong. By this time the crew's been in training for a year and has gotten to know each other fairly well. Even then I was surprised when one of the astronauts mentioned an upcoming car show they couldn't attend because they would be in quarantine that weekend! Mike would have loved to attend the show but went into quarantine as planned.

STS-89 was launched on January 22, 1998. Endeavour reentered the Earth's atmosphere on January 31, and like the final sim, everything went perfectly. Terry Wilcutt took manual control of Endeavour and guided it to a picture-perfect landing.

A couple of hours later Mike told me, "From my perspective the entry was a lot like we trained; the simulators do a pretty good job." He added the only surprises were the G-forces and a transsonic buffet [some bouncing when slowing down from supersonic to subsonic speed], which aren't simulated."

The most intense training sessions are the "Joint Integrated Simulations," where everybody participates. The Mission Control flight controllers and payload teams are at their consoles and the astronauts in the flight simulator; they go through sessions where everything's simulated, especially things going wrong. A typical simulation starts with a situation where there's a slow leak on the shuttle that cannot be stopped. It isn't a safety concern, but it will cut the mission short. Mission Control and the payload teams have to decide how to re-plan the mission to maximize the amount of science they can accomplish before the shuttle has to come home.

Kalpana Chawla talked about one extended 24-hour simulation: "We were in a simulator. Around 1 A.M. we are told by the control team to pretend. [Ilan] was told to pretend that he has a cramp in his leg while he's on the ergometer for the European ARMS experiment. So he got on the radio and made up this, 'I just got on the ergometer and have a cramp.' The ground told him, 'If you can, can you try it again, because the timeline is falling behind and they're trying to figure out what to do? You may want to talk to Laurel,' the physician on our shift."

The sim sup wanted to get the payload engineers and scientists to think together as a team—would they decide the leg cramp was serious enough to have Laurel stop whatever she was doing to examine Ilan? Should the team decide to let Ilan rest for a while or reschedule his activity on the ergometer? Either decision would result in re-planing the rest of the shift's activities.

K.C. continued, "Then when [Ilan] called ground again for the second try, the [simulation] team told him to suggest that he has a cramp again. I was in this one corner of the module working on my experiment, so I had not heard he was told to pretend again. I hear this 'ARGH' and I go, 'Oh my god, what happened, Ilan,' and he said, 'I have a cramp again.' So he called the ground and told them, 'I'm sorry I have a cramp again.'" An audio clip of K.C.'s comment is on this book's website.

If the payload team decided to ask Ilan to consult with Laurel, the simulation team would have given Ilan instructions for how much of a cramp he had in his leg to enable Laurel to respond. Based on her response, the sim sup would give Mission Control additional information for how to handle the situation.

The training cycle is designed so the crew is at their peak performance just before launch. If there's a long delay, the training level is ramped down, then ramped up again to reach optimum performance for the next launch date. Rick Husband said that his crew would have been ready and fully trained for a July 2001 launch if Columbia had remained on its original schedule. With the many delays to their mission and changes to the planned launch date, the crew redid much of their training. They must have felt like they were on a roller coaster, with intense training leading up to a planned launch date, but with each slip a corresponding reduction in the intensity of their training before it increased again.

There was, however, some benefits to the delays. Because the STS-107 astronauts spent so much time in training, they had the opportunity for some optional training that is considered desirable but not absolutely necessary. For example, they participated in a camping trip at the National Outdoor Leadership School (NOLS, pronounced "knowles"). NOLS caters to businessmen who like the outdoors and want to have a bonding experience. Many previous astronauts enjoyed NOLS and recommended it as a potential training tool. Before the trip, Husband told me, "That'll be a really great opportunity for us to get to know each other, get used to each other working in close conditions and working together as a team well in advance of going flying."

The STS-107 crew visited the NOLS site in the Wind River Range of Wyoming, from August 20 to 31, in 2001. Husband said, "It went really well.

The STS-107 crew during their trip to NOLS.

Ilan Ramon and Mike Anderson during NOLS. Both photos courtesy of NOLS.

We went there in August, nine nights out in the mountains in Wyoming. Two different tent groups—red shift and blue shift. We covered about 35 miles over nine days, which included climbing Wind River Peak, which is about 13,200 feet. It was a real good opportunity. The instructors from NOLS did a real super job facilitating discussions on leadership and communications skills, self-care—how to take care of each other and yourself. I was pleasantly surprised at how many parallels there are to a trip like that and a space mission. You're busy trying to adapt to a new environment— even brushing your teeth is different out there than it is at home. You've got to take care of yourself, pull your own weight, help other folks out if they're having a hard time, or other folks helping you out. The teamwork aspect is good. You're keeping an eye on the fact that you're a team and operating as a team, making sure all goes well."

Husband added, "We had talked to other folks who had gone, and they all said when they came back after these trips, 'Man if we could only now get assigned together as a crew.' We had the benefit of already being assigned as a crew, going out there and going through the experience and getting to know each other on a level most crews experience when they go on their missions. We had that benefit that other crews haven't had up to this point. It was a great experience with group dynamics, operating as a leader, also as a follower and also how to take care of yourself, which is an important thing being a member of a team as well."

Mike noted that Ilan really shined at NOLS, "Like an Eagle Scout twice over. This was a piece of cake for him." Because they're in such an isolated area they have to rely on each other as a team. No cell phones, no e-mail, they're pretty isolated from the outside world. The NOLS team said the astronauts were excellent participants; they were enthusiastic, interested in learning new skills and excited about their experience.

Dave's dog, Duggins stayed with neighbor Cindy Swindells while he was at NOLS. Duggins was already fairly old and in bad health. Swindells recalled, "We couldn't reach Dave by telephone. If anything went wrong it would be up to us to take care of Duggins. By the time Dave got back, Duggins was in heart failure." Brown put Duggins in his car and drove him to Texas A&M's veterinary school in College Station, a two-hour drive, for open-heart surgery. Later Dave wanted to bring the Swindells up to visit Duggins, knowing it would make him feel better. He gave Cindy and her daughter Jill a ride in his airplane to the hospital. Swindells said, "We showed up with balloons and treats and the next day he turned around and started getting better."

Like everybody in the United States the STS-107 crew was deeply affected by September 11th. Rick Husband was at JSC's flight medicine clinic. He said, "I had been inside and heard an airplane ran into the World Trade Center. That's all I heard, just an airplane. I thought some small airplane on a sightseeing joyriding ride hit the building. But then I went into the lobby where there was a TV going and heard it was a big airliner. The first thing I thought was a terrorist attack, that was before the second airplane hit. I thought that was a horrendous thing to have happened. Terrible."

Ilan Ramon was in Germany, the country of his father's birth. He said, "I was making a trip with my father and brother touring my father's birth city, Berlin. I was in Berlin on September 11 and I was as astonished as anybody. I got a call from my wife; she was watching in real-time what happened. Since then we were watching television as everybody." I asked him if he felt the 9/11 tragedy increased the bonds between the United States and Israel and he agreed— "I feel like the people in United States and the people in Israel got closer unfortunately because of these circumstances." A recording of my question and Ramon's answer are on this book's website.

Since Kalpana Chawla has dark skin and speaks with an accent, I asked her if she had encountered any prejudice since 9/11 from people who mistook her for an Arab or a Muslim even though she's a Hindu. She said, "I haven't encountered anything personally, although I do know people who talk about it. I guess in Houston [in] the area we live is culturally very diverse, so I wouldn't expect such a thing there. People I know who did experience it were in small towns, suburbia."

A month before 9/11, I talked to NASA Administrator Dan Goldin about STS-107 and Ilan Ramon's presence on the crew. Goldin told me, "We flew a Saudi Prince, it was okay. We have flown a lot of times, we'll do whatever's necessary to make sure it's safe." Saudi Sultan Salman Abdulaziz Al-Saud flew on the shuttle in 1985 as a VIP passenger in exchange for the Saudi government purchasing a satellite launch from the shuttle. But unlike Ramon, he was just a passenger without any flight responsibilities.

In the weeks after 9/11 there were rumors that NASA was considering removing Ilan from the crew because having an Israeli on board might serve as a magnet for terrorists. But removing Ramon would be a very good example of giving in to the terrorists—letting them affect how a major civilian government agency did its day-to-day work. It would also

Rick Husband and Ilan Ramon learn how to use MEIDEX under the watchful eyes of trainer Lisa Anderson and flight activities officer Terri Schneider. Photo by Philip Chien.

hurt international relations—a true example of if we change our lifestyles then the terrorists have won. But the rumors were just rumors fueled by speculation and post 9/11 paranoia. Husband told me, "I never heard anything like that—it was probably just rumors."

The crew spent a week in Florida in December 2001 training at the Spacehab facility. Rick Husband was kind enough to invite me to sit in on a training session on December 12. One thing that really surprised me was the large size of the team there to support the astronauts' training. About 20 engineers and scientists patiently sat on the sidelines ready to assist or answer questions whenever they were needed.

Trainer Lisa Anderson and flight activities officer Terri Schneider monitored Husband and Ilan Ramon as they studied how to operate MEIDEX. Husband and Ramon used a laptop computer to send commands to operate the experiment's cameras. Husband said, "Say we're trying to monitor a dust storm, if we need to make real-time adjustments where the camera is pointing we can do that. That's all being recorded while we're doing the run."

Rick gave me a tour of the training module, a full-size mockup of the actual Research Double Module that was sitting across the room. The training module doesn't look like the real thing. A large opening to pass items in and out is a good giveaway. We had to get on our hands and knees to crawl in the hatch. In space, Rick would just float through. The front half of the module had an area on the left side with a workbench where the blood draws would take place. On the opposite side was the ESA ergometer for the ARMS experiment. Husband proudly showed me the color-coded cardboard signs the crew would use as a "calendar" to indicate the flight days. His red shift's signs were black numbers on a red background—the colors of his alma mater Texas Tech. The blue shift signs were yellow on navy blue for Navy astronaut Willie McCool. The back half of the module included mockups of the experiments. The back wall and intermediate wall were filled with microwave oven-size lockers.

Rick Husband during a tour of the Spacehab training module with the author. Photo by Philip Chien.

Some of the experiments occupied two locker spaces. Each locker contained a facility, like a drug-mixing machine, freezer, or growth chamber for different experiments.

The training module was a busy place, with instructors showing the astronauts how to operate Spacehab's systems and the experiments. Husband and I left the module and he showed me some of the experiments he was responsible for.

As McCool arrived, Husband called to him, "Blue shift leader" and McCool replied, "Red shift leader" — clearly these were people who were very comfortable with each other, using inside nicknames.

The launch was on schedule for July 19, 2002, and the crew completed most of their training in June. It's traditional to have a "cake cutting day," where the astronauts invite their families inside Mission Control to see a bit of the behind-the-scenes activities. Dave Brown's brother Doug and girlfriend couldn't make it, so he invited neighbor Al Saylor. Saylor, a private pilot, recalls, "I was thrilled. Flying the simulator at NASA as his guest." Saylor got an opportunity to sit in the motion-based simulator for launch and landing. He flew with Willie's family when Dave was called away. "Willie was in the right seat, where he would normally sit, and each of the kids sat in the commander's seat. And then Willie invited me to sit up front and fly a launch and recovery as well. As a pilot, having that dream come true, there's no way to describe how wonderful I felt about taking that slot and representing Dave's family."

On July 26, managers made the official decision to slip STS-107 behind the two ISS flights because of the flowliner problems. Ron Dittemore that said Husband's reaction to the news was "what you might imagine from somebody who's been training for a number of years and gotten to within a couple of weeks of his launch date, only to find out that he's now four or five months from his launch date. I would say there's some personal disappointment on his part, but professionally he understands exactly what we're doing." Columbia's crew got another unplanned vacation.

Spacehab's Pete Paceley said, "As you get pumped up and you get ready and you're so close and you get excited—it's like letting the air out of a balloon. It's just real demoralizing." More so than all of the previous delays, this one occurred close to the launch date, as everybody's becoming more excited and starting to make final plans. Family members and friends were already making arrangements to take time off from work and made airline and hotel reservations and were looking forward to going to Florida for the launch. Many decided to combine the launch with a family vacation or extended family reunion. But suddenly everything was put on indefinite hold—anywhere from a couple of weeks to five months. If the airplane tickets and vacations were already paid for, the families and friends would have to decide whether or not to use them anyway even without the launch.

Crew secretary Roz Hobgood noted, "It was crazy, we didn't know when we were going to go again. You have to start all over [with the invitations]. Think about having seven kids all getting married on the same day. And all of the sudden the grooms says, 'I can't do it'—it's not a very pleasant thing."

The delay affected far more than just the astronauts and their guests. Laboratories scheduled to support the mission stocked up on supplies and had already ordered the various chemicals and special materials that would be needed. In some cases those chemicals had an extremely short shelf life, so replacement chemicals would have to be ordered for the next launch date.

Scientists who needed to be in Florida to prepare their experiments or participate in the payload control centers in Texas and Maryland had to juggle their schedules to take time away from their day-to-day classes and research so they could support the mission. In many cases their families arranged vacations so they could see the launch in person. Many students with experiments were anxious to fly to Florida for the launch.

The MEIDEX team had arranged for a specially equipped aircraft to be on call to fly underneath dust storms and had to put those preparations on hold. Belgian scientists were scheduled to view the Sun simultaneously from Columbia and other spacecraft. The SOLSE team was scheduled to make observations simultaneously with high-altitude research balloons and other spacecraft. But shuttle schedule changes are a fact of life, and all of the teams realized the mission would eventually fly, and far better to fly after technical problems were fixed.

Once the decision was made to slip behind STS-112 and STS-113, everybody involved with STS-107 could work toward the new launch date and re-plan their lives. Dave decided to put the unexpected delay to good use. He went to film school to improve his filmmaking skills. Rick flew to Florida anyway. But instead of flying in a NASA T-38 training jet, he took a commercial flight with his wife and children for a family vacation.

K.C. prepares for emergency parachute training. Dave Brown can be seen in the background with his camcorder, videotaping the training session.

Willie McCool in his life-raft during training.

The Columbia astronauts reentered training, targeting a launch in the December-January timeframe. By this time they were one of the most thoroughly trained shuttle crews ever because of the many delays and most of their training was just repeating their training as a refresher.

There are many activities that everybody hopes will never be needed but the astronauts are trained for just in case. After the Challenger accident, NASA was instructed to add some limited escape capabilities. It was realized that there was no practical way to make an escape system that would operate under all circumstances, but it would be possible to provide something.

If the shuttle cannot make it to a runway the crew can jump out and parachute into the ocean. The only way this system will work is if the shuttle is in a level flight and below the speed of sound, so it's got a very limited time during the mission where it's usable.

The orange launch and entry suits, Advanced Crew Escape Suits (ACES — pronounced "aces"), are full pressure suits, similar to ones used by high altitude test pilots. The suits provide protection from high altitudes, the cold, and water. The parachute harness includes an oxygen supply, an one person life-raft and survival kit with a signaling mirror and emergency radio. The suits are orange to make them as visible as possible to help search and rescue workers to find the astronauts.

Astronauts practice bailing out of the shuttle and getting into their life-rafts. Columbia's crew had already completed the training., but because of the delays they took the opportunity to do it again on November 13, 2002 as a refresher.

Columbia finally made it to its launch pad on December 9, 2002. The next major activity was the crew's dress rehearsal with the launch team. It had been scheduled for early July but was pushed back because of the flowliner delays. On December 17 the Columbia astronauts flew to Florida. Officially NASA calls the dress rehearsal the "Terminal Countdown Demonstration Test" (TCDT). The astronauts suit up and go through the activities they do for an actual launch. The simulated countdown ends in a simulated emergency where the engines ignite but there's a technical problem and they're shutdown. The astronauts go through the steps to shut down the engines and practice an emergency exit from the shuttle if that becomes necessary.

The astronauts also inspect the shuttle, their equipment, have fit checks of the orange launch and entry suits, and learn how to drive an M-113 armored personnel carrier — basically a tank without weapons. The M-113 sits at the base of the launch pad. If there's a major emergency, the crew will exit the shuttle and jump into slide wire baskets that whisk them to the base of the launch pad. They will either enter an emergency bunker, or get in the M-113 and drive as far away from the pad as quickly as possible — making a hole in the fence on the way out. As one astronaut put it — probably the second most exciting way to leave a launch pad —

K.C. learns how to drive an M-113 armored personnel carrier during the TCDT.

the first being a launch of course. Learning how to drive a tank is certainly one of the more unusual forms of astronaut training!

After the TCDT the crew flew back to their homes in Houston. Their final five weeks before the launch would be spent doing final refresher training to ensure that they were in peak condition, and also time off for a Christmas-New Year's vacation.

With all of the delays, the STS-107 crew ended up spending 136 weeks training for their mission—74 weeks longer than a typical crew. They spent 3,506.6 hours training with Spacehab, and 15 weeks at various experimenter facilities. It was one of the most thoroughly trained shuttle crews ever.

The STS-107 crew poses with a T-38 aircraft.

CHAPTER 14   P R E S E N T I N G   N A S A
T O   T H E   W O R L D

Besides flying in space, astronauts are also NASA's most important faces to the public. The job description for astronaut includes being a public affairs goodwill representative, and all astronauts are expected to make public appearances, give interviews, and appear at press conferences. Some of the astronauts are more comfortable with this responsibility than others, and some do a better job of communicating NASA's goals and their personal thoughts. As a normal part of the training for their mission, the STS-107 crew had a press conference before the mission. The preflight press conferences were scheduled for June 28, 2002, as Columbia's schedule was slipping again, this time due to flowliner cracks. Husband said, "It's just one of those things. When we found out [about the delay], there's a letdown because you're geared toward a certain date. But if they found something with the hardware which needs to be fixed, that's definitely the right thing to do. We're more than willing to let them do their work and looking forward to going and flying a good vehicle whenever they're done."

NASA decided to have the preflight press conferences anyway, even though the launch date was uncertain. Unlike most preflight press conferences, which are attended by only a handful of reporters, there was a fair amount of media attention on STS-107 because of Ilan Ramon. Many of the press who covered the briefings were only interested that there was an Israeli, and whether that made the shuttle more of a target for terrorists. The press were required to go through metal detectors before entering the public affairs building.

The first briefing was jointly presented by lead flight director Kelly Beck, who gave an overview of the shuttle's activities during the mission, and by mission scientist John Charles, who talked about the experiments. Beck started, "Our mission is unique compared to the flights we've been flying recently to the International Space Station, in that it's a multidiscipline research mission involving space sciences, life sciences, physical sciences, as well as the educational arena."

She noted, "It is very unique. Our launch window is different. The main science driver is to get sufficient lighted opportunities over the areas of interest for dust storm observations." Because of the security rules, she didn't mention the launch time, and if asked would have been required to explain that security rules did not permit the public release of the launch time until the day before liftoff. Beck acknowledged the delays in the flight and the uncertainty about when Columbia would fly. She said, "It does have some history. We are at the point now where our teams are really ready to do it. We're trained and ready to go fly. We just need to wait for when Columbia's ready." Charles gave a broad-brush overview of many of the different STS-107 science experiments.

Next a series of briefings were held for each group of experiments. Since there were so many experiments on STS-107, the payloads were grouped in categories—the microgravity experiments in one briefing, life sciences in another, Earth observations in the third, and education in the final payload briefing. Even with all of the briefings, many of the payloads were not covered, and in most cases a manager gave a general briefing that covered a broad range of payloads. Some of the scientists and managers who gave the briefings did a better job than others, explaining the details for their projects. These briefings were monitored by just a handful of reporters. The final briefing of the day was the one the reporters were waiting for—the seven astronauts.

A public affairs person introduced commander Rick Husband, and he introduced each of his crew. The crew dressed in matching polo shirts with a stylized version of their logo. Laurel had a reputation of having an entire wardrobe with the crew's logo, and she wore STS-107 earrings to the press conference. Each of the astronauts talked briefly about his or her responsibilities, and Mike gave a fairly comprehensive description of the science.

The preflight press conference for the
STS-107 crew in June 2002.

Husband said, "Our mission will be conducting research that will help to enhance life for all of us on Earth and also learn more about the way we live in space as well." Brown said, "One day I'll get to do atmospheric research, the next day combustion science, the next day physiology research." Chawla said, "Feel free to address me as K.C., it's a lot easier. On this flight I'm very excited to serve as the flight engineer. It took me until age 14 or 15 to say I wanted to become a flight engineer and pursue flight design."

Mike Anderson talked generally about the wide variety of science activities on the mission. He said, "This flight will be a research flight and that's quite different. It kind of harkens back to the old days of the Spacelab missions we so successfully flew in the 1980s and 1990s. The [scientists] have really packed this mission with some outstanding science. We're going to do a little bit of everything on this flight. We're going to do some physical science. In the area of physical science we're going to cover everything from fluid physics and geophysics to combustion physics. On this flight we have the combustion module. I think that's one experiment everybody's looking forward to. Everybody likes to watch things burn. When you burn things in space they burn quite differently and you see something that's really quite interesting. On this flight we're going to start off by lighting two of the largest rocket boosters, we're going to ride this huge flame ball into space producing 7 million pounds of thrust to take us up into orbit. When we get there and we shut down those engines and go the back and open up our Spacehab, we're going to try to burn some of the smallest flames that we've ever burned—we're going to create a very lean mixture, we're going to try to ignite it. And if everything works just right we're going to get some very small flame balls just floating there in space and burning. Very, very small flames. If we do this successfully, we'll be able to study this lean burning process and with the information we learn from that make more efficient and more environmentally friendly burners and engines back here on Earth. So we're really looking forward to getting up there and working on our combustion science.

"We're also going to be doing a lot of life sciences. As you see, we've got two doctors on the flight, they're going to be very important in helping us out with those experiments. In life science we're going to continue do what we've always done—we're going to use space as a place to learn more about the human body and biological organisms and how they work in space and how they adapt to the things that are different in space and try to take that information and bring it back on Earth and find some way of helping things out here on Earth.

"One of the more exciting things we're going to do up there in terms of life sciences is an experiment from Europe called ARMS. It stands for "Advanced Respiratory Monitoring System." If you look at the human body, it's really a fantastic creation. And it's made of a lot of very complex control loops that automatically control everything from your blood pressure, your heart rate, to how fast you breathe. And you really have no control over those things but they do a great job of taking care of everything by themselves. We really don't understand them, and in this experiment we're going to try to understand them a little bit better. We're going to measure all these control loops and then we're going to stimulate them by taking them into space and take away something that they're all very used to—gravity. We're going to take away gravity and see how those control loops respond to that. And that's going to give us a great opportunity to really investigate these automatic control loops in our body and how it really works.

"We're also going to do some Earth science while we're up there. We're going to take advantage of the fact that the space shuttle gives us a great platform where we can look back at the planet Earth and also look out in to space. We're going to study this planet. We're going to look at dust storms and see how they affect the environment and the hydrology of the planet and we're going to take a look at the ozone layer. There's been a lot of talk about the ozone layer and whether or not it's being depleted. We're going to study the ozone layer. We're going to do it a little bit differently. In the past we've always looked directly at the Earth to measure the ozone layer. But with an instrument called SOLSE we're going to look at the limb of the Earth. By doing this we hope to get a better understanding of what's really happening to the different layers of ozone in the atmosphere, where the changes are taking place. We're really excited by the Earth science we have on board this mission.

"If you look at this flight, it's really jam-packed with a lot of science, we're really excited about getting up there and making everything work. We can't say enough about the experimenters and engineers that put this flight together."

Ramon said, "I'll be taking part in a variety of experiments on STS-107, including the MEIDEX experiment, which will hopefully enable us to better understand Earth climate and global warming and cooling."

Not surprisingly, most of the media attention was on Ilan Ramon and what the other members of the crew thought of him. Ramon stated, "I think my presence on this wonderful crew is just another crewmember. As you know we have several Israeli experiments on board. And there's no talk [about security] other than the usual risks after 9/11."

Husband said, "We're confident with all of the measures, which have been taken with hardware inspections [on the flowliners] as well as we are with security. And we have every confidence in all the people who are working on those issues and they all have them very well in hand."

McCool said, "We as a crew have embraced Ilan because he's such a wonderful individual regardless of nationality. He's just been a true pleasure to have. He's a hard charger, wonderfully warm and personable. We've come to embrace and love Ilan."

Kalpana Chawla was asked about Ilan's sense of humor. She said, "In one of our sims there's a bicycle where you're supposed to exercise and some data measurements are taken. In our simulator we do not have the real bike. Somebody for fun brought a kid's tricycle and Ilan was there sitting on this tricycle with some of the [experiment] gear on his nose and face—he was just following the procedures, simulating it. I just find that extremely impressive as a person."

Ramon said, "Every time you're the first it's meaningful. To be the first Israeli astronaut is meaningful to a lot of people, I believe all the people of Israel. It's also meaningful to a lot of the Jewish community around the world. I invited my family and my wife's family and all my friends. Hopefully anybody that can make it will be there. I'm sure my father and brother will be there. Brothers, sisters, cousins, and friends—everybody."

Husband said, "We're going to go up there and have a great time, and do the things that folks back home have trained us to do and make the folks back home proud." Yes, he really said that, and, yes, he really meant it.

Normally NASA accommodates about five to ten reporters who ask for individual interviews with each of the astronauts after the astronaut press conference. But for STS-107 there were 19 organizations. Three were Israeli television stations and newspapers, seven were reporters who normally cover the space program, and nine were reporters who were only interested in Ilan and what the rest of the crew thought of him. Some didn't even bother to interview the rest of the crew; others only asked the rest of the crew about their interactions with Ramon.

During the crew's TCDT dress rehearsal, there's an opportunity for the press in Florida to meet the astronauts. As part of post-9/11 security, the media were instructed not to tell anybody outside of the press site

Ilan Ramon sits on a child's tricycle in the simulator, pretending that it was the ARMS ergometer. Photo from the author's collection.

Ilan Ramon training on the actual ARMS ergometer.

when the astronauts would be present and live transmissions were prohibited. NASA security officer and K-9 handler Ken Cox came over with his dog, Blitz, who sniffed the press cameras and other equipment and inspected the area. The crew bus arrived, escorted by three SWAT guards. One guard left and the other two kept a careful watch.

I asked Rick about his thoughts on the many delays, and Ilan about what he'd like to tell the people of Israel.

Husband said, "To be honest whenever we have launch delays initially there may be letdowns or disappointment, but then we realize we've been given a gift of more time to prepare. We've made really good use of our time; we were able to take some time off, so that was really good for us and our families morale-wise. We're doing great now and really looking forward to the launch. We feel like we're as well trained as we can possibly be."

Ilan said, "Of course everybody's waiting for the launch. The crew, the family, and back in Israel and we're ready to go. And I'm sure everybody is excited to go, at least myself and my family. My family had to sacrifice a lot. Even just to go backwards in July [because of] the delays. So I'm happy to be able to go next month."

I moved off to the side where I could take photos of the crew. I asked one of the SWAT guards if I could go into the field behind the crew to shoot a "reverse" shot showing the crew talking to the media, and got permission.

The astronauts were asked to talk about their favorite experiments.

McCool said, "One of the experiments I'm working on is SOLSE. It's to evaluate a new algorithm to measure the ozone levels in an altitude profile, rather than looking straight down from a nadir view. I think the world as a whole is interested in global warming and the ozone layer and this will certainly help in understanding the ozone and its depletion."

Ramon said, "The complement to this experiment is the MEIDEX from Israel. It researches the dust particles and their impact on global warming and precipitation. I'd also like to talk about the education experi-

The STS-107 astronauts talk to the press at the Kennedy Space Center. Photo by Philip Chien.

ment we have on board—S*T*A*R*S. This is a payload with six experiments from six countries like China, Japan, Australia, Luxembourg [sic], Israel, and the United States. I'm excited to be a part of this experiment for the excitement of the students and I know they're excited."

Brown said, "We have some that study osteoporosis, calcium loss. That's something that's a problem for everybody as we get older, particularly women. It turns out that astronauts—when we go to space and we no longer have the stress on our skeletons—we lose calcium. We're going to be studied for that and we're actually a very good model for a very accelerated model for what happens to people over many many many years, so it's useful to study us to know how to slow or prevent osteoporosis here on Earth. We also have quite a few locker experiments that actually have bone cells in them to study the same metabolism for why bone cells gain or lose calcium.

Anderson cited the many different types of experiments but said that one of his favorites was the Combustion Module. He said, "These experiments will be studying the combustion experiment. Here on Earth we get about 85 percent of our power from the combustion process—burning petroleum or some process like that. With the combustion module we hope to investigate the combustion process and find ways to make it more efficient, to make it cleaner, and also to make it safer."

Clark discussed the European ARMS experiment. "Being a physician, that's one that deals with physiology—both heart and lung physiology— we're going to be doing a complicated set of breathing maneuvers. There are eight different experiments with dozens of investigators, primarily from Europe. They're looking at the changes in the oxygen exchange in the lungs as the blood flow changes with different body positions. They're looking specifically at the changes in respiratory or lung function when patients have to be on their backs for extended periods of time. In fact there's some good evidence that patients in intensive care units would be better off on their stomachs as opposed to their backs. This involves a huge amount of overhead to take care of them that way, but certainly if it's much better oxygen exchange and lung function it's worth that overhead. They've been studying us in different positions—on our stomachs and on our backs and studying the air exchange preflight. They'll look at it inflight. Microgravity simulates that since the blood doesn't get pulled down into the legs in the same way. And we'll be studied again immediately postflight and for several weeks after we return home."

Chawla said, "We are growing lots of crystals on this flight—physiological a lot of protein crystal experiments, in fact there are hundreds of them on this flight, and materials. We are growing crystals. One experiment that really interests me is zeolite crystals. Of course we grow crystals on orbit because we can grow them bigger and then look at their structure and get to understand the structure better. And hope that perhaps we can come up with a new drug if it has been a cancer-related protein crystal or some material-related advancements. I'd like to highlight the zeolite crystal growth experiment that we are doing. We carry precursor solutions in tubes that we mix on orbit. Once the solutions are mixed we put these tubes in a furnace and let the crystals grow. Zeolites are materials which have a lot of holes in them, much like a sponge. But they attract materials of unique characteristics toward them. The applications are very wide-ranging and very interesting. One of them is, for example, dyes that hold better to the paper like printing media. Another one is alternative fuels like hydrogen as opposed to gasoline. It's hard to store hydrogen at room temperatures. You have to compress it and then it becomes a hazardous thing to deal with. One of the things the [scientist] is doing is come up with a material that can have hydrogen embedded in the zeolite material so you can store it at room temperatures."

Husband was asked about whether or not he felt the presence of an Israeli on the crew put them at more risk because of the potential for terrorist attacks. He said, "I don't think we feel there's any particular increased risk. Spaceflight is a risky business. Certainly since September 11, we've heard in the media that it's not only Israelis but Americans that are being targeted in different ways. We see ourselves as one crew united, just as we see ourselves united with the rest of our fellow countrymen in the situation we find ourselves in now."

The rookie astronauts were certainly the most anxious. McCool said, "A lot of excitement. Every time we do a new event like this it's a new milestone. It's a reality check is what I like to tell people. It brings us to the reality that we're ever closer to a launch."

Dave Brown said, "It's been six and a half years since I came to NASA, and I was really hoping I was going to go sooner than this. I got the most excited today when we pulled out of crew quarters in the astrovan and I saw the flashing lights on the police cars and the helicopter following us. I really felt like a little kid. I remember getting to the vehicle and thinking I should really be a lot more serious, paying attention to the detail. I'm sure it will be the same and even more on launch day. It was pretty neat. I was pretty excited even though there was no fuel on the vehicle and we weren't going anywhere today except back here."

After the question and answer session was completed the crew agreed to pose for a group shot out of the shadow. As the crew posed I teased them by shouting out, "I've heard some rumors about some painful needle jabs" and Husband shouted back, "You've heard nothing!" McCool added, "You know Phil, we do need more volunteers for the blood draws," and I wondered whether or not I would get "volunteered" on the spot for their next training session! The crew then got in their bus to go to their next activity. It was the last time I saw Columbia's crew in person.

CHAPTER 15

P R E Q U E L:
S T S - 1 1 2 A N D
S T S - 1 1 3

While the Columbia astronauts were training for their mission, events were happening with the other shuttles which at the time seemed unimportant—but ultimately would lead to Columbia's destruction. STS-112, with the repaired flowliners, launched on October 7, 2002. It made it to orbit safely, but engineers were surprised to find a dent on one of the solid rocket boosters when it was recovered. An investigation revealed that a piece of foam had fallen off the ET's bipod and hit the booster with enough force to bend the metal. NASA was aware that pieces of bipod foam had come off before, on STS-7 in June 1983, STS-32 in January 1990, and STS-50 in June 1992. It had been a decade since the most recent case where NASA was aware of foam coming off of the bipod region. Engineers thought they had solved the problem after STS-50, so the loss of bipod foam on STS-112 was a surprise.

The STS-112 crew was informed about the incident only after Atlantis landed because the incident didn't affect the rest of their mission. Several months after the accident, crewmember Dave Wolf noted that he wasn't very concerned when he was informed because—at least from his personal perspective—it hadn't harmed him or his mission.

The STS-112 incident was addressed at the Flight Readiness Review (FRR) for STS-113, on October 31, 2002, the mission that immediately preceded STS-107. The FRR is the meeting where managers of each of the

A dramatic view of the STS-112 launch from a video camera mounted on the External Tank.

shuttle's sub-systems address any problems which occurred on previous flights, describe any changes to their systems, and determine whether everything's ready for launch. The FRR is co-chaired by NASA's highest shuttle managers—the Associate Administrator for Safety and Mission Assurance, and the Associate Administrator for Spaceflight. The individuals in those positions were Bryan O'Connor, an ex-astronaut with two shuttle flights, and William Readdy, an ex-astronaut with three shuttle flights.

Normally it's desirable for organizations to note that everything is on schedule and approved, along with an update on any changes or corrections to problems on previous missions. But occasionally an organization will note that paperwork has not been closed out or engineering work is still underway, along with a promise things are proceeding on schedule and the issues will be closed out before the final reviews two days before launch. There's a strong desire not to be the one who comes to the review with a problem that could delay the launch.

Jerry Smelser, the External Tank program manager, made a presentation that noted, "External Tank Thermal Protection System foam loss over the life of the shuttle program has never been a 'safety of flight' issue." In bold type, his PowerPoint chart stated, "The ET is safe to fly with no new concerns (and no added risk)." Essentially, the External

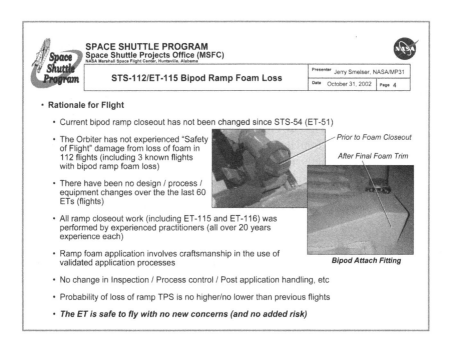

The Marshall Spaceflight Center claimed that falling bipod foam was not a safety issue. Photo credit—CAIB

Tank team said that the falling bipod foam had never been a problem before, so they didn't think it would be a problem in the future.

That simple statement was treated with disbelief by many in attendance. Boeing deputy orbiter program director Mike Burghardt recalls that he looked at a colleague and thought, "There's no way we'd try to pull off something like that," but there was an unwritten rule for management meetings — just brief the managers on what you're responsible for, don't criticize somebody else's system.

Safety head Bryan O'Connor reacted to Smelser's simple statement that the bipod foam was not an issue. Shuttle engineering office manager Paul Shack e-mailed colleagues about O'Connor's reaction at the. FRR, "They [the ET team] were severely wirebrushed over this and Bryan O'Connor (AA for Safety) asked for a hazard assessment for the loss of foam." One would assume it's not desirable to get wirebrushed. . . .

Several months after the accident O'Connor explained, "It wasn't clear to me what [Smelser] had said was all there. I asked him to expand on that and I even asked for the hazard analysis report, and somebody actually brought it to me during the meeting. So it was a clarification. He did make a comment that it was not a flight safety issue, and my concern was whether or not he was talking specifically about that event [on STS-112] or in general, whether it had been taken off the books as an issue at all. The hazard analysis report showed that it was still a generic problem that we have to deal with. The program assured me at the time that they were working it — not as a constraint to flight, though, but offline." In simpler terms O'Connor said he was asking Smelser for a clarification whether the situation was a non-issue or something which needed to be worked between shuttle flights, but without stopping flights.

NASA Associate Administrator for Spaceflight William Readdy added, "We both remarked on that. I think the exchange was something to the effect of, 'Is there any credible aerodynamic transport mechanism that could get foam from that area to impact the orbiter?' And the answer that we were given at the time was 'No.' But clearly, in retrospect, there's some stuff about the aerodynamics that we don't know. I'm sure that as a result of the accident we'll go back to the wind tunnels, we'll go do further analysis."

O'Connor and Readdy's comments are on this book's website.

Instead of classifying the STS-112 foam loss as an "In-Flight Anomaly," — a serious condition which must be corrected before the next flight — NASA decided to assign it as an "action" for the External Tank team, which was to determine the root cause of the foam loss and propose a corrective action. Again, this meant something which needs to be corrected but isn't important enough to shut down flights while the corrections are being made. The Columbia Accident Investigation Board (CAIB — pronounced "Kaeb") noted that this was inconsistent with the previous known cases of bipod foam loss from a decade and more before, which had been considered In-Flight Anomalies. The ET team was initially given a deadline of December 5 for their report, which was eventually pushed back to early February. With that move, managers made the conscious decision to fly STS-113 and STS-107 as-is, without any changes and without understanding why the STS-112 incident took place.

The CAIB believes the managers made that decision because of the pressure to keep the shuttle flying so ISS could be completed on schedule, in February 2004. CAIB chairman Hal Gehman, Jr., a retired Navy admiral, said, "We were alarmed by the way the bureaucracy handled the knowledge of the debris hit on STS-112—as an in-flight anomaly—and how it was treated as a hazard. It was treated as something to look at. We were concerned that this was done on purpose, knowing full well that if it was categorized as a hazard or In-Flight Anomaly, it would have delayed the follow-on flights. We considered it part of the evidence that schedule pressure had invaded every part of the system."

Did the Marshall engineers do any analysis to justify their conclusion that the falling bipod foam was not a flight safety risk? When I asked about it during a press conference in September 2003, External Tank manager Neil Otte avoided the question. Gehman said, "To my knowledge, they did an administrative review of foam coming off and concluded that they were at the same place they were 20 years ago, and we've been doing all of this flying for 20 years and so why should we consider foam to be any more of a problem today than it was before? They took no steps to truly understand what was happening."

After the accident, the CAIB concluded, "With no engineering analysis, shuttle managers used past success as justification for future flights." The accident board chastised NASA managers for failing to recognize the unexpected foam loss on STS-112 as a safety concern and accepting the recommendation to continue flying the shuttle even with the unexplained problem. Astronaut and manager James Halsell notes the STS-113 FRR was the event that, more than anything else, sealed Columbia's fate. Engineers made a fatal mistake by assuming that the falling bipod foam could not hit the shuttle. (Of course all of these conclusions were only reached after the accident occurred—with hindsight.}

The Marshall Spaceflight Center External Tank team said STS-112 was only the fourth out of 111 shuttle missions where bipod foam had been lost. But what they ignored was that many missions were launched at night and had no camera views of the tank, and there were cases where problems resulted in no film showing whether or not the bipod was intact. It wasn't 4 cases out of 111 missions but 4 *known* cases out of the 73 flights where there was adequate photographic documentation of the bipod region.

A close reexamination of the film from every shuttle flight after the Columbia accident revealed two additional launches where bipod foam was lost. In those cases, the foam loss was extremely small, so it was missed during the normal examination of the flight film, but after the Columbia accident anything that could have contributed was reexamined with a fine-toothed comb. Statistically, bipod foam had been shed on the average of 1 in 11 flights where there was adequate photography. Instead of what was passed off as a rare event—occurring only once every 27 missions—the bipod foam loss occurred more than twice as often. It would only be a matter of time before NASA's luck ran out and the bipod foam damaged something important.

Coincidentally, 5 of the 7 cases where bipod foam was lost, including STS-107, were on Columbia. This confused the CAIB and they wondered whether or not there might be something unusual about Columbia that caused this statistical anomaly. Columbia was the first operational shuttle, built before Challenger, Discovery, Atlantis, and Endeavour, so it was different in many ways. Could that account for the different statistics? The CAIB discovered one reason—Columbia had an engineering camera in its belly far longer than the other shuttles, so it had more opportunities to take photos of the ET as it dropped away after being jettisoned. Moreover, because Columbia could not launch as much payload as the other shuttles, it was assigned primarily to science missions, where in most cases the launches were in daylight. (Only two of Columbia's 28 missions launched at night.) As a result, there were more opportunities to see whether or not bipod foam was missing.

Still, the question remains: Why were there so few cases of bipod foam loss on the other shuttles? The two non-Columbia bipod losses were on Challenger on STS-7 and Atlantis on STS-112. No cases of bipod foam loss were ever discovered on Endeavour or Discovery. It's possible there may have been something unusual about Columbia, like its heavier construction, or some facet of the types of missions which Columbia flew (primarily payloads that remained within the shuttle's cargo bay) which accounted for this unusual bias in the statistics. Or it may have been just a coincidence.

The decision by the managers to continue flying after the STS-112 bipod incident, almost three months and one mission before STS-107, would have dire consequences. William Readdy and Bryan O'Connor signed the Certificate of Flight Readiness for STS-113's launch.

STS-113 launched November 23, 2002, at 7:50 P.M., after sunset. As the shuttle arched over the Atlantic Ocean, it flew into deeper darkness. It was the 39th launch without any way of knowing whether or not bipod foam had been shed. Statistically, on 3 of those 39 flights, bipod foam was shed but went unnoticed because of the lack of adequate photography.

Daytime launches and landings are more desirable for safety. There are more visual cues for the pilots, especially important if an emergency landing is required. Also, in the daytime, tracking cameras can show more detail. Since the beginning of the shuttle program, an image analysis team goes over video and film for each shuttle launch to see if anything unusual happened that could affect the shuttle. With night launches, far less data is available.

Daylight was not, however, a requirement. Early in the shuttle program, NASA made the decision that while it is desirable to launch and land in daylight, it may be necessary to do so at night, accepting the added risks when it's necessary to accomplish the mission's goals. Flights with night launches and landings have more conservative weather rules to compensate. On flights to ISS, the launch window moves about 22 to 23 minutes earlier each day. Over 65 days, it moves around the clock — from daytime to sunrise to night to sunset and back to daytime.

Before STS-107, NASA had launched 84 shuttle missions during the day and 28 at night. But exactly when daylight and nighttime occurs is a bit fuzzy, with many launches occurring in twilight.

When a mission doesn't require a night launch or landing to accomplish its goals, it's designed with a daytime launch and daytime landing. STS-107's launch window was specifically constrained by the desire to have a daytime launch and landing, whenever they occurred — even if the mission had to be cut short for some reason. Believe it or not, the

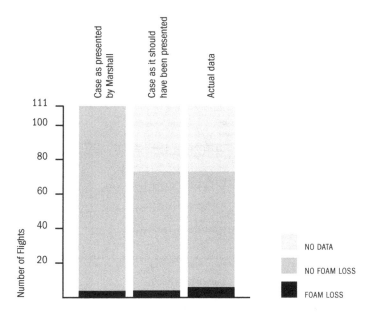

The foam break-off data as presented by Marshall (4 known cases), as they should have presented it (allowing for the missions where no photography was available), and the real-world data unknown at that time (two additional cases). Chart by Philip Chien

opening of the launch window was actually determined by sunrise two days after the planned landing day at the backup landing site at Edwards AFB in California! With those rules in place, even if the mission had to be extended two days due to bad weather or technical problems and had to go to the backup landing site, it would land just after dawn. If the mission was cut short for any reason, each day it was cut short would move the landing time later in the morning. The only case for a nighttime landing would be if there was a couple of hours delay in the launch followed by a major emergency a couple of minutes into the flight, which would require an emergency landing in Spain—an extremely unlikely set of circumstances.

The other important factor in determining the launch window was the MEIDEX experiment. It needed good lighting over its primary region of interest, the Mediterranean Sea.

With all of those constraints STS-107's window for a January 16 launch opened at 10:39 A.M. EST. Because Columbia didn't have to rendezvous with a moving target in space, the window would last two hours and 30 minutes—plenty of time to wait for bad weather to clear or to take care of technical problems.

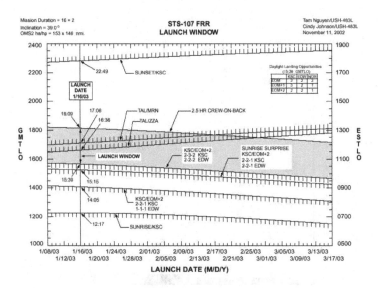

Launch window charts show the allowable launch time for Columbia. The launch time was designed to permit a daylight launch and daylight landing under all possible conditions. For security reasons, NASA never released this chart, which shows the allowable launch time each day from January through March 2003. Photo from the author's collection.

CHAPTER 16

PREPARING FOR LAUNCH

There was one minor advantage to the mission's many delays: the crew had more time to spend with their families and were able to go on vacations that would otherwise have been missed. Rick Husband said, "Last summer, when we found out there was going to be a delay, we saw a tremendous opportunity to take some summer vacation that we wouldn't have otherwise had. We were able to take some time at Thanksgiving and Christmas as well."

Dave Brown spent the holidays with his parents and older brother, Doug. Doug asked his brother one of the more natural questions astronauts gets asked by their closest family: "Dave, it's pretty risky. What do you want me to say if you don't come back?" Dave replied, "Well, I would want the program to go on," and he added that he fully accepted the risks he took in life since the time he joined the Navy—an honest question to a family member and an honest reply. Little did Doug Brown know at the time how often he'd be repeating his brother's thoughts.

After coming back from the holidays, the crew had an extra press conference on January 3. This set of refresher briefings was added because of the five-month delay due to the flowliners. Flight director Kelly Beck and mission scientist John Charles gave an updated mission overview, then the astronauts talked again about their planned activities. Once again, much of the attention was on Ilan Ramon, and once again the security question was asked—even though it had already been answered several times.

Was Husband concerned about the cracks on the BSTRAs causing another delay to the mission? Husband told me, "This was a great find [by the technicians] and those folks did a great job. When we first heard about it, it sounded to me like the initial characteristic was, in the end they didn't think it would be a problem. As it turns has out it looks like they will come up with sufficient rationale for it to be perfectly fine for us to fly [as is]. … It looks like, fortunately this time, we'll be able to launch on the 16th of January and we're looking forward to it."

Ramon said, "I have a lot of patience. To be with this magnificent crew it's a pleasure. I don't want to be delayed again, but I'm sure that we'll have a wonderful time together as we've had in the last two and a half years of training."

One of the VIPs on Ilan's invitation list was Ezer Weizman, a former Israeli president who was also one of the first pilots in the Israeli Air Force. But Weizman's age and health prevented him from flying to Florida for the launch. Ramon told me, "I got a phone call from him two days ago. He wished me all the best, and he's very, very angry with envy. He would like to go instead of me."

Six days later, the astronauts entered quarantine. The isolation is for medical purposes, to prevent them from catching a communicable disease. Young children are especially prone to sniffles and colds, so the last time they get hugged by their parents is just before they enter isolation; they cannot have physical contact again until the mission is over — unless there's a major launch delay. (Spouses can get health certificates signed by a NASA doctor which permit them to visit the astronauts in quarantine.)

The day the crew enters quarantine they spend as much time as they can with their families. Even if everything goes perfectly this will be the last time they can hug their children for three weeks, and if there are delays it can be much longer. And in the back of every astronaut-parent's mind is the thought, "If there is an accident this will be the very last time I get to hug my kids."

Dave left Duggins with his neighbors. He had already asked Cindy Swindells if she would become Duggins's permanent owner if something happened to him. By this point Duggins was 14, old for a Labrador, and in failing health. Dave discussed with Cindy what he wanted if things took a turn for the worse.

The Columbia crew started their quarantine on the evening of January 9. The only people in close contact with the crew would be those who were medically cleared. The astronauts could go anywhere, but stayed at least 10 feet away from people who weren't medically cleared. The crew moved to a set of dormitory-like buildings at JSC, where they could concentrate on their final preparations for the mission. Families could visit, but the children were supposed to stay at a distance.

The astronauts' final days in Houston were spent going over their flight plans and final simulations with the Mission Control team. The crew also started to adjust their sleep cycles. Since STS-107 was a two-shift mission to maximize science around the clock, the crew split into the red team of Rick Husband, Kalpana Chawla, Laurel Clark, and Ilan Ramon, and the blue team of Willie McCool, Mike Anderson and Dave Brown. Each team started to shift their wake-sleep cycles to match the times they would be awake during the mission. During their "days" they spent time in extremely brightly lit rooms, at "night" they wore dark sunglasses to help to encourage their bodies to adapt to their sleep cycles for the mission.

The day before the astronauts were scheduled to fly to Florida, Jonathan Clark took his son to the quarantine area. He recalled, "I took Iain in to see Laurel before they flew to the Cape, and she hugged him.

That was a deliberate thing [to break the health quarantine rules]. I wouldn't have given that up for anything."

STS-107's flight readiness review was held on the same day the crew entered quarantine. Each portion of the shuttle program presented the status of their systems, any problems they were working, and the progress solving those problems in preparing for the January 16 launch.

Anything that was an issue on Columbia's previous mission or an issue with another shuttle that could affect STS-107 was addressed at the FRR, and generally in a fair amount of detail.

The BSTRA cracks were one of the major topics. Engineers presented their rationale for why it was safe to fly without any changes, or even inspecting Columbia. Engineers had put a BSTRA with a crack through the equivalent of many flights and concluded that the cracks wouldn't get any larger. Some engineers questioned whether or not it was justifiable to fly as-is without even looking at Columbia's BSTRAs. Engineers and managers decided to proceed with the preparations for the launch while continuing to analyze the BSTRAs . A final report on the BSTRAs would be made at the launch readiness review two days before the launch.

There were seven In-Flight Anomalies (IFAs) on Columbia's previous mission, most notably the clog in the Freon loop. Each was addressed, including how the problem was corrected.

Marshall Spaceflight Center engineer Terry Greenwood gave the presentation for the ET team. The briefing mentioned that STS-107 was the first flight of the new Block II main engines with the older lightweight External Tank. It also mentioned an engineering analysis because of a change in oxygen pressure in a supply line. However, the lost bipod foam problem from two launches ago on STS-112 was not mentioned, and no one at the meeting asked about the status of the efforts to determine why the bipod foam had fallen on STS-112 and the status of the efforts to prevent it from happening again.

At the end of the FRR, managers decided to proceed with the January 16 date for Columbia's launch, pending the final resolution on the BSTRAs. For copies of the presentations, go to this book's website.

While most of Columbia's payloads were loaded months, or in some cases almost a year in advance, others had to be prepared close to launch. Many biological payloads, like rats and plants, needed to be at the proper phase of their growing cycle. The professional scientists prepared their payloads at NASA facilities at KSC, the Spacehab facility in Cape Canaveral, and the Florida Institute of Technology (FIT), a university about 35 miles south of the space center.

Fortunately for FIT, the preparations took place during the winter break. The European scientists were assigned laboratories to prepare their experiments along with whatever materials they had requested. By the time the students came back to school, the scientists were ready to fly and they only needed the two laboratories that would serve as their control center during the mission.

Students from six countries flew their own experiments in Spacehab's "S*T*A*R*S" Bootes (Space Technology and Research Students) program. Bootes indicated it was the second experiment in the S*T*A*R*S series, but for the most part the experiment was just known as S*T*A*R*S. The S*T*A*R*S were commercial educational payloads, with each group paying Spacehab $60,000 for the opportunity. Each of the experiments was about the size of a paperback book. Israeli students flew a "chemical garden" that grew blue and white colored crystals. Students from Australia flew orb weaver spiders to study how webs form in microgravity. Chinese students flew a silkworm experiment, students from Liechtenstein flew bees, a graduate student from Japan flew a closed biological system with medaka fish, and students from the United States flew an ant farm.

The students prepared their experiments at Astrotech, in Titusville, Florida to the west of KSC. Each group had assistance from engineers and laboratory technicians. It was the second time the students prepared their experiments; they had participated in a mission simulation in April 2002.

The Australian spiders traveled halfway around the world in a Styrofoam box to make their spaceflight. The students had to fill out the paperwork to transport animals internationally. The other animals, including the silkworms, bees, and ants, came from companies that supply animals for laboratories and high school biology classes in the USA.

The diet for the spiders consisted of fruit flies. But they weren't imported—they were just plain fruit flies collected in Florida. The students devised an innovative system to attract the flies: a rotten apple was the bait, and a set of plastic tubes were used to capture the eggs from the flies in an agar gel. As the flies hatched in space, they would be captured by whatever webs the spiders wove. The students created a chamber with seven

The balsa wood hive for the Spice Bees experiment with its pre-carved channels.

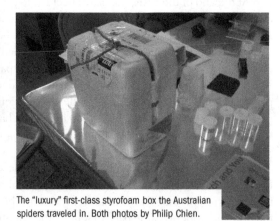

The "luxury" first-class styrofoam box the Australian spiders traveled in. Both photos by Philip Chien.

subchambers with a spare spider in each. If the primary spider was balky or didn't want to weave webs, an astronaut could flip a knob to let a backup spider into the main chamber. The species selected weaves a new web each day, eating the old web before building a new one. The students hoped to watch several webs as they were created over the two weeks in space. Besides watching the spider in action, the students were interested in testing the web's strength. Their chamber had a screw-on top where the astronaut could insert a stick. The astronaut would move the stick around to try to collect some webbing material, much as you could use a broomstick to clear webs from a difficult to reach corner. The stick with its clinging web material would be stored for analysis by the students.

The Israeli "chemical garden" had eight 50-milliliter clear containers (about the size of a 35-mm film canister) for crystal growth. Cobalt Chloride produced blue crystals against a white background and Calcium Chloride produced white crystals against a blue background. Students from Yonathan Netanyahu Ort Motzkin High School in Haifa chose those chemicals to match the colors of Israel's flag. Two pairs of containers were on the front of the experiment and two pairs on the back. Having four crystals with each color ensured that even if there were mechanical or electrical problems the students would have a good chance of getting video of at least one of each colored crystal. During the mission, Ilan Ramon would remove the experiment and turn it around to expose the second set of vials to repeat the experiment. The students would monitor the crystals growing on the Internet. Ramon had worked with the S*T*A*R*S team for several years, even before he was formally assigned to STS-107, and met many of the Israeli students. Five of the students flew to Florida to prepare their experiment.

The US experiment, from Fowler High School, in Syracuse, New York, was the simplest—an ant farm with 15 harvester ants. A yellow-colored gel was used instead of soil. The gel was the food for the ants as well as the medium they would live in and bore tunnels through. To prevent

Students from Israel prepare their "chemical garden" experiment. Both photo by Philip Chien.

Students from Fowler High School in Syracuse, NY, pose with the space ants experiment.

the ants from tunneling before launch the students devised a plunger-like mechanism with a door to prevent the ants from entering the "farm." The first time the astronauts accessed S*T*A*R*S, they would push down the plunger to let the ants in.

The ants, spiders, and chemical garden were the only experiments the astronauts would operate. The rest would run on their own, although if the astronauts wished, they could peek inside and see how the experiments were doing.

The Chinese students couldn't make it to Florida, so a Los Angeles high school acted as their representatives. Congresswoman Diane Watson wanted to get a Los Angeles school involved in the space program, so the Dorsey High School found local businesses that paid for the students' travel expenses so they could participate. The moths were prepared in their larval stage. The students ground up mulberry leaves into a paste that they put on sponges in plastic containers to provide the moths with a source of food after they hatched.

The Liechtenstein bees experiment featured a balsa wood block. The students used an X-acto knife to carve channels to give the bees a start. Ordinary carpenter bees would munch on the wood during the mission. The block was weighed before launch, and the students planned to weigh it after landing to see how much the bees had eaten during the mission. They also planned to put the wood block on display to show what they had accomplished. The experiment was named the "Spice bees" because the students were fans of the "Spice girls" band. It was the first space experiment ever for the tiny country of Liechtenstein and garnered a fair amount of interest. The students even gave a presentation to His Serene Highness Hans-Adam II, Prince of Liechtenstein, and gave him a sweater and cap with their experiment's logo. The country, never one to pass up an opportunity for more income from stamp collectors, issued a 90-centime postage stamp in March 2002, worth about 60 cents.

The most sophisticated student experiment was a closed ecosystem by Maki Niihori, a 22-year-old Japanese graduate student. Her experiment grew tiny medaka fish, snails, and small plants. She had run the experiment up to three weeks on the ground with a stable ecosystem and wanted to see how well it would work in space.

Even as the astronauts prepared to fly to Florida on January 12, engineers continued to debate the BSTRA issue. The astronauts were instructed to wait until a management meeting was completed where there would be at least some confidence that the BSTRAs would get cleared. After the management meeting was completed, the "go" was given to start the countdown. Over the next three days technicians would perform the final tasks to prepare Columbia for its launch.

That evening, crew secretary Roz Hobgood visited the quarantine area to finish some paperwork and was invited to join the crew for a meal.

She noted, "Some were eating breakfast and some were eating dinner. I got to sit between Rick and Laurel. It was just a great family time—we laughed and cracked jokes. We knew enough about each other that we could find the faults in the person and tease them. It was just a wonderful time. I remember when they left it was just a really big goodbye. They got their bags and gave big hugs and got in their vehicles to go to Ellington (the airport)." Hobgood had expected to see them in Florida before launch but was too busy coordinating all of the events, so that was the last time she saw her crew. She noted, "There's several hundred guests and when you're their contact, they're [calling your] number." The astronauts flew from Houston to KSC. After scores of flights to Florida for training or to inspect Columbia or their payloads, this time it was for real. Normally they would fly in their T-38 training jets, but because of the bad weather they flew in a business jet, which is less sensitive to high winds. When Husband stepped off the aircraft he was greeted by a handful of NASA workers and managers who support the crew in Florida. He joked, "This is a more civilized arrival." (As opposed to climbing out of the cockpit of the T-38.) Each of the seven crewmembers stepped up to a microphone and talked about their excitement.

Husband said, "We're absolutely thrilled to be here. We know we've got a good vehicle to fly. The STS-107 crew is looking forward to flying a great 16-day science mission. We want to express our gratitude and appreciation to all of the folks who worked so long and hard on this mission— whether it be on the vehicle, on the paperwork that goes along with the vehicle, the checklists that we do, the experiments, and all of the training we've gotten as well. We thank everybody so very much; we're looking forward to flying a great mission. Thank you very much."

McCool said, "I'm very excited to get up and go. The karma is feeling right, the weather's good. So let's get up and do it."

Brown mentioned his previous experiences helping to prepare other astronauts for launch—"Worked here at the Cape before coming down here and really thankful for all of the hard work by a lot of people coming up to this and still working really hard for us. Real excited to go do the science. Really happy to be here."

Chawla noted, "It has been a long road for us, but, like Willie said, today though even it started as a rainy gloomy day, but it was almost like the sun has finally shone at dusk time. I'm very, very happy to be here. I think we're going to go."

Anderson started, "I'd like to thank everybody for showing up, I really didn't expect such a big turnout tonight." He was joking—because of post 9/11 security measures the only ones in attendance were NASA photographers, a couple of managers, and the aircraft support people. Anderson continued, "This flight has been a long time coming, some of us have been in training for about two and a half years. I'd really just like to thank the sci-

entific community for really being patient with us on this flight and working hard with us to continue to bring the payloads up to date and to work
with us and train us [and get us] ready for this flight. We've been training
for about two and a half years, and we're going to go up there and do a really
good job of 16 days of science and bring back some really good results and
hopefully that will keep the scientists busy for years to come. I just want to
thank everybody for your support on this flight."

Clark realized all the good comments were already taken by her
crewmates, so she just said, "The same as everybody else, absolutely
thrilled to be here. Most of all I'd like to again thank our terrific team
that's on the ground. Thrilled to go do a lot of work, see some incredible
things and spend some more time with this great group of people I'm
with today. Thank you everybody, take care."

Ilan Ramon said, "I'm happy to be here, finally. It was a pleasure to go
the long way—two and a half years—because of the great team, great crew,
great trainers, great flight directors, great engineers. The route to the target is
more important than the target. We're going to go for the target, but we
enjoy the route as well." It was a long route for all of the astronauts, but especially Ilan. Six years had passed since he was selected to become Israel's first
astronaut, and four years had passed since he was originally scheduled to fly.

As part of NASA's post 9/11 security policies, there were certain
changes in pre-launch activities. The news media, for example, were not
permitted to see the astronaut arrival, the crew's locations were not
revealed, and activities were not confirmed until they moved on to their
next location. But the astronauts performed the same activities on the
same timeline as missions before 9/11.

The astronauts' immediate families were also affected. On all of the
previous shuttle flights, even after 9/11, the astronauts' immediate families stayed in condos in Cape Canaveral. (NASA only pays for the travel
expenses for the spouse and dependent children, but the condos are
fairly large, and often extended families will stay there too.) But because

The STS-107 crew arrives
in Florida for launch.

of the perceived terrorist risks with Ilan onboard, the STS-107 crew's immediate families stayed at Patrick AFB in the visiting officers' quarters.

Rick and Evelyn Husband visited the Patrick officers' quarters in July 2002 after the mission was delayed, and the family went to Florida anyway for vacation. Evelyn said, "One of the things we did was check out the facilities at Patrick, just to see if that would be a place the families would be comfortable in, but we would be far more protected. At that point Rick made the decision that he thought Patrick would be the best location for us. For what we needed it was perfect. It was real easy for them to have us all in one place and protect us." As a career Air Force family they were very familiar with living on Air Force bases, but certainly not accustomed to the amount of security that Evelyn and the children would have before launch. Jonathan Clark said, "It was very self-contained where we were. They had patrol boats, the whole area cordoned off, it was incredibly tight security for launch."

Security personnel followed the families wherever they went in public. Evelyn Husband said, "You couldn't do anything easily, all of us had family members and friends who watched our children while we spent time with the crew. None of them could go anywhere unaccompanied, without a special agent and people watching them. We didn't know what the threat was, and they were trying to take good care of us. Even just to go to lunch was a huge undertaking. All of these special agents and police and undercover agents that would sit around and [try to] act totally natural and be totally out of place."

Evelyn recalled an incident when her daughter Laura went to a local tourist trap, Ron Jons: "You had all of these men in suits looking at teenage girls necklaces and clothing like they're blending in, in the store, and they didn't blend in at all—they were watching my children. But everybody just took it in stride and we really appreciated all of the protection we had."

Most astronauts throw a big party for their families and friends, but cannot attend because they're in quarantine. Crew secretary Roz Hobgood noted that from her perspective, it was like coordinating seven receptions at the same time. She said, "It's madness. You survive on candy and chocolate and sheer energy. You have people calling you needing to change their seating, asking where to stay. It's seven different people—and all of their families, and their friends. It was always fun—they would come and tell me the stories about their friends and how they met them." In the case of international astronauts like Ilan, the prelaunch party is often a major reception sponsored by the country's space agency.

Dave's extended family stayed at the Wakulla motel in Cocoa Beach. They decided to have a family reunion, even though the key family member they were honoring couldn't attend!

Before the mission NASA had an informal press briefing with its top security official, former Secret Service agent David Saleeba. He said there had been no credible threats against the space shuttle since September 11, including STS-107. "We always have concerns about these missions because the shuttle is a fairly prominent symbol of American technology."

Saleeba said he was ready to call up even more protection, if necessary. Saleeba acknowledged that NASA was very aware of the additional threat potential because of Ilan Ramon's presence on the crew. He said, "Our antennas are up more than usual. Based on the world climate today, I would be foolish to try to make you believe that the presence of an Israeli astronaut on this flight would not make it a higher profile flight. It's unfortunate that that is true, it shouldn't be. He's first and foremost an astronaut and a scientist. And on a secondary note, he's Israeli. We fly international people, people from all over the world on many of our flights. But because of the fact that he is Israeli, and what's going on in the world today, it's a natural assumption that this may be a higher profile flight."

Saleeba was misleading or misinformed in his statement that Ramon was a scientist. While Ramon was performing science experiments for scientists, he was a military fighter pilot by training. First and foremost he was a combat veteran of two major wars with Arab nations, in addition to his rumored participation in the 1981 attack on an Iraqi nuclear reactor.

Saleeba said, "You're not going to see that much difference in the area of security for this launch than you did for the last several launches (after 9/11). We're trying to still maintain an openness within the agency while still maintaining an appropriate level of security." He was lying. Besides the changes to the activities for the astronaut families, additional security precautions were going into effect at the hotel where the VIPs were staying, and there were also changes that would affect the public, including the public in Florida with no interest in the launch.

The Hilton hotel where the Israeli delegation was staying had a roadblock at the entrance, with police checking cars for explosives and ensuring that their occupants had a legitimate reason to go to the hotel. The sudden increase in security surprised hotel guests with no connection to STS-107! Local police from surrounding communities were asked to help support the launch, supplying officers and police cars.

Security escorts were also provided for high-level Israeli officials. Israel Space Agency director general Aby Har-Even said, "When I went to a restaurant I have two policemen with me and I'm not used to this kind of attention."

Several reporters noted this was the mission that was going to get the most intense interest before launch (because of an Israeli crewmember and the potential for a terrorist attack) and absolutely no interest after launch (because it's a microgravity mission without any dynamic activities like spacewalks or a rendezvous with ISS).

In contrast to the intense training for the astronauts up until the time they flew to Florida, the period at KSC was far more relaxed. There were daily briefings on the shuttle's status, the payloads, and the weather forecast. The astronauts unwound by doing acrobatics in their T-38 training jets. Husband and McCool flew the Shuttle Training Aircraft, a Gulfstream II business jet modified to fly like the shuttle. It would be their last realistic training of the approach to the runway they would land on at the end of the mission.

Even though the crew was in quarantine, they could have phone and e-mail contact with the outside world. Not surprisingly, there were lots of VIP phone calls, especially for Ilan. Israel was extremely excited about STS-107. Three hundred VIPs from Israel and major Israeli news organizations came to Florida to watch the launch. In Israel, school children received 25 hours of lessons about the mission.

Dave's dog Duggins took a turn for the worse the day the crew flew to Florida, and Cindy Swindells wondered whether or not Duggins would make it through the night. Cindy and her daughter Jill slept on Dave's couch while Duggins laid on the floor. He was too weak to get up, but would spin around to follow them as they moved around.

The next morning, Cindy talked to Dave and he asked her to take Duggins to the dog's cardiologist. Cindy recalls that Jill grabbed Dave's wristwatch so Duggins could have something with Dave's smell during his last moments. Brown called the cardiologist to inform him about the situation and asked that Duggins should be put to sleep if the situation was too bad so Cindy wouldn't have the burden of having to make that difficult decision. An ultrasound verified that Duggins was dying and Cindy agreed with Dave's wishes. She said, "Dave and I talked Tuesday evening. I told him I felt bad that he couldn't have been there. He felt bad because we had to do it, but even if he was here we would have gone [with him to the veterinarian] anyway. We talked about the Duggins." Cindy's son, Adam, had a basketball game that evening and she held off telling him about what they had to do until after the game.

Cindy recalled, "Wednesday morning Dave called me—he wanted to know the outcome of the basketball game—and wanted to know how we were all doing and wanted to make sure we were still coming for the launch. He had the opportunity to tell me how much he appreciated everything we had done over the years and how we had adopted him and the dog and how he never worried about leaving the dog—because the dog was with family." Wednesday evening the Swindells flew to Florida for Dave's launch.

On Tuesday, January 14, managers held the Launch Readiness Review meeting, where all of the outstanding issues were addressed. The only open topic was the BSTRAs. Everybody was confident the BSTRAs were not going to delay the launch, but the team was given an additional day to put together their presentations and rationale.

At the standard prelaunch press conference, shuttle program manager Ron Dittemore acknowledged the risks for spaceflight and why NASA takes those risks. He said, "It takes a lot of work, and some risk, to get this vehicle into space, and we do it for a purpose. Because we've taken the risk and prepared for it, we want to accomplish every one of our objectives."

Could Israel look forward to future space travelers? Har-Even was realistic; his space agency is an extremely small organization in a small country. Israel's space agency is part of the ministry of science, which includes the science directorate, culture directorate, and sports directorate. Har-Even said, "We're working on other ideas, part of them for the space station. But the space station is a joint venture of a lot of countries who have contributed a lot of money. We know, for example, that Canada has invested more than $900 million. Brazil has budgeted about $40 million per year. The budget of our space agency is very limited—it's a few minutes of NASA's yearly budget. At the moment we don't have any idea to invest in a large instrument. We have suggested to NASA a few small instruments we can donate, like a very interesting instrument for measuring the weight of the bones. But at the moment it's still in discussion."

The next day the BSTRA team made their full presentation to the managers for why it was acceptable to fly as-is, without inspecting Columbia's BSTRAs, and it was accepted. Managers gave the go-ahead for STS-107.

The astronauts hold a barbecue on the beach for their closest family and friends. Each crewmember selects just five guests, and there are backup family members in case somebody selected for the party catches a cold or doesn't pass the health check. It's a final opportunity for the crew to share some private time with their families and closest friends. They all have their own private thoughts—aware that if anything goes wrong it may be the last time they see their loved ones. But it's tempered by realizing that your loved one is about to experience the greatest event of their careers.

Laurel invited her brothers Jon and Dan, sister Lynne Salton, and her mother Marge Brown. Dave asked his brother Doug, but did it in such a casual manner that Doug didn't think it was a big deal. So Doug decided to spend time with his friends instead and missed out on the unique opportunity because Dave just said, "How'd you like to join me for dinner the day before launch?" Dave ended up selecting his parents,

The STS-107 crew at the prelaunch beach party for their families. Photo from the author's collection.

Dave Brown and his "wife" Jeff "Goldy" Goldfinger during the pad spouse's tour. Photo courtesy of Jeff Goldfinger.

Rick and Evelyn Husband during the pad spouse's tour. Photo courtesy of Evelyn Husband.

friend Ann Micklos, and long-time Navy buddy Jeff "Goldy" Goldfinger as his guests. Goldy noted Dave knew he was a space buff and would really appreciate the unique opportunity. As the other guests boarded the bus to leave, Brown told Goldy to stay behind. In effect, Goldy became Dave's "wife" for the tour the astronauts give to their spouses. They joined Mike and Sandy Anderson and Willie and Lani McCool for an up-close personal tour of Columbia. The red shift and their spouses had done their tour earlier because of the different sleep cycles.

A much larger party was held at the Hilton hotel for the Israeli VIPs. The center of attention was Ilan's wife Rona, and their four children. Of course Ilan couldn't be there himself.

One experiment made it on board at the last moment—almost literally. On New Year's Eve, just 16 days before launch, managers approved the late addition of six canisters filled with tiny worms. The purpose of the experiment was to see how well a synthetic worm food would work in space. If it worked as anticipated worm experiments lasting several months could be performed on ISS. There was extra space underneath a moss experiment which was already scheduled, and managers were loath to waste the space. The worms and canisters had previously flown and passed all of the flight safety reviews, but the approval for the new food didn't arrive until the end of December. Safety people had to be assured that the new worm food was not hazardous in any way. One of the scientists offered to drink a glass of the worm food if that would satisfy the safety panels, but fortunately the safety people didn't ask for that demonstration! No crew activities were required; the worms would grow on their own in space and would be studied afterwards by the scientists. The worm team, especially post-doctoral student Nate Szewczyk, worked overtime to prepare the canisters in time for the launch. It set the record for the most rapid preparation of a shuttle experiment for flight.

Starting three days before launch, technicians loaded about 2,000 pounds of payloads, including S*T*A*R*S, protein crystals, rodents, moss,

worms, cell cultures, and other time-critical experiments. Spacehab manager Pete Paceley said, "It went flawlessly. That was a huge logistics effort of
vans, experimenters, and technicians all working to get this stuff loaded. In
our wildest dreams we couldn't hope for something to go as well as that
went." The last items were put on board less than a day before launch.

The blue shift—Mike, Dave, and Willie—got up Wednesday evening
for launch. It would be half a day before the launch and another several
hours before they would go to sleep in space for the first time. With plenty
of free time before the rest of the crew woke up, the astronauts could do
what they wanted. Willie called crew secretary Roz to ask her to make a
last-second change for his guests. She said, "Willie called me about 10 P.M.
and wanted me to find a couple and change their seating. They were hand-
icapped and needed some assistance, so I spent a couple of hours trying to
find them at their hotel to get them a different setup." She recalled that that
was the last time she spoke to one of her crew.

Dave decided to call neighbor Al Saylor, a mortgage broker, who
flew down to Florida for the launch. Saylor said, "This voice came on the
line and said 'Hello, I'm interested in getting a mortgage for a house in
Houston,' and I said, 'Dave Brown, what are you doing calling me up
talking about mortgages?' And that's how he started the conversation. He
went on to tell me that he wanted me to know that Duggins had passed
away. I expressed my sympathies for him and he appreciated that. And
then he asked me—and this is indicative of Dave Brown—'Did you go to
the reception tonight?' I said, 'We saw your video and you looked great
and you're quite the actor besides the producer of films.' He asked, 'Did
you get the chance to meet my mom and dad—aren't they great?' That's
what he was he was actually asking me. We had talked about his family
and I kind of felt like I knew (his brother) Doug and the Judge (Dave's
father) and his mom from a distance."

Saylor and several of Dave's neighbors had flown to Florida in a small
aircraft, landing at nearby Space Coast Regional Airport. That airport was
one of several affected by the additional STS-107 security measures. The
9/11 terrorist attacks made Americans very conscious of the fact that an airplane could easily be turned into a guided missile simply by putting a terrorist who is willing to give up his life at the controls. The space shuttle,
loaded with half a million gallons of extremely energetic rocket propellants
is a significant target. NASA and the military instituted intense security precautions, including shutting down general aviation in the area surrounding
KSC. Only scheduled commercial airliners flying their normal flight corridors under radar and transponder contact with a control center, emergency aircraft like police and rescue helicopters, and military aircraft are
permitted within a 30 nautical mile radius. Five general aviation airports
are shut down six hours before launch by these restrictions, with the airports shut down an additional three hours for STS-107.

CHAPTER 17

L A U N C H
D A Y

January 16, 2003, was an excellent day for the launch. The weather was mild, and good viewing conditions were expected. Managers held a meeting during the night and gave the approval to load 500,000 gallons of liquid hydrogen and oxygen into Columbia's External Tank. The red shift was awakened at 5:30 A.M.

Some of the workers were a few minutes late because of the additional security checks. Launch director Mike Leinbach said, "We dealt with it—no big deal." There were fewer tourists than usual. Members of the public had to purchase $51 tickets to get on a bus to go to the viewing site, and if there was a scrub, they'd have to pay again for the next attempt. Those who did purchase tickets parked their cars off site, where they boarded buses to the viewing locations.

The astronauts posed at their dining table with their mission logo for NASA's photographers—more of a photo-op than an actual meal. While Public Affairs called it breakfast, it was actually a late lunch for the blue shift. The crew was all smiles, looking forward to a good launch; they gave the camera a thumbs up, while Rick and Laurel held up both hands.

The pilots were given a weather briefing with the forecasts for the launch and the handful of runways where they could land if an emergency required a rapid landing. During the first two minutes of a shuttle launch, while the solid rocket boosters are burning, there's nothing the crew can do but hang on. After the boosters separate, Columbia could make a U-turn and come back to KSC if there was an emergency like an engine shutdown. That would be an extremely challenging flying task for Husband and McCool, flying a completely loaded shuttle with a full mission's worth of supplies on board and a mostly full tank. With all of the momentum from the liftoff, the shuttle would continue to fly forward but gradually slow down. After the shuttle reached zero speed, the tank would be ejected and the shuttle would turn around and aim itself back toward Florida. In the simulations it works well. In real life, everybody's relieved that no one has ever had to actually fly such a difficult maneuver. At 3 minutes, 50 seconds after launch, Columbia would have so much forward velocity it could not make it back to Florida and would have to continue across the Atlantic.

If one engine failed between 2.5 and 7.7 minutes after launch, Columbia could have made a safe landing at either Moron (pronounced "More own") or Zaragosa in Spain. Overlapping part of that period, Columbia could have done a very rapid trip around the world, landing at either Edwards AFB in California or White Sands Space Harbor in New Mexico. After 5.5 minutes, Columbia would have enough energy to reach a safe orbit even if an engine had a premature shutdown.

Astronauts tend to take an extra long shower on launch day, and nobody's going to complain if they use a lot of hot water. It's the last real chance to be clean for the next two weeks. In space, they are limited to sponge baths.

And one of the least publicized features of launch day: When you see the astronauts walk smartly out to board the van to go to the launch pad, they're wearing adult-size diapers. There's a chance that they'll spend several hours lying on their backs, plus launch and additional time before the shuttle's toilet can be activated in space. And if you've gotta go, you've gotta go. So astronauts wear diapers for launch and landing and during spacewalks. Most astronauts take the precaution "to go" in advance to minimize the chances that they'll need to use the diaper.

As the astronauts left their living area and entered the elevator down to the street level, Dave Brown could be seen aiming his camcorder at the NASA photographers who were shooting him, still documenting everything he could.

The astronauts walked out of the Operations and Checkout building to the "astrovan," a modified motor home that drove them to the launch pad. There have always been armed SWAT guards on alert—on the ground watching the press and workers who have shown up to wish the astronauts luck, on the tops of buildings, and in security vehicles in the

The STS-107 astronauts walk out of their living quarters to the astrovan on their way to launch.

convoy. After 9/11, the security was increased even more. Astronauts have talked about how surreal it is to look through the astrovan's front window and see a helicopter flying just above the road with a SWAT team member on the helicopter's skid, making sure everything is clear.

Security closes off roads before the astrovan starts its journey. The most open areas are the main cafeteria and press site, where many people have access. The astrovan stopped at the Launch Control Center to let off managers who accompany the astronauts. A worker decided to show his own support for the STS-107 mission's international flair by waving an Israeli flag in front of the astrovan.

The media could view the astrovan as it traveled past the press site and dropped off the managers. Many filmed the flag incident, but NASA had a security blackout on official information until the crew arrived at the launch pad. NASA waited until then before broadcasting the prerecorded video of their prelaunch meal, suiting up for flight, and leaving the crew quarters.

In contrast to a normal workday, the pad is almost abandoned. Before the ET is loaded with propellants, the pad is cleared of all non-essential personnel. Unlike on other days, the shuttle feels "alive." The ET makes noises as the metal contracts due to the supercold propellants inside, and it looks different, with a tiny sheen of ice and vapors coming off its surface. The ice inspection team had checked the tank earlier to ensure that no ice was forming that could fall off during launch and damage the spacecraft. When the astronauts arrive, the only people at the pad are the "closeout team," which helps them inside. More than one astronaut has joked that, on launch day, the pad's too dangerous for anyone to be there—except them of course.

When the shuttle's on the pad, the seats are lying on the "floor," with the astronaut almost lying down. The cabin is fairly small, so it's difficult to maneuver everybody inside, and especially difficult for the pilots, who have to do a pull-up to climb into their seats. Technicians and fellow astronaut Doug Hurley helped the astronauts into their seats and strapped them in. It was a process with which Dave was very familiar; he was one of the support astronauts who helped strap the astronauts in on three flights before he was assigned to STS-107. The process is carefully choreographed, with one astronaut getting helped into a seat on the flight deck and another astronaut into the middeck at the same time. Husband was the first to enter Columbia, at 7:53 A.M., climbing up to his seat on the flight deck. The next was Ilan. Before entering Columbia, he held up a "boarding pass," a stylized version of the STS-107 logo with wings marked "Global Village Launch Pass." He entered the middeck and was strapped into the furthest seat from the hatch.

As Willie entered the shuttle, Dave used his camcorder to document the astronauts entering Columbia. While Willie was helped into his seat on the flight deck another technician helped Laurel into her seat on the middeck.

Ilan Ramon holds up a "boarding pass"
before entering Columbia's crew cabin.

Dave Brown videotapes Willie McCool
as he enters Columbia's crew cabin.

When it was Dave's turn to enter, he handed the camcorder to one of
the members of the closeout team, who returned it to crew quarters to
await Dave's return at the end of the mission. McCool was in the right
front seat, and Brown was in the seat behind him. As technicians helped
Brown get into his seat on the flight deck, Anderson was helped into his
seat on Columbia's middeck.

The astronauts always enter the shuttle in a specific order because of
the lack of space and the maneuvering required. Kalpana Chawla had to
be the last one to enter because her seat was used as a platform and step
for Husband and McCool to reach their seats. After everyone else was in
place, Chawla entered, at 8:45 A.M. and was helped into her seat. Each of
the astronauts performed communications checks with the Launch Control Center and Mission Control. The closeout crew removed protective
covers from control panels and other equipment; these covers are put in
place to ensure that switches aren't accidentally flipped while the crew is
being seated. The astronauts put on their helmets and gloves. After wishing the crew a good mission, the closeout crew left the crew cabin and
closed Columbia's hatch at 9:17 A.M. After a pressure check to ensure that
the hatch is properly sealed and the shuttle is airtight, the closeout crew
left the launch pad. From this point on, the astronauts are on their own;
the only other people within 3.5 miles are a rescue team in an M-113
armored personnel carrier a mile from the pad.

The pilots have a few switches to flip, but for the most part the crew is
on their own as they listen to the launch team go through the final preparations. It's time for them to reflect on their lives and everything it took to
get to this point. There's rarely any nervousness or fear—the time for that
was long ago, if at all. There may be some adrenaline, especially in the
rookies, but everyone realizes that a scrub (canceling a launch because of
weather or a technical delay) is just as likely as a launch. Statistically, half
of the launch attempts are scrubbed. But as the countdown proceeds
toward zero, the anxiety and excitement grow. Ilan had told crew secre-

tary Roz Hobgood, "I'm not going to be excited until I'm in my seat, buckled up and taking off."

Nothing can prepare a rookie astronaut for what spaceflight is like. No matter how good the simulators are, and no matter how much the veteran astronauts describe what the experience is like, no written or verbal descriptions can match the real thing. It's like imagining what it's like to climb a mountain, jump out of an airplane, or do any other incredibly thrilling activity for the first time. No matter how well prepared you are and how much you're looking forward to the experience, the actual event will always be more than you could have ever imagined, and include some surprises.

In contrast to the many program and technical delays before the mission, the STS-107 countdown went like clockwork. Everything proceeded on schedule, with only extremely minor problems, what NASA calls "nits." As the managers went through their approvals, the anticipation could be heard in Husband's voice as he announced "CDR is go."

The tension increases as the countdown comes out of its final hold, nine minutes before launch. Five minutes before launch, pilot Willie McCool was asked to turn on the auxiliary power units. The running joke is, a pilot's only "official" tasks are to turn on the auxiliary power units before launch and lower the gear just before landing. In reality, they have many more tasks where they are not the "official" crewmember in charge.

Launch director Mike Leinbach, noting the mission's many delays, said, "If there was ever a time to use the phrase 'all good things come to people who wait,' this is the one time. For you and your crew, best of luck on your mission, and from the many, many people who put this mission together, good luck and Godspeed." Husband replied, "We appreciate it, Mike. The Lord has blessed us with a beautiful day here, and we're going to have a great mission, we appreciate all of the great hard work everybody's put into this, and we're ready to go." An audio clip of Leinbach and Husband's exchange is on this book's website.

Columbia's 28th and final launch took place at the opening of the window, at 10:39 A.M. EST on January 16, 2003. As usual, it was incredibly spectacular. The weather was perfect, with mild temperatures and a brilliant blue sky with a few clouds. The three main engines ignited in sequence six seconds before launch. From viewing sites, the crowds saw white "smoke" (actually water vapor) coming out around the pad silently—it would take several seconds before the sound would arrive. When the countdown reached zero, the twin SRBs ignited. Columbia almost jumped off Pad 39A, with over 7 million pounds of thrust, rapidly clearing the pad and quickly building up acceleration.

Even daytime launches make the sky brighter, and then the vibrations arrive at the viewing sites. A little at first, when the vibrations from the main

Columbia's final
launch. Author
Philip Chien
can be seen in
the foreground
of this NASA
photo.

Columbia's final launch. Author Philip Chien can be seen in the foreground of this NASA photo.

engines arrive, then they build into a crescendo as the full force from the combination of the main engines and SRBs hits you. It's an incredible experience, especially if you're at one of the closer viewing areas.

Thousands of cameras photographed the launch, from $3 disposables to sophisticated tracking cameras. Among those cameras was the E208 tracking station, 18 miles to the south, in Cocoa Beach. The camera is located next to the beach, between a condo and a hotel. Many of the STS-107 guests, including Dave Brown's family, stayed at that hotel. The E208 camera was not perfectly in focus—soft-focus was the term the technicians used—but having that camera in action wasn't considered mandatory, just desirable. With so many shuttle missions launching at night, and with a total of 112 previous flights, the tracking cameras were no longer considered mandatory, just desirable. But since the camera did function, albeit out of focus, it tracked Columbia's launch.

At 33 seconds after launch, the thrust from the main engines automatically decreases to reduce the maximum amount of acceleration on the astronauts and everything else inside the shuttle. At 45 seconds after launch, capcom Charlie Hobaugh radioed, "Columbia, Houston. You're go at throttle up." That call indicates the point when the main engines build back up to full thrust. Husband replied, "We copy—go at throttle up." That was the point in flight where the Challenger accident took place, and many, seventeen years after the catastrophe, breathed a sigh of relief when Columbia passed that milestone without incident.

At 81 seconds after launch, the E208 camera saw a 1.67-pound chunk of bipod foam fall off the ET. Unlike all of the small pieces of foam which had fallen in the past, even the previous known chunks of bipod foam, this particular piece was gigantic — about the size of a small suitcase, almost twice the size of the largest previous piece of debris. Over 90 percent of the bipod foam had broken loose. The foam was an incredibly light object and quickly slowed down, much as a piece of paper tossed out of a moving vehicle will quickly come to a stop. Columbia continued to accelerate. In previous cases where bipod foam was lost, it missed hitting any critical place. But a combination of random vibrations, upper level winds, wind shear, and bad luck resulted in the foam hitting Columbia's left wing. The launch videos and films were analyzed the next day. It would be 17 days before the true consequences of that impact were known. As far as anyone knew on launch day, everything had gone perfectly.

At 2 minutes, 4 seconds after launch, the SRBs completed their burn and separated. Parachutes deployed to slow the SRBs' descent into the Atlantic Ocean. Specially equipped recovery ships towed them back for refurbishment and reuse on future flights.

Traditionally, the commander makes an in-cabin announcement about 3.5 minutes after launch for the rookie astronauts as they climb through 50 nautical miles — the altitude where they officially become astronauts. There's usually a cheer in the cabin, and occasionally the ex-rookies will clasp hands.

Around Mach 13, the engines swiveled to flip Columbia right side up. The main computer determines whether it's a clockwise or counter-clockwise spin, and it's a 50/50 chance. McCool noted ahead of time, "Rick and I are placing bets whether I get the view because the roll is in my direction or in Rick's direction so he gets the view."

The engines shut down 8.5 minutes after launch, and half a minute later the ET was jettisoned. A pair of cameras mounted in Columbia's belly took still photos and movie film of the ET as the shuttle moved away.

The instant the engines shut down, the astronauts became weightless. For the rookies, it was their first real exposure to the microgravity they would become accustomed to for the next 16 days. Rookies Willie McCool and Dave Brown were on the flight deck and saw their first spectacular views of space as Columbia flew over the Atlantic Ocean and reached Europe. On the middeck, rookies Laurel Clark and Ilan Ramon were certainly thinking about floating up to the flight deck for their first views of the Earth. Before the launch Ramon had said, "Since I'm a pilot — when you get into a fighter you anticipate how you feel when you pull some Gs (acceleration.) Here it's the contrary — I'm anticipating how does it feel to float." Of course even the three space veterans were thrilled to be back in space, too.

While everybody would certainly be relieved that the most dangerous portion of the mission was over, they were certainly aware that they wouldn't be completely safe until Columbia landed. While those thoughts went through their heads, the astronauts were also busy concentrating on their activities and following the flight plan.

After the ET was jettisoned, it continued to fly around the world with Columbia. Both were on a suborbital trajectory. The tank would reenter the Earth's atmosphere, where most of it would burn up, with any remaining pieces splashing down in the Pacific Ocean to the southeast of Hawaii. Columbia, by contrast, would fire its engines to put itself into a stable orbit 41 minutes after launch.

Hobaugh radioed Columbia, "You are go for the ET photo maneuver," and Husband replied, "We copy, Houston, go for the ET photo maneuver. Thanks." It was a normal go-ahead to fire the thrusters to flip Columbia over so its windows faced the tank. Those photos are routinely taken on every mission for post-mission engineering documentation and analysis.

Mike got out of his seat and opened a locker where a Nikon F-5 35-mm camera with a 400-mm telephoto lens and Sony PD-100 DVCAM video camcorder were stored. As Anderson floated up to the flight deck with both cameras Dave got out of his seat and looked at the Earth through the overhead windows. Anderson gave the camcorder to Brown and used the Nikon camera to take photos of the ET. Brown's video was transmitted to the ground the next day. The video showed that the tank had rolled over, exposing its back and bottom. After the accident, there was an intense search for the film from those cameras and the automatic camera in Columbia's belly, which could have shown how much foam had fallen off the ET, but they were never found.

After the Orbital Maneuvering System (OMS—pronounced "ohms") engines circularized Columbia's orbit, Hobaugh said, "And to have a sensitive moment we hope your wait to space was worth it. We saw a flawless ascent and insertion for your now veteran crew of astronauts. And especially a big welcome to Ilan as [Israel] joins the international community of human spaceflight." Husband replied, "We thank you very

Dave Brown gets out of his seat
for the first time in space.

much, Houston. We made sure we welcomed everybody to space, and they're all doing great—thanks a million."

On orbit, the crew quickly got to work. The "red shift" (Husband, Chawla, Clark, and Ramon) went to work immediately, while the "blue shift" (McCool, Brown, and Anderson) prepared for bed after completing a few tasks.

The most important task when the shuttle arrives on orbit is to open up the payload bay doors. A set of radiators is on the inside of each door, and they are critical for the removal of heat generated by the shuttle's electronics and crew. If there was a problem opening the doors, Columbia would have to return to Earth within a day. Brown's neighbor Jeff Kling was the Mission Control engineer responsible for the mechanical systems. Brown had told Kling, "When we get on orbit, I'm going to open up the payload bay doors, and I hope I don't screw it up and make a fool of myself in front of you." Kling thought, "Here's this guy who's ultra-overqualified for everything messing up the payload bay doors—I don't think so." Once the doors were opened, Mission Control gave Rick approval to reconfigure Columbia's systems for its stay in space. Husband replied, "And Houston, we copy—go for orbit ops. Thanks."

Later in the mission, Willie described one of his first incredible sights, about an hour after the launch: "Rick said, 'Willie, come look.' I came to the back window and it was dark outside, but the tail of the orbiter we could see, and it was glowing red, to orange, yellow to white, all in about 15 seconds. I saw my first sunrise across the tail of the orbiter."

The astronauts helped each other climb out of their bulky orange suits. Laurel, Ilan, and Mike "converted" the shuttle from a rocketship (seven seats for the crew, with all of their equipment tucked away in lockers and storage locations) into a spaceship (seats folded and put away, and equipment installed all over the place). It's an extremely busy period, and the crew carefully choreographs which tasks each person performs.

Some automated experiments were activated quickly. Dave flipped the four switches to activate FREESTAR. Willie radioed, "Tell the CEBAS [Closed Equilibrated Biological Aquatic System—pronounced "see bas"] folks I got the tape installed at MET [Mission Elapsed Time] 1 hour 48 minutes, so our activation is complete." CEBAS included plants, snails, 16 swordtails, and 50 small yellow-belly cichlids in a 2.6-gallon aquarium.

Just five hours after launch, McCool, Anderson, and Brown were sent to bed. They had been up for half a day before launch, setting up their sleep cycles opposite the "red team." But even though they had been up for almost 18 hours, the adrenaline rush from the launch certainly had to make falling asleep difficult. If they felt the need, they could have taken a strong sleeping pill with the approval of the flight surgeon. By the time they went to sleep, the rookies already had more experience than John Glenn had on his historic Mercury mission in 1962.

Clark's sorority roommate Martha Wilson (the elementary school teacher) said, "I had my whole class [in Wisconsin] watching the launch on a big screen. I was so nervous because I was thinking of the Challenger accident. A couple of hours later, we could hear the shuttle talking to Mission Control and we heard Laurel talking to a woman [capcom Stephanie Wilson]. I could just hear the smile in her voice and I just knew she was so incredibly happy. She was just in her element, this was the coolest thing she had ever done in her life."

When Clark entered Spacehab, about 4.5 hours after launch, she said, "We're thrilled to be in this beautiful new laboratory and anxious to get going on 16 days just packed full of science from many, many different disciplines. We're thankful for all of the hard work of the Spacehab team." Wilson replied, "Thanks for those words, you've got big smiles down here and we're all looking forward to working with you."

Paceley said, "I had babied this module through its development phase, working on it for five years. When you get up there in the flight and see all of these photos and video coming down—it just tickles you pink. There's no better feeling than to see those folks up there working on these payloads like we practiced, and things going well."

The shuttle has an internal intercom, so the crew can talk among themselves from different locations. One of the two channels worked properly, but on the second channel, the astronauts inside Spacehab could hear the crew cabin but not vice versa. It was a minor annoyance; what NASA calls a "nit."

The red team had their first meal—a lunch. Before the mission, Rick had selected peanut butter, grape jelly, tortillas, pears, almonds, and tropical punch. Obviously he knew he was going to be busy and eating on the run, so he decided a peanut butter and jelly sandwich was the best meal for the first day. Kalpana chose black bean soup with rice, tomatoes and eggplant, tortillas, cherry blueberry cobbler, and a grapefruit drink. Laurel wanted a shrimp cocktail, tortillas, pineapple, cashews, and an orange-grapefruit drink—light snacks chosen by somebody who's uncertain how her body is going to react to spaceflight. Ilan selected kosher chicken and noodles, green beans with mushrooms, crackers, strawberries, trail mix, a brownie, and orange juice.

After lunch, the red shift continued setting up Spacehab. Within a couple of hours Spacehab was running, and its first science activities were underway. Paceley said, "The module activation went very, very well. I remember being extremely relieved [there weren't any problems]." Laurel checked on the rats, and Ilan got on the ARMS ergometer for the first of several sessions.

One important variable that can't be predicted is whether any of the crew will get sick, especially the rookies. Space Adaptation Syndrome, or "space sickness" as it's commonly known, affects about half of the astronauts—and more so rookies than experienced astronauts. It's similar to motion sickness, but it is hard to predict who's going to get sick and who isn't. In one famous case, a scientist didn't get sick but the ex-acrobatic jet fighter pilot did. It's not surprising that when the body is put in an extremely different environment it's going to react, and different people will adapt better than others. Many astronauts who experience space sickness on their first missions have less severe reactions on their later flights because their bodies now know what to expect.

In one case, space sickness almost cancelled a mission's key goals. Apollo 9 astronaut Rusty Schweickart was supposed to test out the Apollo spacesuit, and it would not be safe to get sick while wearing a sealed helmet. (The flight rules required the spacewalk before the Lunar Module could fly on its own. If there was a problem redocking the Lunar Module, then its two astronauts would have to do an emergency spacewalk to get back home.) Fortunately, Schweickart overcame his sickness the next day, and the rest of the mission was accomplished.

Astronauts who experience space sickness just push ahead and do their tasks, even if they have a queasy stomach or bloated feeling, but it can reduce their productivity. As a rule, flight planners try to schedule fairly relaxed activities for the first day or two to allow time for astronauts to adapt, and in most cases that works well. Space sickness is an extremely private medical issue that NASA won't discuss publicly, although some of the astronauts are more candid than others.

Laurel compared her shuttle experience to the first few days living on a submarine. Later in the mission she said, "I feel wonderful now. The first couple of days you adjust to the fluid shifting, different environments, how to fly through space without hitting things or anybody else. But after a couple of days you get in the groove, and it's just an incredibly magical place." Willie also acknowledged that adjusting to spaceflight made it difficult for him to concentrate on his hand-intensive tasks the day after the launch.

Paceley said, "We had a lot of concern about the early part of the flight because it was busy. We had things like crew riding on the ergometer, we didn't know if they'd be sick or not. We were really kind of worried early on that we'd be able to accomplish everything we needed to accomplish, especially the first couple of days. Everything went just flawlessly, things were going extremely well."

WATCHING THE LAUNCH

The crew's spouses and children watched the launch from the roof of the Launch Control Center. Many astronaut spouses have talked about how lonely it is to be standing there and watching their husband or wife launch into space, even though you've got your children with you. No one else can truly appreciate what a spouse goes through—hoping that everything's going right and worried about everything that can go wrong. Astronaut Jim Wetherbee, the commander of the flight before STS-107, said, "The one it's probably toughest on is my spouse, Robin. I think it would be incredibly tough to watch a loved one climb on a rocket. It's very easy for me to do, because I live and die as a consequence of my actions. As Robin explained it, it's as alone as you can possibly feel, standing there on the building watching a loved one launch into space. There's nothing anyone can say or do to help you in that kind of situation. So I feel really fortunate that she's allowed me to do this six times now. I think she's a lot stronger than anybody I've ever met."

The astronauts' extended families, including parents, siblings, and close friends, watched from a VIP site about two miles north of the Launch Control Center. Other family and friends watched from other viewing sites at greater distances from the pad.

Dave Brown's neighbor, Cindy Swindells, held Duggins's dog collar as she watched the launch with his family. She recalled, "At least Duggins was there in spirit for Dave's launch."

I met with Laurel Clark's mother and siblings an hour after the launch. Laurel's mother, Marge Brown, told me, "It was very nice. I hope all of it goes as well as it did today." Were they nervous while watching the launch? Jon Salton, an engineer who had worked on the shuttle program, said, "Anyone who has watched [video of the] Challenger [accident] can't even hardly bear going through [the point where the Challenger happened]. After that point, you can relax." Lynne added, "When we saw the Solid Rocket Boosters drop away, everything was still fine, my heart lifted a little, and then they got to Main Engine Cutoff." Dan noted, "It seemed after launch it hit me—I was holding my breath for 10 minutes."

The person who missed Laurel the most was certainly her son, Iain. Marge said her grandson told her, "Why can't any of you go, then my mom won't have to go?" Lynne Salton added, "He wanted to see the

Judge Paul and Dorothy Brown
watch their son launch into
space. Photo courtesy of the
Brown family.

launch, but he wanted one of us to go instead of his mom." Like any other nine-year-old he just missed being away from his mother. Iain watched the launch from the Launch Control Center's roof with his father, Jonathan Clark. Clark said, "Iain had a very bad feeling about this mission. He had not wanted her to go. I've asked him about [whether or not he thinks he had a premonition] and he said yes. There's a lot of unresolved anger and guilt."

Many believe Ilan Ramon's daughter Noa and other children had premonitions about their parent's death. According to the Israeli press, "In a haunting prophecy, Noa Ramon said, "I lost my daddy." Of course those claims were only made after the accident. Prophecy is always easier when it's hindsight.

But premonitions about the Columbia accident were just examples of selective memory and the natural tendency, especially for young children, to be worried when their parents are away. Every astronaut's child feels the same thoughts in the days before every shuttle launch, culminating on launch day—it's just an extreme version of what any child feels when a parent goes away for a long trip.

Consider the situation for Noa Ramon—it's a couple of weeks before launch and you're a five-year-old girl living in a foreign country. You're getting out of school and everybody's anxious to go to Florida for daddy's big adventure. From a child's perspective the most important activities are getting out of school and going to Disneyworld! And Ilan Ramon said as much before the mission. Two weeks before the launch he told me, "My girl is looking forward to Disney. The [older] boys are looking forward to both [Disneyworld and the launch] I think. And they are excited."

You can't see your father because he's in quarantine. Your mother and brothers are all excited about what's happening, and your mother is busy with all the preparations for the trip to Florida. The night before launch, there's a giant party and your family is the center of attention. There are some extremely important VIPs at the party and everybody from the Israeli Ambassador to your next door neighbor is asking you, "Are you nervous?," "Are you excited?," "Are you worried about your father?," and especially "Are you worried your father won't be coming back?"

The next day you go on top of a building to watch a rocket sitting on the horizon with your father on board. When it launches there's an incredible amount of light and noise—more than you've ever felt in your life. Then the vibration comes and it is intense—"Like being in an earthquake, but without the bad parts," was how one scientist described watching her first launch in person—and she was an adult who knew what to expect. So who wouldn't be nervous under those circumstances?

A delegation of Israeli VIPs showed up at the press site, with a security escort of seven police cars. The Israeli ambassador to the United States, Daniel Ayalon, said, "We are very privileged to join this very pres-

Israeli ambassador Daniel
Ayalon talks to the media.
Photo by Philip Chien.

tigious club of countries which have had astronauts in space." Israel
became the twenty-ninth country to have a citizen fly in space. Ayalon
said about the launch, "That was so moving, so touching. I was thinking
the skies were colored blue and white—our national colors. We had deep
beautiful blue skies and with this smoke (from the SRBs) coming it was
very, very moving."

NASA Administrator Sean O'Keefe talked to the press after the
launch. O'Keefe said, "It couldn't have been a more picture-perfect
launch—flawless in its character," and of course at that point that was all
anybody knew.

O'Keefe was asked about keeping the shuttle flying safely. He mentioned NASA's culture, citing the upcoming anniversary of the Challenger accident: "Not many people in this agency have forgotten about it.
All Americans, and I'd venture to say a lot of people around the globe,
remember where they were that day, and it really had a searing effect on
all of us. I think it's changed forever the approach this agency—those
who have committed themselves to exploring on behalf of the American
people—that you do that without being reckless and do it in a way that's
responsible and absolutely respects the fact that there are people
involved in this."

O'Keefe said he had just left the families of the Columbia crew. He
noted that he met with the families before each flight to remind everybody
the astronauts were real people—"These are people who really depend on
everyone who's engaged in this." O'Keefe added, "I don't see this becoming a case of withdrawing from that. I don't think it's ever going to get to a
stage where anybody becomes complacent, because you can really put a
name, face, and personality on human beings behind the consequences
of what we do every day, and we're all very mindful of that."

The next time O'Keefe walked in that auditorium was 17 days later,
in a much more somber mood, to officially inform the world Columbia
was destroyed and its crew had died.

CHAPTER 18

BEHIND
THE SCENES

When the engineers examined the videos from the long-range cameras trained on STS-107, they were shocked to see something fall off the ET and hit Columbia's left wing.

Two of the cameras saw the debris. One was a sharply focused camera at the Cape Canaveral Air Force Station with an oblique view. The other was the "soft focused" E208 camera in Cocoa Beach, which had a better angle.

Based on where it appeared to come from and its color, the engineers concluded that what had broken off was a piece of foam from the bipod, and not a piece of ice. Unlike the previous cases where a chunk of bipod foam had fallen off in this case the debris hit the shuttle, shattering into dust. The questions engineers asked over the next ten days were: Where did the foam hit the wing? With how much force? And most important, did the impact cause any damage — and if so, how much?

In increasing order of importance, the questions NASA managers would need to answer were: Was the incident of enough concern to halt shuttle flights until the reason it came off again was understood? Would it cause enough damage to require repairs to Columbia before its next

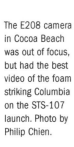

The E208 camera in Cocoa Beach was out of focus, but had the best video of the foam striking Columbia on the STS-107 launch. Photo by Philip Chien.

flight? Could it cause enough damage to affect Columbia's reentry? And, most important, could there be enough damage to result in the loss of Columbia and its crew?

The Debris Assessment Team had to determine four important factors—the speed of the debris, how much it weighed, where it hit the wing, and at what angle.

The speed was the easiest to determine. By studying the video frame-by-frame, the engineers could observe the piece come off the bipod and measure how far it traveled from frame-to-frame. Once the foam fell, it quickly slowed down from 1,568 m.p.h. to about 1,022 m.p.h. in just 0.161 seconds, just as a piece of paper tossed out of the window of a moving car quickly slows down when it leaves the car. In effect, the wing "ran into" the slowing foam. During the mission the engineers calculated 511 m.p.h. as their best estimate for the speed of the foam hitting the wing.

How much the debris weighed could be estimated from the known density of the foam and the size of the piece. If the camera was in sharp focus, its dimensions would have been easier to determine, but the analysis team had to make do with what they had. The length and width could be estimated from the video, but the depth would have to be extrapolated by studying the video and how the piece tumbled. The engineering teams came up with two possible estimates—20 inches by 20 inches by 2 inches, or 20 inches by 16 inches by 6 inches. Unlike the previous cases, in which a chunk of the foam came loose, on STS-107 almost the entire bipod foam had broken free.

The Columbia Accident Investigation Board (CAIB) later noted the estimates for the size and velocity that were determined within a couple of days were remarkably close to the best information derived after the accident, with several months of research and much more analysis.

The final factor—where the piece hit—was the most difficult to determine. The out-of-focus video could only isolate it to the left wing, fairly close to where the wing attaches to Columbia's fuselage. The video ran at 30 frames per second, so the foam traveled several feet in each video frame. The video showed the foam smashing into the wing and a shower of powdered foam coming from behind the bottom of the wing with nothing coming over the top. This led engineers to believe that the chunk of foam hit the bottom of the wing and not its leading edge (the front of the wing). That was the key incorrect assumption in their analysis. The estimate of the impact's location was only a couple of feet from the actual impact point, but it made an important difference. A head-on hit would result in all of the energy impacting the wing, but a glancing blow would result in very little force. Because of the shape of the leading edge, the precise location of the impact was extremely important: a one-foot difference would mean the difference between a head-on collision and a fender-bending sideswipe.

The engineers estimated that the foam hit the wing at an angle of 20 degrees. The theoretical models gave 13 possible locations: 12 that hit the black tiles on the bottom of the wing and one that hit the gray reinforced carbon-carbon (RCC) on the leading edge. The engineers examined the critical systems in those areas, primarily the left main landing gear. A heavily damaged tile could result in excess reentry heat getting in the wheel well and destroying the landing gear.

Based on the weight of the foam and the speed it was traveling when it hit Columbia's wing and the angle, the engineers were able to calculate the impact force, or kinetic energy. Energy is very dependent on the speed an object is traveling. Doubling the mass doubles the amount of energy, but if you double the velocity you quadruple the energy. This is why a pencil can be propelled into a tree trunk during a tornado—the pencil may be very light but it's traveling at an incredibly high velocity. With the larger estimate for the size of the debris, it worked out to a kinetic energy of 7,992 foot-pounds. In real-world terms that's the equivalent of a compact car hitting a brick wall head-on at 9.4 m.p.h.

The wing's leading edge is covered with 22 custom-shaped RCC panels. They're numbered from 1 (closest to the body) to 22 (tip of the wing). From the crew cabin windows, the astronauts can view only the outer portions of the wing. The engineers were certain that the strike took place fairly close to the fuselage, out of view of the windows. It would be pointless to ask an astronaut to try to take a photo of that area.

If the engineers were correct in their assumption that the foam hit the bottom of the wing, it would have been in an area that was extremely difficult, if not impossible, to access on a spacewalk. The bottom of the shuttle is an extremely smooth surface, with no handrails or way for an astronaut to hold himself in place.

The most important tool available to the analysis team was a fairly simple computer model named Crater. Originally developed during the

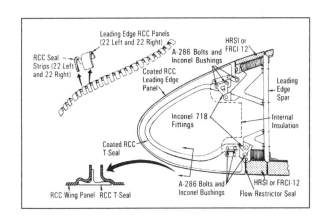

A cross-view of the front of the RCC, which covers the leading edge of the shuttle's wing.

Apollo program, it was designed to calculate the impact from small particles. The engineers extrapolated the results from Crater and came to the
conclusion that there would be some damage to the tiles, but not enough
to cause a safety of flight concern.

Many critics have complained that Crater was unsuitable for analyzing impacts from large pieces of foam, since it was only designed for small
particles. They're right in pointing out that Crater was used in an "uncerti-
fied" manner. But that's only part of the story. The laws of physics are the
same for large and small pieces. Large pieces have a bigger area, and that
can actually reduce the amount of impact to a particular point because
the impact is spread out over a larger area. And in any case, even if it was
not certified for such a large extrapolation, Crater was available and there
was no time to come up with something better. It made sense to use something that could provide at least some reasonable information.

Crater was misused unintentionally. The engineers believed the
foam hit the tiles on the underside, not the RCC on the leading edge.
Crater was intended to analyze damage to the tiles, not RCC. While
Crater could predict the amount of force with which the foam hit RCC,
it could not predict how much damage it would cause.

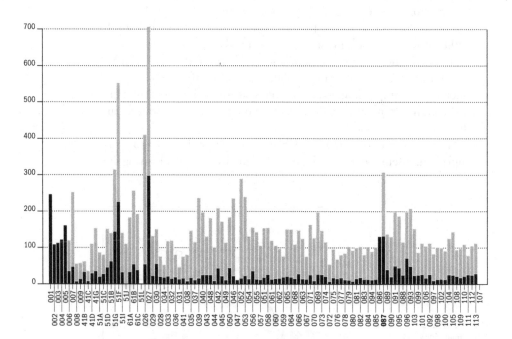

The number of damaged tiles on each shuttle mission. The total is indicated by the light gray bars.
Black bars show damaged areas greater than 1 inch. Because of the 51-L Challenger and STS-107
Columbia accidents, there's no data for those missions. The STS-87 flight had over 300 damaged tiles
(see photo on facing page) because of changes to the formula for how the foam is sprayed on the ET.

Managers, astronauts, and engineers look at the damage to many tiles on Columbia after the STS-87 mission. (The damaged tile areas appear as white blotches in this photograph.)

The shuttle's tiles are extremely delicate — about as strong as a piece of stale bread. You can easily push a pin through a tile; they are extremely fragile if they're improperly handled. In contrast, the RCC panels are made from stronger material, similar to fiberglass. Neither is designed to withstand impacts — the shuttle's specifications called for nothing to hit the tiles or the RCC. But in practice, small pieces of foam fell off the ET and hit the tiles on every shuttle mission, and NASA considered it a routine maintenance issue, not a flight safety concern. After the accident, manager Ron Dittemore noted that there were 11 shuttle flights with substantial damage to the shuttle tiles, and all had landed safely.

What was not really appreciated until after the accident was how the mechanical strength of the RCC varied widely. Since there were no requirements to test the RCC against impacts, the mechanical strength varied from panel to panel. The RCC was designed to protect the shuttle from heat — not impact. The specifications called for the RCC to withstand just 0.006 foot-pound of energy — the equivalent of dropping a penny eight inches. That was less than one millionth of the force of the foam that hit Columbia's wing. This book's website includes the presentations the Boeing engineers generated from their analysis for the Mission Evaluation Room (MER).

The story of how engineers reacted to the video of the foam falling off the tank and hitting Columbia's wing is detailed very thoroughly in Chapter 6.3 of the accident report which is available on this book's website. But there is information that was not addressed in the CAIB's analysis.

A handful of reporters were aware that engineers were examining how much Columbia could have been damaged by debris during the launch. They were informed that the foam impacting the wing was a non-issue and the engineers were not concerned about the shuttle's safety. Mission Control manager Phil Engelhauf noted in the January 24 Mission Management Team (MMT) meeting reporters in Florida had asked about the lost foam. To ensure that the crew wouldn't be surprised by a question during an inflight press interview, Rick Husband and Willie McCool were sent an e-mail informing them about the foam issue. Other than that, there was no need for the crew to be concerned, since there was nothing they could have done to help with the analysis. If the reporters hadn't asked about the situation, there would have been no reason to inform the crew at all about a situation they couldn't control or assist in—that's the standard rule that has always applied. For the bipod foam loss on STS-112 in 2002, a problem with the SRBs on STS-70 in 1995, and similar cases, there were no press inquires, and the crews were not informed until the post-mission debriefings.

On the other hand, on STS-95 in 1998, the parachute door fell off during the launch, hitting one of the main engines. It didn't damage the engines, and Discovery made it to orbit safely, but there were concerns that the parachute might be damaged by heat during the reentry. Managers had to make the decision whether or not to use the parachute during Discovery's landing. The astronauts were informed of the problem during the mission because the situation did affect how the pilots would fly Discovery during its landing.

The e-mail from flight director Steve Stich warning Columbia's crew that they might get a question about the foam, along with Husband's replies, were released after the accident. They're on this book's website.

After the accident many questioned whether or not the crew should have been involved in the analysis, including Senator Bill Nelson, who had flown on Columbia on a political junket. That answer is and should be a resounding absolute no. The crew has one responsibility from the time they enter the shuttle before launch until they exit it after landing—their mission. Anything else is a distraction. If they can provide on-the-spot information of use to the engineers, then they should be contacted. But other than that, they should do their job without the distractions of information they don't need.

There are hundreds of engineers, including scores of experienced astronauts, who work problems during a mission. Astronaut Mike Lopez-Alegria noted, "We've got seven sets of eyes up there, they've got 700 sets of eyes studying the problem on the ground."

Also brought up after the accident was whether NASA administrator Sean O'Keefe should have been kept informed. O'Keefe was what's commonly known as a "bean counter"—a financial person brought in to get NASA's infamous budget overruns under control. O'Keefe does not have any engineering knowledge and has always acknowledged that. Certainly, if enough people felt something was wrong, the NASA administrator could help to "rally the troops" and encourage them, but that situation didn't occur. O'Keefe did have highly qualified engineering personnel working for him—Associate Administrator for Spaceflight Bill Readdy and Associate Administrator for Safety and Quality Assurance

Bryan O'Connor, both former test pilots and astronauts. They were involved in the preparations for STS-107 and informed about what was happening during the mission.

Spacehab's Pete Paceley was in Mission Control, but not involved in any of the discussions about the foam. He recalled, "I sat in that control center every day for hours and hours and hours. Aside from seeing the little video clip [of the foam] a few days into the flight, and aside from the MMT discussions, I had no idea [they were discussing the foam issue.]" He added, "Looking back on it, I remember seeing the Mission Evaluation Room jammed full of people on occasion, and now I can say—oh, they were probably talking about the debris strike."

SPY SATELLITES AND COLUMBIA

Many engineers analyzing the foam strike wondered if a spy satellite image could give them additional information for how much the wing was damaged.

The CIA headquarters in Langley, Virginia, has a small museum with various historical artifacts. One display case has a shuttle tile with a small model of the space shuttle—but no caption or explanation for why it's on display. Rumors claimed the KH 11-3 Keyhole satellite was used to take photos of Columbia on STS-2 in November 1981 to see if any tiles were missing, but the images were inconclusive. When the parachute door fell off on STS-95, NASA asked the military for assistance, but the images were of little use because they didn't have enough resolution to show whether or not there was any concern for damage in that area.

The problem is distance: The greater the distance, the less detail you're going to see, and any satellite camera that could take a photo of the shuttle would certainly be at a great distance—hundreds of miles above. Another problem is the rapid motion of both spacecraft—traveling at 17,500 m.p.h. in different directions. Spy satellites are designed to track targets on Earth, not a rapidly moving target in space, so there are a very limited number of times when spy satellites are in the proper place to take an image of the shuttle.

Spy satellite images are highly classified, but there are procedures to give permission to people who need temporary access. Each federal government organization has personnel assigned as "Scientific and Technical Intelligence Liaison Officers" (STILO—pronounced "stylo") who act as intermediaries with the military for obtaining temporary clearances.

The CAIB report analyzes the engineers' requests for spy satellite images and why mid-level managers turned down those requests. Amazingly, at one point an informal request for spy satellite images was placed by engineers studying the situation but countermanded by mid-level NASA managers. The CAIB analyzed many of the reasons for why the requests were turned down and criticized the lack of open communications between the engineers and the MMT. It's important to remember that the managers were aware of the fact that previous attempts to take images of the shuttle did not provide very useful information because of their low resolution, and that could have affected their thoughts as to whether or not the requests would do any good. From a "big picture" view, the military was gearing up for the war with Iraq. so all of their intelligence assets, including spy satellites, were presumably busy concentrating on gathering whatever strategic data would be useful. If a request did come from NASA, it would have to be an extremely high priority request and would disrupt the scheduling of one of the spy satellites.

On February 3, two days after the accident, manager Bill Readdy wrote what he called a "memorandum for the record." It started, "Wednesday, January 29, 2003, in the early afternoon, Mr. Michael Card, NASA Headquarters Safety and Mission Assurance, visited me in my office. Mr. Card and an individual from another agency had been discussing the external tank (ET) debris issue during STS107 ascent. He wanted to discuss an "offer of support" from the other agency with respect to observing Space Shuttle Columbia on orbit." Clearly Readdy wrote this post-accident memo realizing it would be publicly released, and that his actions during that meeting five days earlier, where he turned down the offer for military assistance, would be questioned. As a career military officer he was circumspect with his phrasing—"an individual from another agency"—without specify that the agency was the National Imaging and Mapping Agency, the military organization that controls the day-to-day tasking for spy satellites. In the memo, Readdy cites the analysis by the engineers which concluded the debris was not a safety of flight issue, therefore spy satellite photos would not be needed. A copy of Readdy's entire memo is on this book's website.

If the requests for spy satellite photos were approved, it would have been fairly easy to calculate when Columbia passed underneath a spy satellite's orbit. The flight plan would have been adjusted to rotate Columbia to face its belly toward the satellite in its far higher orbit. The mission planners may have had to change the science activities to accommodate the unplanned maneuver, but it wouldn't have had a significant impact. At most, an hour of science would have had to be rescheduled, and another piece of science that wouldn't be affected by the maneuver would be put in its place if possible. Most of the staff at Mission Control, and certainly the public, wouldn't have been aware of the additional maneuver, and only a few managers and engineers would need to know the actual reason for the unexpected change. Even with all of the attention on STS-95 with John Glenn on board, NASA was able to coordinate with the military and obtain in-orbit photos of Discovery without it getting revealed to the public.

But the question remains: Would spy satellite images have had enough resolution to determine the extent of the damage? After the accident, Congress asked O'Keefe in a public session about the possibility for spy satellite photos. O'Keefe noted, "Tom Clancy [novels] would have us believe the quality is extraordinary. They may not be as close to that reality as the novelists would have us believe." He wasn't kidding.

While most of the details about spy satellites are classified, there is enough publicly available information to determine whether or not a spy satellite image could have done any good. Spy satellites obey the laws of physics, just like every other satellite. The exact resolution of their images is classified, but a maximum value can be easily calculated. Spy satellites are launched on Titan IV rockets, and that puts a maximum on the size of the spacecraft and its main mirror. The mirror is believed to be 94 inches (2.4 meters) across. Snell's Law, a standard physics formula, places a limit on the maximum theoretical resolution—assuming perfect conditions. There's no way the resolution can be any higher, but just how close the actual resolution approaches the theoretical limit is what's classified.

CAIB chief Hal Gehman said, "It's more complicated than just do you have a camera. It's how far away is the camera and what's the light and what's the angle of incidence? But is it possible a hole [of about 16 inches]—it is within the realm of capability, to take a picture of something that size? Now, once again, you've got to remember: You're looking at a black hole on a black surface. So whether or not it would've been vis-

The launch of a Titan IV rocket on December 20, 1996, from Vandenberg AFB in California. The payload is the USA 129 photo reconnaissance satellite, which could have taken a photo of Columbia in orbit. Photo credit—USAF.

The orbits of USA 129 (moving from North to South) and Columbia (moving from West to East) crossed as the two spacecraft passed over Central Asia. Satellite tracking map by Philip Chien.

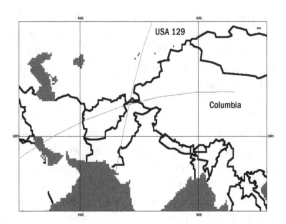

ible or not, and what the angle and the shadows would've shown, very, very—very, very hard to predict."

The US government had two operational imaging spy satellites during STS-107, officially known as USA 129 and USA 161 in unclassified documents. USA 129 was launched on December 20, 1996. USA 161 was launched on October 5, 2001. Amateur satellite observers enthusiastically track these satellites. Satellite tracking analyst Ted Molczan studied their orbits. He found suitable conjunctions, where a spy satellite's orbit intersected Columbia's orbit, on January 18, 20, 25, 29, and 30. The spy satellites could have been instructed to look toward Columbia as it passed underneath and snap a photo. (Columbia would have rotated itself to face its lower left wing towards the satellite at the best Sun angle.)

Molczan determined that the best encounter was on January 20 at 12:18 A.M., with USA 129 passing within 118 miles of Columbia as the two spacecraft traveled over the region where Kazakhstan, China, and Pakistan meet. That "close" approach took place 3 days 13 hours 39 minutes after launch, while the blue team was on duty. McCool has just reoriented Columbia to point its payload bay toward the Sun for the SOLCON (Solar constant) experiment; Anderson was accessing the Laminar Soot Process experiment; and Brown was starting a daily exercise period. The other four crewmembers were asleep.

If the decision was made to have USA 129 image Columbia, McCool would have been given an update in the flight plan to maneuver Columbia for the spy satellite image instead of aiming Columbia towards the Sun for SOLCON. Mission Control wouldn't even need to tell McCool the reason for the maneuver, it would just be phrased as a set of coordinates to input into Columbia's computer. The SOLCON data take would be rescheduled for later in the mission.

The spy satellite resolution for that pass could be as good as 2.5 inches, enough to see a fair amount of detail. That's the smallest feature that could theoretically be detected under the best circumstances. That's hard physics—there's no way anything smaller could have been seen, no matter what. Additional real-world limitations which would limit the actual resolution include the difference in contrast with the material beneath any suspected damaged areas, how rapidly a spy satellite can aim itself and maneuver to follow a moving target, Sun angles, and any "signal noise" that can affect quality. It's also important to note that this particular pass was fairly early in the mission, when the image analysis team was analyzing the launch videos and just starting to discuss whether or not they should even put in a request for spy satellite imagery.

GROUND-BASED TELESCOPES

Besides spy satellites, there were several opportunities each day where Columbia passed over the Starfire Optical Range at Kirtland AFB in New Mexico and AMOS (pronounced "ae mos"—Air Force Maui Optical and Supercomputing Site) on top of a 10,000-foot dormant volcano in Hawaii. These locations have extremely sophisticated telescopes which use lasers to create an artificial "guide star" to compensate for the blurring effects of the Earth's atmosphere. The observatories have far larger telescopes than spy satellites, more opportunities where they can take images of the shuttle, and can be much closer to the shuttle than a spy satellite flying at a far higher altitude above the shuttle's orbit. On the other hand, the spy satellites don't have to take their images through the atmosphere with its blurring effects. Overall, spy satellites can have slightly better resolution than ground-based observatories for taking photos of the shuttle in space.

The AMOS telescope did take photos of Columbia on January 28, but not because of any request from NASA. Like many amateur and professional astronomers, the AMOS team takes satellite images as a challenge to test their system's capabilities.

Two of AMOS's telescopes were used, an infrared camera attached to a 12-foot Advanced Electoptic System telescope, and a visible light camera attached to a 5.2-foot telescope. The smaller telescope can take images of the shuttle as it passes overhead in daylight and in twilight— when the ground is in darkness while the shuttle is still illuminated by the Sun. The larger telescope can only take images when the shuttle passes overhead while the telescope is in twilight, but has a laser guide star adaptive optics system. The telescopes rotated under computer control to track Columbia as it flew over Hawaii, and the digital cameras took images 250 times every second.

With such fast exposures, there isn't much to see in the raw images. AMOS deputy commander Major Kelly Hammett said, "The post-processing is the key, it's a blob without doing that." Computers using sophisticated image-enhancement techniques add together 16 images to create a higher-resolution image of about 11 to 20 inches. The team gave copies of the images to NASA after the accident.

In the AMOS visible images, Columbia is facing the Earth, and its payload bay doors block the portion of the left wing where the damage occurred. The AMOS images were taken four days before Columbia's reentry, coincidentally when Columbia was in its "thermal soak" attitude to prepare for reentry. The images clearly show Columbia's payload bay facing the Earth, with the Sun coming from the starboard side. The belly is facing away from Earth, which exposes the main landing gear doors to deep space. That cools the shuttle's tires to acceptable temperatures for reentry. The radiator on the port payload bay door is in its deployed position, getting rid of the excess heat generated by everything that uses power.

Since the accident, the AMOS team has continued to work with their raw data to see how they can improve the image resolution. Instead of 16 frames, they're using 100 frames to improve the brightness and contrast. This book's website includes the AMOS images that were released soon after the accident, and increased resolution images that were processed later.

The AMOS images bring up an obvious question: What would have happened if NASA had put in a request for military assets to take an image of Columbia? NASA's flight dynamics team in Mission Control and the AMOS team would determine the best time, taking into account lighting conditions and when it would be least disruptive to the science activities. With several daylight passes each day, there would have been plenty to choose from, far more than the limited spy satellite opportunities. The astronauts would be instructed to rotate Columbia to face the damaged area toward Hawaii. Within a couple of minutes, AMOS could generate images like the ones publicly released just after the accident. It

The 5.2-foot visible light telescope which took images of Columbia on January 28, just four days before the accident, and an enhanced-contrast image which was generated several months later. Photo credit—USAF.

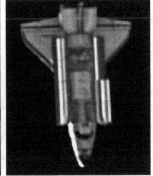

would take a couple days for the AMOS team to create images with about 6- to 8-inch resolution. Hammett said, "The general consensus is, with our viewing capability we would have had the resolution in the ball-park to maybe notice something in the imagery. It's unclear if we would have been able to clearly determine if anything was wrong."

It's important to remember that the engineers studying the damage from the debris, based on all of the information they had, believed the foam hit the bottom of the wing, near the main landing gear doors. That would have been the target for a spy satellite or ground-based telescope. The engineers had no doubt some damage had occurred—the question was the extent of the damage and whether or not it would put Columbia at risk. If a military camera was instructed to take an image, it would have included most of Columbia's bottom and possibly the leading wing, where the damage actually occurred.

Whether or not any damage would show up on a military image is questionable, depending on the exact viewing angle and the material underneath. If there was a dark hole on a dark surface it may not look like anything's wrong. But if the foam ripped off the surface layer of a black tile, its white interior would show. Of course the amount of damage is the most important factor for whether or not the damage is noticeable from a distance. The larger the damage and the greater the contrast, the more clearly visible it would be.

The RCC panels are a light gray shade. The area underneath is olive green, but could be in shadow, depending on the size of the opening and Sun angle. A spy satellite or telescope photo would be black and white, and how much contrast would be seen between the RCC and whatever was exposed underneath is questionable.

If NASA did ask the military to take photos of Columbia, it certainly could have been accomplished. But the quality of those images would be marginal—at best—for determining how much damage had occurred.

ASSESSING THE SITUATION

It is wrong to assume that if NASA had asked for a spy satellite or mountain-top observatory photo it would have provided certain evidence. Even the assumption that spy satellite photos would have provided engineers with more evidence is questionable.

But even if the spy satellite photos would not have done any good in the specific case of STS-107, it doesn't mean that they couldn't have helped in another situation, with damage over a larger surface area and better viewing conditions. Imagine a case where there is a long shallow gouge in the tiles. With the white undersurface exposed, a spy satellite image could have shown the amount of damage. The engineers could have recommended the limited measures that NASA could take to minimize the heat load to that portion of the wing—cold soaking the wing as much as possible, and reducing the altitude and weight as much as possible before reentry. They could have even made the recommendation for a different reentry path that would minimize the heating but would result in the need to have the astronauts abandon the shuttle after it reached the safe bailout altitude.

The question becomes: Was the MMT correct in its decision to turn down the engineers requests to ask the military to take photos of Columbia, even if they believed it wouldn't provide any useful information? And the answer is a resounding "No!" It is important to provide as much information as possible to engineers who are trying to solve a problem, even if you don't think it will be useful. After the accident, NASA signed an agreement with the military where spy satellites will be used to take photos whenever possible in the future, just in case they're needed.

A CLOSEUP EXAMINATION OF THE DAMAGE

If spy satellites and ground-based telescopes wouldn't do any good, what about an up-close personal inspection by the astronauts from just a couple of inches away? Mike and Dave received the minimal 40 hours of spacewalk training that a pair of astronauts are given for every flight, just in case an emergency spacewalk is needed. Normally spacewalkers are confined within the cargo bay or attached to another spacecraft, like the Hubble Space Telescope or ISS. There are plenty of handrails and locations for safety tethers.

NASA had only one set of emergency procedures for one spacewalk outside of the cargo bay—a procedure it hopes it will never have to do. There's a set of doors on the bottom of the shuttle, which have to close after the ET is ejected. The umbilical doors seal the openings for the plumbing that takes the propellants from the tank into the shuttle, where they're routed to the main engines. If the umbilical doors don't close, there's an emergency spacewalk procedure to close the doors manually. A couple of months before the launch Mike said, "That's one we hope to never use, but it's one we always train for." The spacewalkers would take out a laundry bag filled with dirty clothing and tied to a tether. They would keep tossing the bag towards the back of the shuttle's wing, aiming for the gap between the inboard and outboard elevons (the flaps on the wings which control the shuttle while it's gliding in for landing). Hopefully they would eventually get lucky and the bag would lodge itself in the opening. "That's the hard part," Anderson acknowledged. After the tether's tightened between the cargo bay and the back of the wing, the spacewalkers would use it as a handrail to get access to the bottom of the shuttle. It's considered extremely dangerous and only on the books because an entry, with the open umbilical doors would certainly result in the shuttle's destruction.

It would have been possible for Anderson to hang over the side of Columbia's cargo bay and hold Brown by his ankles. Both would remain tethered so there would be no chance that an astronaut could float off by accident. It's never been tried, has not had safety engineers examining it from dozens of different perspectives to try to discover any potential problems, and would not be something to be undertaken lightly. But it's feasible, and could have been done during STS-107.

Dave would be fairly close to the area where engineers believed the foam hit the shuttle's left wing. The question is whether or not there would be enough damage for him to see with his naked eye from within a spacesuit helmet. If the damage was on the bottom, where the engineers believed the foam hit, it would have been in a place the astronauts couldn't reach. The task would roughly be the equivalent of leaning over the side of a building to see if a window on the floor beneath the top floor has been damaged while your buddy is holding on to your feet to ensure that you don't fall off the building. If the window has been completely broken—you'd probably see it. But if it's only a crack—you might not be able to lean over far enough to see any noticeable damage. On the other hand, if there was a large hole close by, it would have been easily visible to the spacewalker.

While a spacewalk to inspect for potential damage would have been possible, it would be tricky. It would also disrupt the science for about two days. The crew would have to temporarily shut down Spacehab, and it would take about a day to prepare for the spacewalk, plus the day the spacewalk occurred. Some automated science could continue, but most crew-related activities would have to be put on hold.

STS-107 didn't even have a spacewalk-compatible video camera. If NASA did go ahead with an inspection spacewalk, the findings would be limited to whatever verbal description the astronaut could provide. A large gaping hole would be obvious, but a smaller amount of damage may not have been something a spacewalker could recognize or describe with sufficient detail. And, of course, even if the spacewalkers did see a

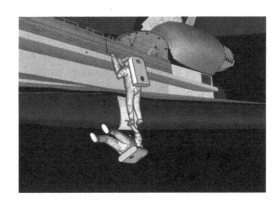

A computer-generated graphic shows a hypothetical inspection spacewalk. Photo credit—CAIB.

damaged area, what could be done to repair it? An inspection spacewalk is one thing, but to attempt a repair in such an awkward location with no handholds or way of anchoring yourself would be far more difficult.

Even with all of these concerns, engineers did suggest the possibility of a spacewalk. But not surprisingly, there were many objections because of the risks and limited amount of information that could be obtained. In addition, managers were resistant to stopping the science for two days to perform a spacewalk which may or may not provide more information. CAIB chair Hal Gehman criticized this decision, saying, "They did not have the prove it safe attitude, they had an attitude where you had to prove it was *unsafe* before they would take any action."

Two months after the accident, after an intense analysis, it was determined that the foam hit the bottom of RCC #8, next to the front of the wing where it would have been visible to a spacewalker. If NASA had decided to go ahead with an inspection spacewalk, they would have had access to the damaged area. Whether or not an astronaut could have seen something from within a spacesuit helmet is another question. If the damage was just a dislodged T-seal, it might not be recognizable to someone who wasn't trained to realize it wasn't normal. The astronauts do inspect the shuttle fairly closely, but they aren't specialists in what's acceptable for flight and what isn't. But if the damage was a large hole, then it would have been obvious.

Without any data from spy satellites or other sources of information, the team analyzing the foam hit did what they could with the resources they had—the video and film views of the launch and the mathematical predictions from the Crater program.

The team was divided as to whether or not it was safe. Gehman said, "There were a lot of engineers that felt the foam event on Columbia was serious and they could not tell whether there was damage or not. But there was insufficient evidence to tell you it was safe—there was insufficient evidence to tell you that it was unsafe. The engineers who felt it was unsafe all felt that it was unsafe for a number of reasons, none which turned out to be true—damage to the wheel well, wheel well door seals, and things like that. No one predicted that it was a hole in the wing."

With the limited information they had, the debris assessment team concluded that the foam had certainly damaged the outer layer of the black tiles, and if it hit the RCC, the damage would be limited to the coating. But while there could be structural damage requiring repairs before the next time Columbia flew, it would be able to land intact with that amount of damage. Their conclusions were announced on January 28, in the flight day 12 Mission Evaluation Room report. Note that the report has a typo, substituting "radial" for "reinforced".

STS-107

TWELFTH DAILY REPORT

028:12:00 G.m.t.

The STS-107 mission is progressing nominally and all Orbiter subsystems are performing satisfactorily. No Orbiter issues have been reported in the previous 24 hours. The Orbiter consumables remaining are above the levels required for completion of the planned mission.

Regarding the debris hit on the left wing last discussed in the Fourth Daily Report; systems integration personnel performed a debris trajectory analysis to estimate the debris impact conditions and locations. This analysis was performed utilizing the reported observations from the ascent video and film. It was assumed that the debris was foam from the external tank. Based on the results of the trajectory analysis, an impact analysis was performed to assess the potential damage to the tile and radial carbon carbon (RCC). The impact analysis indicates the potential for a large damage area to the tile. Damage to the RCC should be limited to coating only and have no mission impact. Additionally, thermal analyses were performed for different locations and damage conditions. The damage conditions included one tile missing down to the densified layer of the tile and multiple tiles missing over an area of about 7 in by 30 in. These thermal analyses indicate possible localized structural damage but no burn-through, and no safety of flight issue.

[signature] 028:12:51 Gmt.

Don L. McCormack, Jr.
STS-107 Lead MER Manager

That's the best information anybody had before Columbia reentered, and the engineers involved in the analysis accepted its conclusions.

However ... During the January 21, 2003, MMT meeting, five days after launch, MMT chair Linda Ham said, "And really, I don't think there is much we can do, so you know it's not really a factor during the flight 'cause there isn't much we can do about it." It was a simple matter-of-fact statement—not a callous comment disregarding the risks. An audio clip of Ham's comment is on this book's website. The logic was, there was no way to make repairs to Columbia's exterior in space, any more than one could climb on an airplane wing during flight to patch up bullet holes from a combat situation.

Many people have claimed that a statement like Ham's is an instruction from management to shut down an activity and not pursue it any further. But the logic of that hypothesis falls flat on its face. Why would managers want the engineers not to continue to examine the issue—even if they thought Columbia was safe and the engineers were wasting time? Ham brought up the issue in the MMT meeting; if she wanted to keep it quiet, why would she be the one to bring it up? More important, if the hypothesis that Ham wanted to shut down the activities by the engineers was true, it would indicate that Ham—with the limited amount of information she had about the situation—believed that Columbia was fatally doomed, so the engineers shouldn't pursue it anymore and just ignore the situation, or that the engineers were just wasting their time analyzing it any further. These assumptions are absurd on the face of it and there's no reason to believe that there's any truth to them. Ham unintentionally chose words that could be interpreted by the engineers as an instruction not to pursue the matter any further.

The transcripts and audio recordings of the MMT meetings were not released until July 22, almost six months after the accident. Linda Ham explained her comments to NASA-selected members of the news media: "When I made a statement about what we could or couldn't do during the flight, when I first was alerted to that, I couldn't even recall making that statement. But, of course, I did go back, re-read the transcript, and listen to the tapes, and sure enough, I did say that.

"Now, if you put that in context to what the MER [Mission Evaluation Room] manager was talking to me about, and the things that I was thinking, the way I recall this is I was thinking out loud. I do know that we do not have TPS [thermal protection system] repair, tile repair, or RCC repair capability.

"The other thing that I was thinking about was having the engineering community go back and get a flight rationale from STS-112. If you recall, two flights prior to 107 [on STS-112], we had the foam come off, a pretty big chunk of foam from the same area that we were thinking came off the 107, the bipod ramp, and it struck the SRB."

Ham is extremely open acknowledging that there were problems with how the MMT operated during the STS-107 mission, in particular communications. She told me, "There's all kinds of things that should have been done differently. There were obviously some communications failures in a lot of different links in the communications chain. There were people in the environment that were concerned and the program management never knowingly heard any of that concern. I never got any of that knowledge of somebody having that big of a concern. We should have been able to communicate more clearly."

Wayne Hale was scheduled to become the launch integration manager in February 2003. Had STS-107 been delayed an additional couple of weeks, everything would have happened when he was in charge. He told me, "If I had been sitting in the MMT team chair I probably would have made the same decisions. I have no defense to that. I find it a little disheartening that Linda Ham was tarred-and-feathered over this, when if I had been sitting in the [MMT] chair I probably would have made the very same decisions for very similar reasons. Linda was not the ultimate culprit in all this."

Ham said, "Based on what I knew, I made the best decisions I could at the time. We were all trying to do the right thing. All along, we were basing our decisions on the best information that we had at the time.

Nobody wanted to hurt the crew. These people are our friends. They are our neighbors. We run with them, work out in the gym with them. I think we all take some personal responsibility for this, and I certainly feel accountable for the MMT. So it has been very difficult through this."

Mission Control manager Phil Engelhauf added, "In the end, we lost the crew and we lost the vehicle, and we can't escape that. And nobody feels worse about that than every one of us who has our hands on these missions every day, but it is not because of lack of good intent or lack of effort on anybody's part. If the system fell down, we will fix the system."

Many within and outside the space program were upset that while many managers acknowledged their mistakes there was no one person to blame, no individual who stood up and said, "It's my fault." This was mentioned to Wayne Hale, and he replied in a letter to the shuttle workers, "I cannot speak for others but let me set my record straight: I am at fault. If you need a scapegoat, start with me. I had the opportunity and the information and I failed to make use of it. I don't know what an inquest or a court of law would say, but I stand condemned in the court of my own conscience to be guilty of not preventing the Columbia disaster. We could discuss the particulars: inattention, incompetence, distraction, lack of conviction, lack of understanding, a lack of backbone, laziness. The bottom line is that I failed to understand what I was being told; I failed to stand up and be counted. Therefore look no further; I am guilty of allowing Columbia to crash."

There are three inescapable conclusions to the MMT's actions:

1. Had the MMT been more proactive after launch and done everything possible to address the engineers' concerns about the foam strike on Columbia's wing, the results would have been the same—the destruction of Columbia. There's no realistic repair or rescue technique that would have worked given the exact circumstances during STS-107. All of the rescue and repair scenarios that have been proposed by space program observers, as well as NASA's best engineers, assume far more knowledge, far earlier in the flight, than what would have been plausible on STS-107. Chapter 48 addresses the difficulties for a rescue mission and just how many things would have had to have happened to make that possible. There just wasn't enough time to determine the extent of the damage and do something about it before Columbia ran out of supplies. At most, if there was more effort, NASA would have known that Columbia was almost certainly doomed and the astronauts would have been given the opportunity to talk to their families one last time. NASA certainly would have done everything possible to minimize the reentry heating, but it wouldn't have been enough.

2. The last point where the Columbia accident could have been pre-vented if mistakes were not made was three months before launch, at the STS-113 FRR, when the bipod foam falling off the STS-112 ET was addressed. At that point the shuttle fleet should have been grounded, and an analysis made to determine why the foam fell off and how to prevent it from happening in the future. Or there would have been a redesign of the bipod to eliminate the need to have foam there if the reason for the falling foam was unexplainable. That was when the mistake was really made.

3. Just because there was nothing that could have been done in the specific case of STS-107 does not exonerate the managers. The MMT was not proactive and did not do everything it could have done to provide hard engineering data to the debris analysis team when they needed it. The MMT did not encourage the lower level engineers to speak up, and those engineers who were concerned didn't feel comfortable going to the managers and pushing for additional information that could have helped their analysis. If there was less damage than what actually occurred on STS-107, there could have been a chance to save the crew—but again, only if things were done differently.

This book's website has all of the Mission Evaluation Room reports and transcripts of the MMT meetings and the press conference that NASA had on July 23, 2003, after the recordings and transcripts were released.

It's important to always remember that when Columbia was ready to reenter the Earth's atmosphere on February 1, NASA believed the reentry would be a safe one. The first indications of trouble didn't happen until sensors started to fail.

At the pre-landing briefing on January 31, entry flight director LeRoy Cain was asked about the foam issue. He said, "The engineers and ana-lysts took a very thorough look at the situation with the tile on the left wing and we have no concerns whatsoever, and therefore we haven't changed anything with respect to our trajectory design. And there's noth-ing we need to do in that regard. So, nothing different. It will be nominal, standard trajectory. I believe that, at this time, we can't say with great detail the degree of the damage, other than all the analysis suggests it would be very minor, in terms of the amount of tile that might actually be missing or had been removed. All of the analysis says that we have plenty of margin in those areas in that regard, and that the impact could not have been from this particular material significant enough to take out any significant amount of tile. So I can't tell you inches by inches or depth, but I can tell you we think it's going to be very small." An audio clip of Cain's remarks is on this book's website.

CHAPTER 19

L I V I N G
I N   S P A C E

For 16 days, Columbia's seven astronauts lived and worked in space. The shuttle was their home and office. Living on the shuttle is like a camping trip. You bring all of your supplies but don't have all of the luxuries and things you take for granted at home.

HYGIENE IN SPACE

Let's get the most common question out of the way, which astronauts get asked constantly: How do you go to the bathroom in space? The shuttle has an extremely high-tech toilet—the "Waste Collection System" (WCS). It looks like an airliner toilet, however there's a set of handlebars to keep the occupant in place. The WCS is a pretty sophisticated device—imagine the engineering challenge involved in designing a toilet that's mounted on the wall instead of the floor, and operates without the help of gravity. A strong flow of air pulls the waste away from the body. Solid waste goes into the compactor tank, where it's compressed into a canister. When a canister is full, it's replaced. After the shuttle lands, some unlucky person working for a contractor gets to remove the canisters and clean the toilet. Each crewmember has an individual funnel for urine (imagine urinating into a vacuum cleaner), and it's dumped overboard. (Some of the urine on STS-107 was collected for science experiments.) Everybody's responsible for cleaning up after themselves, and the pilot's responsible for keeping the toilet in working order. If the toilet malfunctions, there are plastic bags. Husband showed off the toilet area during a guided video tour during the mission, explaining, "Behind this door we have the WCS, otherwise known as the bathroom. This is where we go for our daily hygiene, and you can see there the WCS [seat] with towels hanging from the door and the different cabinets for tissues and different things like that."

There's a small hand cleaner, but no way to shower. Large wet wipes are available for sponge baths, and a rinseless shampoo keeps the hair clean. Brushing your teeth is fairly simple—just spit your used toothpaste in a dirty towel. Shaving is with ordinary shaving cream and razors. Each astronaut has their own personal toiletry kit with whatever items they

Ilan Ramon brushes his teeth.

want. During the mission Rick Husband showed off his kit including Crest toothpaste, a comb, shampoo, Keri, razors, Speed Stick deodorant, chapstick, and shaving cream.

The crew also broadcast video of Dave brushing his teeth and Willie shaving. In a video recovered after the accident, McCool videotaped Brown shaving, and after Dave moved away from the mirror Willie took the opportunity to wave at the mirror and say hi to his kids. Another clip showed Rick combing his hair.

One of the recovered videos shows Ramon vigorously brushing his teeth. He put so much muscle into moving his toothbrush back and forth he was actually vibrating himself. Laurel kidded, "The whole orbiter is starting to shake," and Mike added, "Yeah, you're disrupting microgravity." This got a big grin from Ilan. Anderson kidded, "We're going to have a 'no brush' period before [certain experiments]," suggesting Ilan's motions were so strong they could affect the mission's most sensitive microgravity experiments! Go to this book's website for a video clip.

SLEEPING ON THE SHUTTLE

The flight deck has most of the windows, and astronauts try to spend as much of their spare time as possible looking at Earth. Quite often the crew will have their meals on the flight deck while looking out the windows. Going to sleep on time is always a challenge, but the astronauts realize they must stay on a regular schedule to keep on the mission's timeline. The astronauts tried to spend as much time as possible admiring the Earth before bedtime. At one point capcom Charlie Hobaugh told Willie, "You can let Laurel and Rick know their curfew is almost up." McCool replied, "They know, they're enjoying the view of Hawaii, we just passed over the Hawaiian Islands." Certainly after the reminder from Mission Control Laurel and Rick would have headed to sleep – even if it was reluctantly.

The four sleeping bunks with the red shift. From top to bottom Laurel Clark, Rick Husband, and Kalpana Chawla. Ilan Ramon has already gotten out of his bunk.

Columbia's middeck had four bunks stacked on top of each other, each one the size of an extremely small closet. Certainly you can't be claustrophobic and stay inside the bunk! Before launch McCool said, "It seems fairly roomy to me. I'm a Navy guy so I'm used to being bunked, it's not something particularly new to me." The astronauts share their bunks with somebody on the opposite shift. What's amusing is they shot video of Laurel in one bunk and Willie in another – even though they were on opposite shifts and wouldn't be sleeping at the same time! They even commented about it as they shot the video as a demonstration of the sleeping bunks and how the two-shift mission operated.

During the mission Husband said, "We've been sleeping pretty well, we've got sleep stations in the middeck. They've got sound-suppression padding and it's extremely dark. Some of us wear earplugs as well, so it's pretty quiet and dark in those sleep stations. We've also made an effort to be quiet for the crew that's sleeping while we're in the middeck." Husband mentioned they wear headsets and turn the speaker off to minimize the noise on the middeck while the opposite shift is sleeping, "so it maintains a fairly quiet environment for folks." At one point, engineers asked Willie to try to fix the Biopack experiment by cleaning a clogged filter. Biopack was located fairly close to the bunks so they asked McCool to wrap the vacuum cleaner in towels to muffle its noise.

The two halves of the crew alternated with 12-hour-on, 12-hour-off shifts throughout the mission. Eight hours is scheduled for sleep, and the couple of hours before and after sleep are called "pre-sleep" and "post-sleep." (NASA has a name for everything.) These periods include time for breakfast and dinner, hygiene, reading and sending personal e-mail to family and friends, other personal activities, and reading the flight plan

A typical trash bag located next to the Combustion Module.

Willie McCool uses all of his strength to squeeze some trash into as small a ball as possible.

updates which Mission Control sends each day. The entire crew gets together before and after each shift for a handover meeting, where they discuss the previous day's activities.

On a typical shift, Rick or Willie would be on the flight deck operating the experiments that needed them to control Columbia's attitude. The rest of the crew would normally be inside Spacehab performing their experiments. A few of the experiments were located in the middeck.

TAKING OUT THE TRASH

Clothes and food are packed incredibly tight to minimize their volume. Dirty clothes and used food containers take up more space. Used clothes go inside large laundry bags that get stuffed in a corner. Wet towels and food containers are stored in an airtight sealed container to prevent odors from escaping.

Favorite clothing the crew wants to keep after the mission and other important items are put in a separate "Return to Houston" bag. At one point, Rick asked Laurel if she had seen the specially marked bag. She replied, "The only place I'd ever seen it was in the tunnel, and that was ages ago because I haven't gotten any crew shirts dirty. I'll go look for it." He told her "No, I'll just ask K.C."

There are small trashbags with closeable flaps for the wrappers and other things that accumulate in any laboratory or living space. At one point the crew broadcast video of Willie removing a trash bag liner and using all of his strength to compact it into a ball and wrap the ball with duct tape so it would occupy as little volume as possible.

In one of the recovered videotapes McCool demonstrated how you can create your own fun with a ball of trash. He took the ball and went through various sports—volleyball, football, baseball, soccer, and golf. For soccer he did a forward somersault, hitting the trash ball with his feet. Clark commented, "Wow Willie, that was impressive!" She suggested, "Then there's tennis," and used her arm like a tennis racket to hit the ball of trash.

Laurel Clark cleans
the filter on the CIBX
experiment.

Since the crew cabin's enclosed, there's lots of dust and dirt with no place for it to go. An astronaut is assigned to clean the filters each day, using a vacuum cleaner to clear whatever's collected on them. On Earth, warm air rises and is replaced by cooler air. That's enough cooling for many electronics. But hot air doesn't rise in space, so fans are used to cool electrical equipment. Many experiments have their own filters at their air intakes.

The cleaning procedure is extremely low-tech — an astronaut shuts off the circuit breaker, removes the filter, and uses the sticky side of a piece of duct tape to pick up whatever's collected on it. Then the astronaut puts the duct tape in the trash and reinstalls the filter and turns the experiment back on. While cleaning an experiment's filter, Laurel kidded, "Joys of janitorial duties in space. Yeah! We love taking care of those." She commented, "I am constantly amazed at how much lint there is."

KEEPING HEALTHY IN SPACE

Each day the crew has a "private medical conference" with crew surgeons Dr. Smith Johnston and Dr. Steve Hart. These are extremely private communications not broadcast to anybody else. The crew can discuss any medical issues with their doctors during these daily scheduled sessions. Most of the time it's just chit-chat with the astronaut telling the surgeon he feels fine and then catching up on the news or whatever else is happening. During one session Ilan said he was sick, which naturally made the surgeon concerned. But then Ramon explained he had "ground sickness" (as opposed to space sickness) and needed to spend more time in space!

There is a fairly comprehensive medical kit onboard including medicines, bandages, ointments, injectables, medical diagnostic equipment, surgical tools, equipment for treating injuries, and a microbiological test kit for testing for bacterial infections. Contrary to urban legend, there are no suicide pills.

Ilan Ramon floats
into Spacehab
with his arms
spread wide.

MICROGRAVITY ACROBATICS

The crew clearly enjoyed microgravity, transmitting video of Ilan flying like a superhero, Rick spinning his body and extended his arms to demonstrate a simple physics principle, and former gymnast Dave tumbling in a way he never could on Earth. This book's website includes several video clips of the crew's zero-G activities.

One of the more interesting acrobatics is to travel as far as you can without touching the spacecraft's walls. Columbia had a 15-foot tunnel connecting the middeck to the Spacehab module. Brown said, "The other thing up here that's great fun is when we go from the tunnel coming from our middeck back to our research module. That thing is pretty long, you can get some rolls going—you really can't do any flips but you can do some rolls and spins when you're going down the tunnel and that's a lot of fun."

In a video recovered after the accident Willie tried a roll as he traveled through the tunnel, but he didn't quite make it through before having to reach out and tap the side to keep himself floating the correct direction. Laurel commented, "Oh he touched the side, I give it a, oh..." —she was about to rate him on his microgravity acrobatic performance.

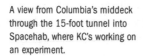

A view from Columbia's middeck
through the 15-foot tunnel into
Spacehab, where KC's working on
an experiment.

Then she exclaimed, "You're not ticklish anymore!" apparently after trying to tickle him. Willie tried again, and Laurel started to announce, "The amazing...," but McCool had to touch the tunnel again. He sighed, "This could take ten tries." That video clip is on this book's website.

Laurel transmitted a video of Dave's acrobatics during the mission. She announced "Live from space. As you know he used to be in the circus, so he's very happy here in space." As Brown did a reverse somersault for the camera, capcom Charlie Hobaugh asked, "Yeah, but can he juggle in space?" and Laurel conceded, "You've got a point there, Scorch." Hobaugh joked, "Unless you're talking about the timeline." Laurel said, "Well, our fantastic timeliners do all the juggling there, we're just following orders up here and having a great time." Hobaugh added, "We'd like to thank Dave for the triple Lindy demonstration." Brown grabbed the microphone and replied, "I don't get no respect." A video clip of Dave's acrobatics is on this book's website.

While Dave Brown was a real ex-gymnast, each of the astronauts could do gymnastics far better than any athlete on Earth. A broadcast showed Laurel and KC hanging upside down and pulling themselves up and down. Technically they may have been pretending they were doing pull-ups, but it would be more accurate to call them pull downs! Then Clark did one-handed pull-downs. She assured Mission Control they were doing their real exercise too.

Laurel tried an impromptu demonstration when she noticed a hinged cue card flapping and wondered what she could do with it. She said, "Let's see if we can make this thing fly (flap back and forth multiple times)" and exclaimed, "Oh! Woo" when she succeeded.

Each astronaut gets a portable CD player and can select several favorite albums. Ilan had asked many of the payload team to loan him CDs to listen to, which he planed to return after the mission.

The CD player is also an excellent science demonstration. A spinning CD will try to hold its orientation like a gyroscope. On Earth it isn't

Laurel reacts as she makes a
checklist flap back and forth.

noticeable because the player's weight is far larger than the CD inside. In microgravity instead of drifting away or tumbling the CD player will wobble in place and try to hold its orientation. This book's website shows Laurel watching her floating CD player. That CD player was recovered along with Columbia's debris with the last CD she listened to still inside, the Irish band "Runrig."

MEALS IN SPACE

Space food is basically camping food or off-the-shelf supermarket products that don't require refrigeration. It is not the freeze-dried astronaut ice cream you find for sale in museums and tourist attractions. Microgravity causes fluids to collect in the head and taste buds can change. Many astronauts report they like different foods.

The food that got the most attention was Ilan's Kosher meals. Ramon chose to eat Kosher food in space in recognition of his religion and its heritage, even though he didn't practice Kosher dietary laws on the ground. NASA food scientist Vicki Klories said, "We have not had a specific request for Kosher food since I've been here (1985). Ilan Ramon is the first. He has included some Kosher items in his menu. The rest of his menu is not necessarily Kosher items but is designed in a Kosher style." NASA contacted the "My Own Meals" company which makes no-refrigeration precooked Kosher meals, primarily for members of the military who want to keep Kosher, and for campers. Contrary to a couple of media accounts the Kosher food was not freeze-dried. Ramon's other foods came from NASA's standard selections, which are not Kosher. For example, the chicken in his chicken teriyaki was not slaughtered by a Kosher butcher, making it unacceptable for a true Kosher meal. Some Jews would consider Ilan's meals acceptable, others would not. There's a saying, "Anybody who is more Kosher than me is a fanatic, anybody who does less is a heretic." At most it could be said some of Ilan's meals were Kosher.

Space food has improved in quality and taste over the last few decades. In the 1960s, the early Mercury astronauts ate mashed up food—almost baby food—out of plastic squeeze pouches. Today's astronauts dine on fancy fare like shrimp cocktails, Mexican scrambled eggs, and cheese tortellini. Most of the food is freeze-dried and just requires the astronaut to inject hot or cold water, wait a bit, and cut open the package—it's the same food eaten by campers. There's lots of thermostabilized food, off-the-shelf supermarket meals that don't require refrigeration. The off-the-shelf packages are cleaned, a color-coded label with a bar code is added, and a piece of Velcro is attached to make the package easy to store on a convenient surface.

The shuttle's "kitchen" can dispense hot and cold water and has a convection oven. The oven's only hot enough to warm precooked foods.

There is no refrigerator or freezer, though. The challenge is to have enough tasty meals for 16 days for seven astronauts, plus a couple of extra days in case the mission has to be extended. That's a lot of meals, and they have to be packed in as small a volume as possible.

In a videotape which survived the accident, Willie demonstrated how to prepare a meal and why a spoon and scissors are an astronaut's eating utensils. He said, "Here's how we eat breakfast. The food is stored in these trays. We look for the colored dots, I want something that has yellow on it." He was searching for yellow because that was the color code for all of his food and personal items. He found one package and read "Granola with blueberries. Mmm, mmm—that sounds good." Willie had picked up the phrase "Mmm, mmm—that sounds good" during the NOLS expedition from Rick. Husband got the phrase during a campout with his daughter.

As McCool prepared the freeze-dried granola, he narrated what he was doing: "Take our dehydrated food over to the galley. We put in our granola with blueberries and there's a needle down here. The needle is going to inject into the package. The package says 2 ounces of water. I'm going to go with cold water and let's see what we have. We mix it up and let it sit for a little while, and after it's sat for a while we open it up and eat it. While we're at it we can get a drink. We'll Velcro [the granola] here so it doesn't go away." He put the granola package in a convenient place and went back to the food storage tray to select his drink. He said, "Let's see—pineapple drink sounds good. I put twelve ounces of water even though it says eight." Willie unclipped the straw and squirted out a blob of the drink and sucked it into his mouth. He said, "Mmm, mmm. Pineapple juice." He then went back to his granola, announcing "Granola· with blueberries—breakfast of champions!" He explained to the camera, "We carry a spoon and scissors with us everywhere we go. We'll start with the scissors, and we'll cut a U-shaped cutout here." He used the spoon to push open the flap he created in the package and scooped out a spoonful of the granola. After putting it in his mouth he

Willie eats a bag of granola as part of a demonstration of how food is prepared aboard the shuttle.

announced, "Good stuff." Then he noted, "One of the fun things to do is to float it," demonstrating how to just send a clump of food floating toward his mouth and chomping on it while it floated in midair. As Willie demonstrated, food in space rarely takes a straight path from the package to the mouth—for astronauts, "Don't play with your food" is a lesson long forgotten.

Astronauts will trade food items with their crewmates—almost like kids in a school cafeteria. Meals are an opportunity to socialize and look out the window, and also playtime. It's common for astronauts to toss M&Ms across the cabin, aiming at a fellow crewmember's mouth. Bread is tortillas because they stay fresh, don't create many crumbs, take little space, and make great flying saucers.

The astronauts can also ask for something special. Maple candies are a tradition among the Canadian astronauts, a pecan pie was sent to ISS as a present for one crew, and on John Glenn's shuttle flight he asked for Metamucil crackers. Rick and Laurel shared Glenn's preference, asking for Cinnamon Spice Metamucil crackers. Rick also asked for Laura Scudder's All Natural Creamy Peanut Butter, and a jar was put on board for him. Laurel also asked for curry paste and a Granny Smith apple. K.C. asked for instant Nile spice low fat black bean soup with rice, lentil soup, and lentil soup with couscous. In addition there's a pantry with a variety of standard shuttle foods in case somebody decides they'd like something different from what they had selected months ago.

Salt and pepper are available in squeeze bottles. Powders would be extremely difficult to handle in space, so the salt is in the form of salt water, and the pepper is pepper oil. Other condiments like ketchup, mayonnaise, and salsa are just normal fast-food packets.

Any drink that can be mixed with water is available, including coffee, tea, cocoa, and a wide variety of imitation fruit drinks – it's not just Tang! While most of the food is "ready to eat" it's possible to put together a meal. When an astronaut asks for peanut butter, jelly, and tortillas, it's pretty obvious he's planning on making a peanut butter and jelly sandwich.

On a typical day Laurel selected macadamia nuts, oatmeal with brown sugar, mocha, and grapefruit drink for breakfast. The nuts and mocha are packed in sealed bags; all she had to do was cut the bag open and enjoy. The oatmeal was freeze-dried and prepared the same way as Willie's granola. The grapefruit drink was imitation grapefruit powdered drink in a metalized plastic pouch with a plastic valve sealed into the pouch. Laurel put a resealable straw into the valve to drink her beverage.

Clark's lunch included a freeze-dried shrimp cocktail which she injected cold water into the package to rehydrate, a foil pouch of tomato basil soup which she just heated before opening, tortillas, what NASA calls "Candy Coated Peanuts" but anybody else would call peanut M&Ms, and a peach-apricot drink.

Dinner included a foil pouch with fiesta chicken, freeze-dried Italian vegetables, more tortillas, a supermarket container of chocolate pudding, a bag of shortbread cookies, and grape drink.

Of course those are the items she selected in advance. On-orbit she could have chosen any of her food items for any meal, the items in the pantry, or traded food with her colleagues.

MORE FUN IN SPACE

On shuttle missions longer than eight days each astronaut gets some time off, two half-day periods for the STS-107 crew. The off-duty period features a private video phone call to their families on the ground. Generally the astronauts will perform acrobatics for their kids and talk to their spouses in a two-way videoconference. The rest of the time off is unscheduled. Normally it's spent looking out the window and admiring the Earth.

Astronauts who are religious can do personal religious activities during their pre-sleep or post-sleep periods but their days off probably won't correspond to any Sabbath on Earth. In the case of Ilan Ramon he noted before launch he was secular so he didn't need any special permission to work on the Sabbath.

Dave videotaped McCool giving a tour of the shuttle's middeck for his movie. At a couple of points Dave asked Willie to start over and redo a scene, clearly intending to splice together the best shots of everything he videotaped.

Shuttle clothing is usually an embroidered shirt with the mission logo and the astronaut's name or nickname. The shirts are in a variety of colors and styles and astronauts will often have color-coordinated days where everybody wears the same type of shirt. The shorts have strips of Velcro sewed on to make it easy to hold things in place. Instead of shoes astronauts wear socks.

KC plucks a floating peanut M&M out of the air.

The STS-107 crew performed a wide variety of science activities. (clockwise from upper left) The Actlight watch for a sleep experiment. A commercial experiment grew roses. Mike Anderson gets on the ARMS ergometer. Kalpana Chawla operates SOFBALL. The entire crew gathers for an educational public service announcement. Laurel operates the Bioreactor. Ilan poses with the Israeli "chemical garden." Rick works on the BRIC experiment.

PART III

SIXTEEN DAYS IN SPACE

What's often ignored in stories about Columbia's final flight was the science that was performed during the mission. The purpose for the flight was to accomplish as much science as possible in 16 days. Mission scientist John Charles came up with the Latin motto "*ex orba siencia*"—from orbit, knowledge. And in terms of collecting science, STS-107 was an incredible success, even though the accident destroyed most of the experiments.

Critics say the science wasn't worth the risks to the astronauts' lives. One example commonly cited is the "ants in space" student experiment. But it's important to note that the ants were just one-sixth of one of the scores of experiments on Columbia. Adding the ants experiment to make use of space which happened to be available did not contribute any added risk to the lives of the astronauts, but it did help inspire students and encourage them to see that science can be fun.

STS-107 had a broad range of different scientific experiments in many fields. Any individual experiment was not the purpose for the mission, but combined they formed an incredible suite of scientific endeavors. Each chapter in this part covers two or three experiments and other activities. In some cases an experiment was operated for several days, in other cases the astronauts just turned on the experiment at the beginning of the mission and let it run on its own. On the surface STS-107 just looked like people working in a laboratory in an unusual location. But deep down it was an extremely complicated and well put together mission.

CHAPTER 20

FLIGHT DAY 2:
MEIDEX, PHAB4, ARMS,
ASCENT VIDEO

The "alarm clock" music on the shuttle is chosen by an astronaut's family or Mission Control. Shortly before a shift was scheduled to wake, Mission Control called one of the astronauts on the other shift and asked him or her to make sure the volume for the speaker next to the sleeping bunks was turned up. (Normally that speaker is turned off to prevent communications with the ground from waking the sleeping crewmembers.) Most of the time the crew is already awake, but it's considered a nice tradition. This book's website has the audios of all of the wakeup calls. The first was "EMA EMA" for astronaut Dave Brown, on Thursday evening at 9:49 P.M. EST. He replied, "Good morning Houston. The blue shift is up and awake and ready to work after that little bit of international flavor, ready for our first full day in space."

The blue shift was responsible for setting up MEIDEX, which was mounted inside a canister in Columbia's cargo bay. Its motorized door would open up before an observation and close afterwards. The MEIDEX camera inside could turn to the left or right to follow a dust cloud or thunderstorm. The astronauts used a laptop computer to control the door, swivel the camera, and turn on the video recorders. Some of MEIDEX's video was transmitted to the ground so scientists could get information on how well the system was working and maybe provide some feedback to improve observations later in the mission. All of the video was stored on recorders within the canister and on backup recorders inside the crew cabin.

Weather forecasters would predict whether or not conditions were good for dust in the "regions of interest." If there was a dust cloud over the primary area, over the Mediterranean, then a specially equipped aircraft could be sent to fly through the dust storm for simultaneous observations. Unfortunately, MEIDEX ended up flying during a season where there's less dust, so the scientists weren't optimistic about getting a dust storm over their primary region during the two weeks Columbia was in space.

Ilan explained, "[The scientists] had to think about solutions for getting improved statistics. They extended the region of interest to [the Atlantic Ocean] west of Africa. In that region usually all over the year there are some dust storms. So in a 16-day mission we are almost sure we will be able to meet some dust storms." With a far larger area there was a greater chance of catching a dust storm. But a full intensity campaign would only be possible in the primary region of interest in the Mediterranean Sea. With a January launch the Israeli scientists could only cross their fingers and hope for an off-season dust storm.

MEIDEX's camera could be used for additional purposes, including looking for upward bolts of lightning. Ramon said, "Even if we are not [successful with the dust storms] there are some secondary objectives for this experiment—Sprites, the clouds to upper atmosphere lightning. The other one is looking at the sea albedo which is connected to the dust storms. I'm pretty sure we'll be able to have some success."

Before the mission Dave said, "Perhaps as a crewmember, one of greatest things about MEIDEX is it requires us to be on the flight deck looking out the window, which I think will be a real pleasure." Ilan said, "The first opportunity [I get will] be to look for dust storms in order to be able to have a mission success for MEIDEX as well as looking through the window and seeing Israel and Jerusalem from space."

Brown set up MEIDEX and on January 17 at 4:46 A.M. He reported, "We do not see any dust streaks or plumes." A couple of minutes later he reported "We now estimate the cloud cover at 50 percent." After the pass was over, Dave passed on his thanks to the scientists and engineers who developed the experiment. He said, "I'd like to say congratulations to the whole MEIDEX science team. They've really opened up a new chapter on dust and aerosols and how they contribute to the Earth's climate and, in particular, global warming. And it's a big day for the Israeli science community. [It's] opened a new chapter now in participating in human spaceflight. I'd like to add my personal congratulations to Yoya (Joachim Joseph), who's come a long way to see this day. So congratulations to MEIDEX."

Later Brown prepared the European ARMS experiment for his first session and radioed, "Spacehab POCC Columbia—Putting ARMS ergometer setup in work." CIC Beth Vann replied, "Okay, Dave, we copy, we know you can't wait to get on that bike."

Ilan got to work with MEIDEX later that day, during a pass over the Middle East. He said, "The MEIDEX science is over, the door is closed. All the best and regards to Israel. We just passed over Jerusalem."

Other payloads activated early in the mission included protein crystal growth, the commercial ITA (Instrumentation Technology Associates) space factory, the bioreactor, Zeolite Crystal Growth, Astroculture, earthquake research, and moss plant experiments. The moss had to be exposed to light for one day to start their growth.

On January 17 at 9:42 A.M., 23 hours after launch, Columbia was flying "backwards" and "upside down." It made an ordinary yaw maneuver, which pointed the right wing downstream. This served two functions—

to align the shuttle's inertial measurement units (IMUs), fancy gyro-scopes which keep track of Columbia's orientation, and to orient the "Miniature Satellite Threat Reporting System" (MSTRS—pronounced "misters") experiment mounted on top of the Spacehab module. During the maneuver, something came off Columbia unnoticed. The identifica-tion of that piece is covered in Chapter 45, on the "Mystery Object."

NASA holds status briefings for the press during shuttle missions on weekdays, and occasionally on the weekend if there are major activities. As anticipated before launch, the interest quickly dropped and only two or three reporters showed up for each of the briefings. It would not be an exaggeration to say that there were only a handful of reporters who actu-ally filed stories on a regular basis during STS-107. The only reporters who showed up for more than one status briefing during the mission were Chris Kridler with *Florida Today*, Marcia Dunn with the Associated Press, and myself, representing the *Amarillo Globe-News, Jerusalem Post*, and a handful of additional newspapers.

At each status briefing, mission scientist John Charles would talk about the mission's progress, and a couple of scientists would describe their experiments. In most cases a shuttle manager would describe what was happening with Columbia. The first status briefing covered human life science experiments.

Biomedical scientists say space is an extremely useful tool for exam-ining how the body behaves. In space, the body's immune systems suffer, the taste buds change, the blood collects in different parts, and the heart shrinks. But the body does function, and after it returns to Earth, most of the functions quickly return to normal. Medical researchers can study an astronaut's body in intense detail before and after the mission, and to a limited degree during the mission. Scientist Arny Ferrando called it "Measuring a stressed state on a healthy individual."

Unfortunately some of NASA's more over-enthusiastic promoters try to push the analogy too far. Is the fact that some of the changes during spaceflight are similar to natural aging in the human body an indication that there's a true scientific connection between aging and spaceflight? So far, no scientist is willing to make that claim with research to back it up, but it was used as the reason to justify flying historic Mercury astro-naut John Glenn on the shuttle.

Four experiments concentrated on the human body's adaptation to spaceflight. The Physiology and Biochemistry Team was nicknamed the "PhAB4"—pronounced "Fab Four." The experiments were Protein Turnover During Spaceflight, Calcium Kinetics During Spaceflight, Renal Stone Risk During Spaceflight, and Incidence of Latent Virus Shedding During Spaceflight. The protein and calcium experiments were limited to the payload crew—Anderson, Clark, Brown, and Ramon. They would be poked with needles many times—before the mission,

John Glenn contributes blood for the Protein Turnover (PTO) experiment on STS-95.

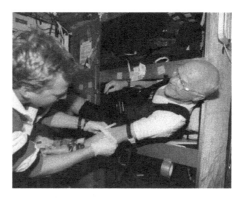

during the mission, and after the landing. All seven crewmembers participated in the urine and virus experiments. The virus shedding research only required saliva samples during the mission, and blood and urine samples preflight and postflight.

The Protein Turnover (PTO) experiment examined why the body doesn't absorb as much protein in space. Dr. Ferrando of the University of Texas said, "What we're looking at here is how the whole body makes and breaks down protein in particular muscles." He explained that based on data collected on STS-95, "It looks like for some reason the body decides it doesn't need to make protein anymore. That whole process dramatically shuts down, such that the body breaks down more than it makes, and as a result over time muscle protein is lost."

The Calcium Kinetics During Spaceflight experiment studied how bones change during spaceflight. Dr. Scott Smith of NASA said, "The loss of bone is a significant health concern for astronauts. The amount of bone that's lost in a short-duration flight, in a few weeks, really is not that significant. It's actually very hard to measure. So what we look at is the change in calcium, things that we can measure in the blood and in the urine that tell us what the bones are doing." Calcium is the main mineral that gives bones their strength. "Our experiment is looking at what we call calcium kinetics, which is the movement of calcium through the body. The way we do that is we use tracers, we give a special form of calcium to the astronauts, they take one of these orally, one is infused intravenously, and we then collect samples and trace the movement of calcium through the body."

The entire crew participated in the Renal Stone experiment. Spaceflight tends to result in dehydration and calcium loss from the bones, factors that increase the chances for kidney stones. Investigators believe that potassium citrate can minimize kidney stones and are performing a long-term study on as many space travelers as they can. Each astronaut takes a pill each day. Nobody involved in the experiment knows which astronauts get the potassium citrate and which ones get placebos. An independent company prepares identical looking placebos and drugs. Each pill has a serial number. When an astronaut urinates a bag collects a sample. After the mission the urine is chemically analyzed for any indications of potential kidney stones. After all of the data is collected over several missions, the "code" is broken and it's revealed which astronauts took the placebo and which ones took potassium citrate. If the astronauts who took the potassium citrate have less tendency to form kidney stones, it would indicate that the procedures are effective.

The Incidence of Latent Virus Shedding During Spaceflight was the simplest life science experiment to perform—the astronauts were asked to contribute some spit each day. Dr. Duane Pierson wanted to test the theory that herpes virus reactivation and shedding into saliva and urine increase during spaceflight. The samples would be examined to see if they contained the viruses after the mission.

Collecting saliva samples was extremely simple. Each morning the astronauts put a cotton pad, similar to one you'd find in a dentist's office, in their mouths and bit the pad for a couple of seconds. The soggy pad was placed in a sealed plastic bag and labeled with the astronaut's name and the time. That activity was videotaped during the mission, and Rick Husband commented, "I hope nobody's eating dinner while watching this." Some of the saliva samples were frozen; others stored at room temperature.

The astronauts kept logs of the food they ate and the amount of urine they generated. Laurel Clark was especially conscientious about the science being performed on her body. At one point she asked if it was okay for her to substitute cocoa for peach-mango drink for her breakfast the next morning. The scientists discussed it on the ground for about a minute before telling her it was okay as long as it was the same amount.

One morning Laurel checked the timeline for what activities were scheduled. She declared, "Willie the man is drawing our blood today!" and shouted to him "Hey, Willie. When you draw our blood you can just put some mocha straight into my veins, okay? That'll work." She was kidding. An audio clip of Clark's comments is on this book's website.

Another experiment into the body's adaptation was Sleep-Wake Actigraphy and Light Exposure During Spaceflight, which evaluated how well the astronauts slept. The theory is a spacecraft in orbit goes through 16 day-night cycles every 24 hours, which messes up the body's circadian rhythm. Understanding how and why the body acts differently in space provides data for why some people sleep better than others. The crewmembers wore "Actilight" wristwatches. They look like regular wristwatches but are actually small computers with a light sensor and motion detector. Obviously if there's little motion and the room is dark, the person wearing the watch is probably sleeping. Each morning the astronauts entered comments in a sleep log on how well they slept.

When Laurel woke up on Flight Day 15, she picked up her crew note-book and declared, "Sleep log." and added, "Yes I'm still sleepy," a moment later. Husband agreed, with an extremely loud yawn. She asked Husband when they went to sleep the previous night and Husband told her 8:45. Clark questioned, "You went to bed the same time as me?" Husband thought about it a couple of seconds and said, "Well okay, you put down 8:40," and Clark started laughing. Husband declared, "You went off floating off back into the Spacehab for a while, we thought we were going to find you sleeping in a corner!" An audio clip is on this book's website.

The Protein Turnover and Calcium Kinetics experiments required blood draws and injections of the calcium and amino acid tracers. The science crew all signed medical waivers agreeing to participate as the guinea pigs, however they always had the right to refuse at any moment for any reason. The astronauts realized the importance of the science and made the decision to participate as willing test subjects. The crew cooperated 100 percent, but they did joke about just how willing they were. At one point the crew transmitted a video showing Mike strapped against the wall of the storage area and Laurel about to insert a needle into Mike's arm. Dave narrated, "This is Laurel, I think she's telling Mike that this isn't going to hurt her a bit." He added, "We decided to use the straps to help restrain the subject and also make sure that Mike can't make a quick getaway from Laurel."

Normally you'd expect the doctors, with the most experience and knowledge, would be the best at sticking a needle into a vein, but that wasn't always the case.

At 9:47 P.M. Friday (January 17) Dave Brown radioed that there was a problem putting the needle in Mike's vein and they had tried twice. Dave said, "We have a good chance on the third try," and added, "I don't mind at all." (Of course he wouldn't—he was on the correct side of the needle!). Beth Vann, in the Payloads Operations Control Center (POCC), replied, "If Mike doesn't mind," and Anderson replied, "No, Beth, I don't mind." Vann confirmed, "Copy that Mike." A couple of minutes later Dave radioed the good news, "POCC, Columbia, third time's the charm, we're complete with PhAB4," and Vann replied, "Good news, Dave. Thanks for letting us know. And a special thanks to Mike too."

In one of the updates to the crew, Mission Control noted, "The PhAB4 vampires couldn't be happier with how smoothly the sample col-lections tracer infusions are going. They really appreciate all of the call-downs to let them know how operations are going. Special thanks to Mike for 'sticking' it out and giving Dave one more chance to get that monovette filled. After FD3 [Flight Day 3], the entire payload crew can truly say—'I gave at the office!'"

Later in the mission Dave admitted to me, "It didn't hurt me a bit. We've trained quite a bit, and [spent] time in hospitals before that. But some days you're right on and some days you're not. Exactly how it felt you'd have to ask Mike, but we're still talking to each other." Brown added, "In fact he has to turn around and draw my blood, so it's pretty good motivation to do your best." Mike didn't seem upset that it took Dave three tries to get it right. An audio clip of Brown's remarks is on this book's website.

The payload crew really gave everything they could for the scientists. Dr. John Charles commented, "They're giving 100 percent, they're there for us 100 percent of the way. And incredibly, they're doing it enthusiastically." Each of the four payload crewmembers was injected with tracer chemicals twice and donated blood ten times during the mission.

The NASA life science experiments concentrated on the blood and the bones. Europe also flew a group of human life science experiments, but they concentrated on the lungs and heart. While they involved heavy exercise and wearing monitoring equipment, at least they didn't involve any needles.

Europe's ARMS (Advanced Respiratory Monitoring System) was an intense research into the body's physiology. Scientists from Denmark, Sweden, Italy, the Netherlands, and Germany sponsored different studies. Each of the payload crew pedaled an ergometer (stationary bicycle). Before the mission Brown said, "We're studied before we go, beginning of the flight, middle of flight, and end of flight. Pulmonary test at rest and while riding a bicycle." While an astronaut pedaled the ergometer, the harness would monitor heartbeat, blood pressure, and blood hemoglobin—and run an EKG. The astronauts breathed through their mouth into a device that measured their respiration and oxygen levels. Brown noted that the tests were "2 hours wired up and focused on what they're doing."

Laurel Clark and Dave Brown
on the ARMS ergometer.

One of the researchers' questions that would have a practical application on Earth was, "Is it less stressful on the lungs for a person to be lying on their back or belly?" A simple question, but important for doctors who are treating injured people. Brown said, "One of the ones that I find very interesting, particularly as a physician, is the question of do lungs work better whether you're lying on your stomach or your back? If you're healthy, this really doesn't matter. But, if you're really sick—say, you're in a hospital, you're seriously ill or you've been in a car wreck or you have a family member who is—the question is, would it be better to roll that person over onto their stomach if you're on a respirator in an intensive care unit? Or, is it not better to do that? Should you leave them on their back? One of the ways that we're going to help contribute to answering that question is through the ARMS experiment. We're studied here, before we fly, lying on our stomachs. They also put us on our backs, and then on our sides. Also in Sweden, where this experiment's sponsored, they've actually had people studied in centrifuge at three times gravity, three-g's, and up to five-g's. And then, the last part is we'll go to space where there's no gravity and see how our lungs work. Once we take that back, the scientists that work this project will have a much better idea of how gravity affects lung function. So now, if you're that physician who has that really sick patient, and you have to decide what's the best position for this person who's really seriously ill, now we have a really solid science foundation to answer that question of what's the best way to proceed with this treatment, and is this a good idea? Or not a good idea?"

An experiment from Denmark was focused in the displacement of the heart in the chest—what does any movement in the heart's position do to circulation? The Dutch group developed a method to predict in advance which astronauts will suffer from poor blood pressure regulation upon return to Earth. Dave's girlfriend Janneke Gisolf was a member of that team and using that research for her PhD thesis. They devised a set of tests where they could evaluate the entire crew's performance.

Mike activated the S*T*A*R*S educational experiments. By pushing down a plunger, he released them into the "ant farm," where they would live and burrow tunnels. About 19 hours after launch, he reported, "Access one is complete, and I'm happy to report that the ants are active, and the spider's active also." CIC Brad Korb replied, "That's excellent news, Mike, thanks for the good work." On the ground, I watched Bioserve engineer Mark Rupert performed the same activities in an identical unit.

The red team's first sleep shift started at 10:39 P.M., 12 hours after launch. They were awakened on Friday January 17 at 5:39 A.M. The wakeup call, dedicated to commander Rick Husband, was "America the Beautiful," sung by the Texas Elementary Honors Choir. Capcom Charlie Hobaugh noted that the choir included "the beautiful Laura Hus-

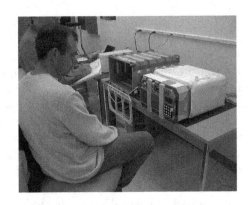

Engineer Mark Rupert prepares to open up the S*T*A*R*S locker in Cape Canaveral with the control experiments. Spacehab's Janet Morgan is in the background. Photo credit: Philip Chien.

band as a featured member." Husband replied, "Thank you very much. It's really great to hear that music, and also to know that Laura was singing with the choir. It's really nice to have a reminder of our family. So I'd like to say hi to Laura and also my son Matthew and my wife Evelyn. We're having a great time up here, we just appreciate all of your support."

When the red shift reported for duty a couple of hours later, Rick said, "We're having a great time and starting to get things squared away where we can move around and really get settled in. So things are going great."

Willie finished his first full day of science on orbit and said, "It was a busy day, an eye-opening day for the first-time fliers, and we had a whole lot of fun. Looking forward to another 15-plus days on orbit."

The astronauts broadcast the video of their launch from a lipstick-size video camera mounted on the shuttle's flight deck. It recorded the launch from the point-of-view of somebody on the flight deck behind the astronauts. For the most part, all that's seen is the back of the helmets of the four astronauts, but if you look carefully through the windows you can see the lighting change as things happen.

One of the astronauts edited the video down to about two minutes. The seats shake gently as the main engines start and then really vibrate back and forth when the SRBs ignite and the shuttle launches. Dave is seen lifting his left arm to see what the launch looked like through the overhead windows with a mirror mounted on his suit's wrist.

Husband narrated, "The vehicle just flew beautifully. We had a great ride, we had brilliant sunshine coming through the windows in the front and it was just a fantastic ride all the way up. Some people have said that sometimes—Columbia, they have to do a lot of work on it on the ground, but when it gets in space it just does wonderful, and we think it's because she just loves to be in space. Because we've got a great vehicle right now, and we're really enjoying it." Husband added, "We were very happy to welcome four new astronauts into the community with Willie McCool, Dave Brown, Laurel Clark, and Ilan Ramon."

The video showed Brown floating out of his seat and looking through the overhead windows to marvel at the Earth from space, and Brown and Anderson taking photos of the External Tank through the overhead windows.

Next was a short clip of Brown's video of the ET after it separated. Husband said, "We've got some clips of the external tank coming up shortly." The video showed the backside of the tank. Husband said, "There's a good shot of the external tank shortly after we pitched up. We're looking at it through the overhead windows. You can see every once in a while ice particles floating across the screen." Dave would have shot the ET for a far longer period of time, but the crew only chose to send down a clip, enough to illustrate what the tank looked like when it was flying along with Columbia, but not so long that it would be boring.

The next video clip showed K.C. studying a checklist and Husband preparing for the engine burn that put Columbia into a stable orbit. During the burn, the video showed Husband holding up a cue card and letting it go. Because the engines were burning, the cue card moved backwards instead of just floating.

That evening marked the first of Ilan Ramon's three Friday evenings in space. But with the intense workload early in the mission, there was no chance to even think about commemorating the Sabbath. Before the flight Ramon said, "I'm planning to make a symbolized Kiddush, a very small prayer, doing with a glass of wine on Friday evening. So I will try if time is available and I'm available, as you know I'll be very busy. If time is available, we're going to go by Houston time. I'll try to do kind of a symbolic Kiddush."

NASA's conservatism has resulted in extremely strict rules about what is and isn't allowed on flights. An American Indian astronaut was prohibited from carrying tobacco, even as a symbol, in a sealed package. Alcoholic beverages are prohibited. In addition, the laws of physics make it impossible to drink anything out of a cup. If you did somehow put some liquid into a cup, surface tension would cause the liquid to spread itself over the inside and outside surfaces of the cup—there is no way to "pour" in microgravity. Ramon noted, "I won't have wine available; we don't carry any wine. We can't exactly drink out of a glass. We'll figure out how to do it."

In space, you adapt. Instead of wine, you use imitation grape drink. And instead of a cup, you drink from foil pouches with a sealable drinking straw. Ramon could squeeze out a blob of grape drink and try to scoop up the floating sphere with the Kiddush cup. He noted, "I'm secular. But I'm going to respect all kinds of Jews all over the world. I will try to make some symbolic traditions in space as already was done by astronauts like Dave Wolf, for instance, who used a dredel. I'll do my best to take any opportunity to [do] several traditions if time permits."

CHAPTER 21

FLIGHT DAY 3:
CALCIUM KINETICS,
PRESS INTERVIEWS

The blue team's wakeup call on flight day three was Pink Floyd's "Coming Back to Life," for Willie McCool, radioed by capcom Ken Ham. Willie replied, "Good morning Ken, there's nothing like waking up to the mesmerizing music of Pink Floyd. Many thanks to my wife Lani, and my three boys Sean, Christopher, and Cameron, for choosing that music for me to listen to. With that, we're ready to get on with another day of science." Ham acknowledged, "Yeah, they have great taste, and we're ready to go."

Dr. Scott Smith of NASA's Johnson Space Center noted that flight day three was the big day for the Calcium Kinetics experiment. Each of the four payload crewmembers swallowed a pill with Calcium-44, a harmless naturally occurring radioisotope of calcium. An hour later, they were injected with another harmless radioisotope, Calcium-42. Dave and Mike took turns injecting each other with the tracers, and Laurel and Ilan injected each other. Later in the day, Willie drew blood samples and K.C. helped out by spinning the samples in the centrifuge. Smith noted, "We're really over the hump for our experiment, the rest of it is watching the tracers as they come out and getting the samples after the landing." Smith's plan was to analyze the samples after the mission for traces of the Calcium-44 and Calcium-42 that would indicate how quickly they were absorbed by the body.

The day also marked "lights out" for the moss experiment. The eight moss canisters were exposed to light for a day to start their growth cycles, then the lights were shut off inside the experiment. For the rest of the flight, the moss would grow without the natural cues from light or gravity. An astronaut would open up the experiment and inject a fixative into one of the canisters every couple of days, so after the flight the scientists would have moss samples that had grown for several different lengths of time.

The red shift's wakeup call was "Space Truckin," by Deep Purple. The song was for K.C., sent up by capcom Linda Godwin. K.C. replied, "We're up and awake. Thanks for the music, we're country music and space truckin today too." K.C. and her husband Jean-Pierre Harrison were fans of Deep Purple. Before the mission, K.C. and her husband and Ilan Ramon and his wife Rona were supposed to go see the band perform. But K.C. and Ilan had to travel to Florida for a "Crew Equipment

Interface Test," so their spouses went without them. K.C. explained, "We were going to go see a rock concert together – 'Deep Purple.' My husband and [Ilan's] wife went, so we've got a lot of funny stories about them meeting the band backstage. I don't think [Ilan Ramon] was a fan. But his wife became a fan coming back from the concert – [Rona said,] 'Wow, this thing is so energetic.'" K.C. added, "I had never been to rock concerts except once. My husband is such a big fan of theirs; we went to Louisiana last year. And I though the same thing – 'my goodness, there's so much energy going on!'"

NASA gave CNN, CBS, and Fox the opportunity to interview the red shift: As Columbia flew over the Pacific Ocean Rick Husband said, "Things are going really great. We're having a great time up here. We had a great ride to orbit, and all the activation of the experiments and the Spacehab went extremely well. And we've really got our space legs up and running."

Ilan said he tried to look for Israel but was disappointed: "It went too fast. It was mostly cloudy so I couldn't see much of Israel, just the north of Israel. But of course I was excited." Ramon had more luck spotting Israel later in the mission.

Ramon also noted, "I think it's a great start and an opening for great science from our nation. Hopefully for our neighbors in the Middle East." He added about his heritage, "An Arab man already flew in the 1980s, so I'm not the first one from [the Middle East]. I feel like I represent first of all the state of Israel and the Jews, but I represent also all our neighbors, and I hope it contributes to the whole world and especially to our Middle East neighbors." Actually two Arabs had flown in space, but they were both VIP passengers with no mission responsibilities. Saudi Arabian prince Salman Al Saud flew in 1985 on a space shuttle, and Syrian Mohammed Faris launched on a Russian Soyuz in 1987.

Chawla said, "[At any time we're doing] two, three, four experiments simultaneously. But it's a lot of fun, stimulating, and we are enjoying it.

A kiddush cup floats between K.C. and Rick as the red team is interviewed by television networks a couple of days after launch.

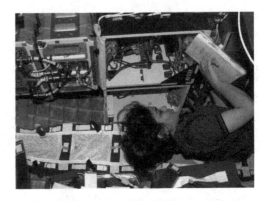

K.C. working at an
experiment rack in the
Spacehab module.

The module is quite big, roomy, and we were able to put it in very good configuration for our work on the very first day, so it's been working out really well."

Ramon did not commemorate the Sabbath the previous day. He said, "It's great to be here in space. We are so busy with all the experiments I didn't even have the chance to think about the Sabbath. As you know I'm secular and didn't get any special permission [to work on the Sabbath]. I'm here with special crewmates and I'm working like them every day. I've got a Kiddush cup but I even missed that on Friday. I hope I will do it next Friday."

Clark said, "The funnest part is definitely swimming through the air, it's just incredible and I'm still enjoying every second of that."

Clark said the biggest surprise for her so far was, "how much the ascent felt just like the simulation. I guess the second one that I noticed immediately was, obviously everything floats, the zippers and all the belts that have D-rings that we hold things down with are always floating and hitting each other and jingling. It makes this beautiful tinkling music in the background all the time. It just caught me off guard, and it was beautiful."

A definitive shot of Dave Brown—
he was almost always with a
camcorder. This photo was taken in
the front half of the Spacehab
module with a digital camera.

CHAPTER 22

The blue shift's flight day 4 wakeup music was "Cultural Exchange," for Dave, radioed by capcom Stephanie Wilson. Brown replied, "The blue team's up, and we're ready to go to work. Thanks a lot."

Brown showed a video of Egypt, Sinai, Jordan, and Saudi Arabia. He said, "It's just been full of beautiful sights up here on STS-107, and the views out the window just don't stop being really, really great." Capcom Ken Ham complimented Dave on his video, saying, "I think you did a great job narrating, I think I can get you a job as a capcom when you come home." The deadpan reply from Columbia? "Ken, I'm not ready to come home." An audio clip of their interchange is on this book's website.

Three separate experiments examined how flames behave in space. On Earth, gravity is a major factor—as air heats up, it expands and gets lighter. Gravity causes the lighter waste products to rise, which pulls in fresh air and keeps the combustion going. On Earth a candle burns in a flame-shape, broad at the bottom and tapered to a point at the top. But in space a candle burns with a spherical flame, and it's smaller and dimmer. A space flame also burns much more slowly, permitting scientists to examine the combustion process in far greater detail. Naturally NASA is

A rare photo shows a 10- by 20-mile shallow extension to the Dead Sea between Israel and Jordan where the hot, arid Middle East climate creates salt from the Dead Sea.

extremely concerned about the safety when something is purposely set on fire in space, even though the flames are extremely small. NASA's Mission Scientist John Charles said, "The very notion of putting a fire on a spacecraft raises the hackles of every safety engineer in the agency." Everything's triple-contained for safety, and there's only a very tiny amount of fuel available.

The first Combustion Module (CM-1) flew on the shuttle in 1997. STS-107 featured an improved CM-2. The three experiments shared six video cameras for visible and infrared imaging, a laser to measure the size of the particles, a gas chromatograph that measured the chemicals that were created, computers, and tape recorders. The experiments took place in a 95-quart (90-liter) chamber, about the size of an office wastepaper basket.

Each experiment had its own apparatus that fits inside the chamber. The Laminar Soot Process (LSP) experiment looked at how soot—the little black particles formed in many flames—is generated. By understanding how soot forms, scientists hope they can make cleaner-burning engines and reduce exhausts. The SOFBALL (Structure Of Flame Balls At Low Lewis-number) experiment looked at the smallest flames ever—less than a watt of energy. They only form under extremely unusual circumstances, like the lack of gravity, but are fundamental to all forms of combustion. LSP and SOFBALL had flown on previous shuttle missions and on STS-107 were being given the opportunity to collect more data. There was also a new combustion experiment—Mist. It examined how a fine mist of water—almost like fog—could be used to put flames out.

The combustion experiments needed pristine microgravity conditions. To get the best microgravity, the shuttle flies with its engines facing the center of the Earth, its nose pointed toward deep space, and its thrusters deactivated. This "gravity gradient free drift attitude" produces the best quality microgravity conditions.

LSP was the first up. Ramon talked to the POCC, going through dozens of settings to get the best possible performance. After the run was completed, the payload team radioed up, "Thanks for the hard work. The LSP science guys have been looking at the data and are extremely happy with what they've seen." Ilan acknowledged, "It's only because of the hard work by the training team."

Later, while showing a video of the soot experiment, Dave commented, "If we can even make a small decrease in the amount of pollution from combustion worldwide, that's a major contribution."

At one point Laurel showed a video of K.C. working on CM-2. She narrated, "We're told the tests came out wonderfully and the investigators on the ground are thrilled with the results they've gotten." The investigator was Dr. Gerard Faeth from the University of Michigan. A couple of days later he said, "Tests conducted on Columbia over the last five days

A tiny flame in microgravity permits scientists to study soot formation in detail.

have provided very valuable results and exceeded all of my dreams for what we could get." Faeth compared it to a prizefighter training for six years and the excitement of winning a championship bout, adding, "If you can observe soot processes in steady gravity-free non-turbulent or laminar flames—flames with smooth surfaces, those flames are appropriate as models for practical flames." Practical flames are those that appear in real life with crackling fires, unexpected pops, tossing off particles, and other hard to predict events. Faeth noted that the video of the LSP flame was "Large, spread out, and stationary, with no flickering, which provides an excellent test bed to study soot."

The team was able to downlink complete sensor and image data from 7 of the 14 tests, along with partial data from the other tests. This book's website includes a movie of a typical LSP run.

The wakeup call for the red team underscored the international nature of the mission. Capcom Linda Godwin radioed up the Israeli folk song "Hatishma Koli," by Hachalonot Hgvohim, for Ilan Ramon.

The shuttle has a couple of methods for returning data to the ground. The S-Band system provides near continuous voice and teleme-

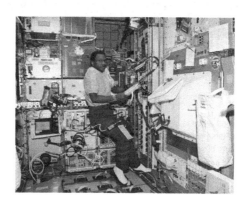

Mike Anderson working on one of the Combustion Module experiments.

try between the shuttle and Mission Control. The key exception is small periods when the shuttle is out of range of NASA's Tracking Data and Relay Satellite System (TDRSS). The Ku-Band (pronounced "Kay You Band") system is only usable when its antenna is extended outside of the payload bay, and when the antenna can be pointed up at a TDRS. Ku is one-way from the shuttle to Mission Control and can be used for video or large amounts of data. S-Band is roughly the equivalent of a dial-up Internet connection, while Ku is a broadband high-speed connection. A backup UHF radio can only be used for voice over certain ground stations, and is used primarily during launch and landing.

There was a problem with the Ku data transmissions. A computer on the ground would have intermittent signal dropouts, and the ground equipment would lock up. While the computer rebooted, a process that took five minutes, scientists couldn't receive any data from their experiments. The data was not lost—it was stored onboard. But the scientists couldn't look at the live data and use it to help plan changes to how they wanted their experiment to run. There was a second payload computer available, but it wouldn't boot properly.

Spacehab's Pete Paceley admitted, "I was sweating bullets through that." Programmers on the ground were able to create a set of software patches for the ground computers to keep the program from crashing when it saw a data hiccup, and they worked with the experimenters to prepare for the hiccups. By flight day 4 the programmers were able to come up with a system where they could predict when the hiccups would occur, and they developed a work-around to ensure that no data was being transmitted while the reboots took place.

The bowl-shaped Ku-Band antenna, seen in the upper left corner of this photo, was used to transmit large amounts of scientific data and video to the ground, including this photo one of the Columbia astronauts took of the Earth covered with clouds.

While Columbia operated flawlessly there was a problem inside Spacehab. On previous Spacehab modules, all of the air was circulated between Spacehab and the shuttle's crew cabin. The shuttle's air scrubbers removed the carbon dioxide and moisture, adjusted the temperature, and added fresh oxygen—and then the rejuvenated air was circulated back to Spacehab. That was fine if there was only one or two people inside Spacehab for any extended period, as when Spacehab was used to transport cargo to the space station. But for STS-107, Spacehab was a full-time laboratory where several astronauts would be working on different experiments all at once: They inhaled and exhaled, their bodies created heat, and their experiments generated additional heat.

Humans generate water vapor—humid breath and evaporated sweat. The shuttle has a dehumidifier for removing the excess water and dumping it overboard. The water separators are similar to a home air conditioner, with a condenser and a drip tray to remove moisture from the air. But since weightless water can't drip, a spinning impeller uses centrifugal force to separate the water from the air. Spacehab also had a pair of separators to remove the additional water created by several crewmembers inside at any given moment.

The shuttle carries "contingency water containers" (CWCs), waterproof bags just in case there's a water spill or there's a problem dumping wastewater overboard. As part of the normal maintenance operations, an astronaut hooked up a CWC to Spacehab's condensate pump and loaded the bag with the collected water. The astronaut carried the bag into Columbia's crew cabin and hooked it up to a valve to dump the water overboard. At least that's the way it was supposed to work.

Paceley said, "I remember distinctly walking into the customer support room, and I made a comment that I was in an upbeat mood that morning." That mood quickly changed when Paceley was informed that Spacehab's impeller was flooded with water; for some reason it wasn't removing the moisture from the air.

One of the astronauts radioed, "Can I stop talking and start mopping now?" About two quarts of water leaked, and the crew employed an extremely low-tech solution—they used towels to mop up the water.

Engineers analyzed the situation and prepared for what if something else went wrong. There's always a desire to remain a step ahead of potential problems. If the second separator also failed, could they try to get the first one working again, or would they have to come up with an alternative solution? And that's exactly what happened, the second impeller also failed. After a couple of hours on the backup impeller, electrical shorts shut it down. The electronics box had vent holes, and engineers believe the water from the leak crept inside.

So there was no way to remove the moisture inside Spacehab. Fortunately there was a second backup—the procedures used on the previous

A rare photo of the
Spacehab water removal
system after it failed.

Spacehab logistics missions—exchange the air in Spacehab with the crew cabin air, and let the shuttle's dehumidifiers remove the humidity. It would require some tweaking to make it work, since STS-107 had more experiments and more people inside. Engineers came up with a solution: by keeping Spacehab at a higher temperature, above the dew point, and Columbia's cabin at a colder temperature, the moisture in the air could be removed by the shuttle's systems. The warmer temperatures in Spacehab would prevent the moisture from condensing on surfaces, and once the air was in the cooler shuttle cabin, the moisture could be removed by the shuttle's condenser.

With the shuttle cabin's air conditioning turned on high, it was "Cool on the flight deck, comfortable in the middeck, and warm in the Spacehab module," according to Rick. If an astronaut felt cold on the flight deck, he or she could use a low-tech solution—put on a sweater. Husband reported that while Ilan and K.C. didn't notice the rise in temperature from 70 to 78 degrees, Laurel did.

Dr. John Charles noted the crew moved several temperature-sensitive items, like the automated blood analyzer and components from the microbial experiment, from the warm Spacehab module to a cooler area. Tracer chemicals to be injected into the science crew later in the mission were put in a refrigerator to keep them fresh.

According to Charles, the crew was still comfortable even with the higher temperature and performing all of their experiments, including exercising on the ARMS ergometer.

There were an additional 13 "occupants" in Spacehab—rats flying to see how they adapt to spaceflight. The mufflers were removed from their cages to increase the airflow, but Charles said the animals did not appear to be heat stressed. The temperature in Spacehab peaked at 84 degrees F: "Tolerable, but not very comfortable," was how the astronauts described it.

Of course comfortable temperature is a relative term. A message to the crew from the FREESTAR team in Maryland indicated that they might have appreciated warmer temperatures. It read, "The folks in the FREESTAR POCC are envious of your 80° temps. The overnight forecast for Greenbelt is a low of 8° F and a wind chill of −5° F. The Hitchhiker Ops Director is threatening to institute nightly Polar Bear runs for the entire POCC! That sure would be timely with your first SOLSE North Polar view today. Good Luck!"

Later in the flight, engineers came up with a system where valves were adjusted to increase the cooling to the experiments without changing the air temperature within the module. On Wednesday January 22 McCool spent much of his time turning the heat exchanger knob to the left or right to adjust the flow. Several times, McCool was asked to travel back to the Spacehab module to turn the knob further in addition to his normal tasks on the shuttle's flight deck. Capcom Charlie Hobaugh asked McCool his location and Willie replied, "I'm on the flight deck but can be wherever you want me to be." Hobaugh told him, "If you can superman down to the Spacehab and give me a call there I've got some more actions for you," and Willie replied, "One of my greatest pleasures is zipping down that tunnel and doing aerobatics en-route, so I'd be glad to do it." Hobaugh said, "We're going to give you a lot of pleasure today to fix it."

Later Hobaugh called Willie and asked him his location. McCool replied, "I'm on the middeck so I can go either way (up to the flight deck or through the tunnel to Spacehab.)" Hobaugh was apparently amused that Willie was halfway between his two work locations and said, "Rats. I'll need you in Spacehab. When you get there we're going for eight, that's five fingers on your left hand and three on your right, or however you will like to count them up. That will also be counterclockwise." McCool chuckled and said, "Okay, I'll do my best to count to eight counterclockwise." Hobaugh replied, "Hopefully we'll just stay with fingers and not have to make you go to toes." Later Hobaugh asked Willie for additional clicks, but never more than eight at a time. For an audio clip of Hobaugh teaching McCool how to count, go to this book's website.

McCool's efforts to turn the heat exchanger knob to the optimum position ensured adequate cooling for the experiments and high enough a temperature to prevent dew from forming without being too uncomfortable for the astronauts.

Once the high heat load experiments were finished, the engineers breathed a sigh of relief—the rest of the mission could continue with a comfortable environment for the astronauts without any additional changes.

CHAPTER 23

Willie's wakeup music on flight day five was "Fake Plastic Trees." It illustrated that shuttle wakeup calls are more for tradition than necessity: It's rare that astronauts are actually awakened by the music. Most of the time, they're already up and starting their day, and in the rare case when they're sleeping really deeply, it's usually a crewmate's knock on their sleeping bunk's door that actually gets them up. This day, McCool was already up and had started his morning hygiene. He apologized, "I was a little late in responding [to the wakeup music] because I was shaving in front of the mirror."

The red shift wakeup call was "Amazing Grace," on bagpipes, for Laurel. She said, "It's great to wake up to the sound of the pipes and inspirational music. That song brings back many wonderful memories for me, and added on another one today. My husband and I got engaged in Scotland, and I send my love to my husband Jon and my son Iain, who has a name originating in Scotland." Clark was based in Scotland as a Diving Medical Officer, where she assisted with medical evacuation from submarines. Laurel loved Scottish culture, even having bagpipes perform at her wedding. Jonathan Clark later noted that bagpipes had performed "Amazing Grace" at their wedding, for Laurel's wakeup call in space, and at her funeral.

The VCD (Vapor Compression Distillation) experiment was a prototype water purifier for use on long-duration spaceflights like ISS or missions to the Moon and Mars. One of the critical supplies for any long spaceflight is water. Besides drinking and hygiene, water can also be used to create breathing oxygen. VCD was contained in a refrigerator-sized rack. It used a partial vacuum to vaporize wastewater and then distill the purified water vapor back into liquid form. The concentrated urine that remains is a waste product that's dumped overboard or pumped into storage containers for disposal. The process removes 97 percent of the urine solids. Manager Cindy Hutchens said, "You could probably get by with drinking it, it wouldn't be all that good." An operational system would use additional purification techniques like multifiltration beds and heating the water again to remove volatiles before the recycled water could be used like fresh water.

If it sounds complicated—it is. VCD used a pump to create its vacuum instead of just hooking up a hose to one of the Spacehab vent ports, where it could have used an extremely efficient vacuum pump– outer space— for free. The combination of pumps, heaters, and motors, made VCD a power-hungry experiment, with a lot of waste heat. It used 155 watts, about as much power as a typical home stereo setup—not very much power on Earth, but lots in space, where every watt counts.

VCD is not a technology developed for the space program. It's already used in resort areas and desert regions where little water is available and it's more cost effective to purify seawater than to ship in fresh water. STS-107's VCD experiment used imitation urine—de-ionized water with some salts added, the equivalent of diluted urine. Hutchens said, "It was basically almost a salt water, we had added chemicals to keep the microbial growth down and that added enough solids to the water [to simulate urine]."

VCD should theoretically work in space, but engineers wanted to check it out under real microgravity conditions before putting it aboard ISS as a major life-support component. It was also important to test the hardware under abnormal conditions—would the system continue to function after a power outage or other situations when something went wrong?

VCD operated by itself for four to five hours each day. Some of the telemetry data was transmitted to the engineers; the rest was stored onboard on a laptop computer. The crew was asked to collect samples of the purified water at preplanned times for analysis. Laurel radioed down that a sample she collected was clear, which was a good indication that the apparatus was purifying the water properly. The telemetry also measured the water's electrical conductivity. Water with contaminants conducts electricity better than pure water, so the low conductivity in the telemetry gave the engineers good indications.

Hutchens said, "We operated successfully on Saturday, Sunday, and Monday [January 18-20]. Our data look very similar to that on the

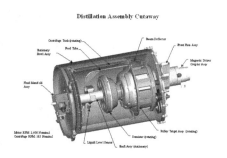

The heart of the VCD is its partial vacuum cylinder, where waste water is vaporized.

ground, so we feel very confident about our hardware. On Sunday, we
did a test to see how it would start up if it lost power, and that appears to
be successful. We're looking forward to getting back our samples and the
recorded data for analysis." Because of the cooling problems in Spacehab, a run scheduled for Tuesday, January 21, had to be rescheduled to
later in the mission.

VCD generated a lot of heat. Imagine running a lot of electronics in a
small room while it's extremely hot and not being able to turn on an air
conditioner. During VCD's last run, Spacehab's cabin temperature rose
to over 80 degrees. The astronauts reported, "It was warm, wasn't unbear-
able but it was warm." The engineers tried to turn off whatever wasn't
needed inside Spacehab to minimize anything else that generated heat.
The thermal engineers were extremely relieved when VCD finished its
six days of operations, since the rest of the flight was experiments that did-
n't generate as much heat. Spacehab's Pete Paceley admits, "I was looking
forward to turning VCD off and getting my module cooled back down!"
Once VCD's final run was completed, the temperatures almost immediately started to come down to a more comfortable 76 degrees F. Paceley
said, "Once we got through all of that—it was a breath of fresh air."

During the mission, the VCD team got some telemetry and the
crew's verbal descriptions. They were supposed to get back their hardware for analysis, the remaining telemetry that was stored onboard, and,
most important, the purified water samples. Hutchens noted, "The
telemetry included a liquid level sensor that showed a high level about 30
minutes into the mission. Was there something wrong with the sensor, or
was it some phenomenon because of microgravity which needed to be
understood?" The team wanted to examine the sensor in a laboratory to
see why it was giving those readings, and if they were true readings and
not a faulty sensor, try to explain why it was occurring. Hutchens added,
"The [astronauts] told us what the water looked like when they took a

Kalpana Chawla
reads a flight plan
inside the
Spacehab Module.
The VCD experiment
is on the left behind
K.C.'s back.

sample. Especially since we didn't get the hardware back, their descriptions helped us a lot with what was going on first hand there."

The VCD team is proceeding with building an operational version for ISS, scheduled for launch around 2008. That unit will include the additional processing to make the water pure enough to drink.

The Microbial Physiology Flight Experiments (MPFE—pronounced "mel fee") studied how bacteria grow in space. Scientists are concerned that bacteria are more resistant to antibiotics in space, a serious issue for long-term astronaut health. The experiment consisted of an incubator and a series of test cards. Each card, which looked a little like a credit-card-size bingo card, contained 30 samples. The samples included three strains of bacteria, five antibiotics, five strains of yeast, and three antifungal agents—chosen to represent a large range of simple organisms. The astronauts inserted the cards inside the AMS 10 (Automated Microbial System) reader/incubator, that would grow the organisms and analyze them. At various points, the astronauts would swap out the cards, and the data from the automated reader was saved on a laptop computer. At least that's the way it was supposed to work.

On January 20, a card tray got jammed in the reader. It took the crew too long to unjam the tray, and the sample was exposed to the cabin temperature for too long so it was considered lost. Another card was jammed two days later and the crew was able to get it unjammed, but it was also considered lost. Five out of six trays were completed, with a total of about eight to ten cards. The setbacks were not as bad as they might have been because many of the samples were duplicated on different cards—just in case samples were ruined for whatever reason. On January 24, one of the crewmembers made what NASA called a "procedural error." A yeast culture was fixed prematurely, ruining that sample. Dozens of samples were processed, and the scientists were anxious to examine the results. Unfortunately, because of the accident all of MPFE's data was lost.

A rare photo of the MPFE experiment with its incubator and card tray holder in the locker on the right and a notebook computer which controls the experiment on the left.

CHAPTER 24

Capcom Stephanie Wilson radioed "Texan 60" for Dave. Brown replied, "It's a beautiful morning looking out the window as we fly over Panama. The blue shift is looking forward to getting up and getting going and getting to work." Later, Rick was awakened by capcom Linda Godwin with "God of Wonders," by Steve Green. He replied, "We appreciate the great words. Boy, looking out the window you really can tell he is a God of wonders. We sure appreciate being able to take a look out and enjoy the view. We're looking forward to another great day in space."

NASA offers the opportunity for international astronauts to talk to high-ranking officials. The VIP videophone call with Israeli Prime Minister Ariel Sharon and Education Minister Limor Livnat was scheduled for January 21. Columbia was connected via satellite links to Mission Control and then ordinary international phone lines to Israel.

In Israel, many took offense because national elections were taking place, and some questioned whether Sharon got unfair publicity talking to Ramon on a very public phone call to space. Ramon said in an earlier interview something that would have ruined an American astronaut's career—he didn't care about the elections: "I haven't voted and will not vote. I will send my father back to vote on behalf of me."

What was amusing is most of the conversation between Sharon and Ramon was in Hebrew. Husband, Chawla, and Clark floated and smiled before the camera, undoubtedly wondering, "Why am I here just taking up space on camera when I could be doing something more useful?" At one point the interview switched to English, and Limor Livnat noted she was very pleased that there were two women on the crew, earning smiles from Clark and Chawla. Sharon invited the entire crew to visit Israel, and Rick Husband replied, "We will certainly be looking forward to that after the flight." Sharon said, "I can assure you that you will find yourself among friends. We are looking forward to meeting you in Jerusalem— the capital of the Jewish people for the past 3,000 years." Husband acknowledged, "If all Israelis are as nice as Ilan and his family are, then I know we can expect a very warm welcome when we come to Israel, and we look forward to when the time comes."

During the interview, Ramon held the tiny Torah that survived the Holocaust. It was on loan from MEIDEX scientist Joachim Joseph.

Israeli Prime Minister Ariel Sharon and Education Minister Limor Livnat talk to the red shift on Columbia.

Ilan Ramon holds up the miniature Torah which was owned by MEIDEX scientist Joachim Joseph. Photo from author's collection.

Ramon concluded the interview by asking every Jew around the world to come to Israel and plant a tree during the coming year, with 40 million trees planted by the anniversary of Columbia's launch.

Ramon probably summarized anything interesting for his crewmates in English.

NASA provided English translation for the NASA TV viewers later that day. It was, all told, the typical "We're so proud of the great job you're doing in space and representing our country" interview with a head of state, which has been done dozens of times on the shuttle, whenever international space travelers fly. Go to this book's website for the original Hebrew and overdubbed English versions of the interview.

NASA's Internet website offers the public an opportunity to send a question to the astronauts. But there's a catch—first the question has to be selected by a NASA public affairs officer, and then the astronauts have to decide whether or not to answer that particular question. The public affairs officers tend to select simple, bland, non-controversial questions. Public affairs will never select a technical question. And strangely, they often select questions about items for which the crew are not responsible, like automated experiments they don't touch during the mission. Public affairs also seems to have a bias toward selecting questions from people with some connection to the crew—relatives of the astronauts, the teachers of the astronauts' kids, and old friends. The questions are emailed to the crew along with their flight plan updates. The astronauts select the questions they want to answer. Willie and Dave answered just 3 of the 30 questions selected by the public affairs personnel, even though they had plenty of time to answer more. The astronauts read the question out loud, along with the name and age of the person asking the question, and then respond with a fairly short answer. NASA puts a transcript of the question and answer and audio clip on its website.

The questions and answers are on this book's website. After Willie answered his two Internet questions on Flight Day 6, capcom Ken Ham thanked him, saying, "Copy that Willie, great job. You're my hero.

CHAPTER 25

Willie's wakeup call on flight day 7 was Peter Paul and Mary's "The Wedding Song," radioed up by capcom Charlie Hobaugh.

NASA's rules call for astronauts to get two half-days off on longer shuttle flights. Typically astronauts use their off time to talk to their families, send e-mail to friends, and spend most of their time looking at the Earth. The first off period was four hours on flight day 7.

One of the more unusual experiments studied earthquakes—not observing them on the ground, but simulating them in a quart-size container of fine sand. The study could also be used to predict how grain acts in a silo, or even how sand castles are made. (How the sand-water mixture acts under pressure can be studied very precisely in space.)

The Mechanics of Granular Materials (MGM) experiment consisted of a latex cylinder with about a quart of fine quartz sand mixed with water. The latex was marked with a grid pattern, and video cameras recorded the experiment from three sides. During the tests, pressure was applied very gradually to the cylinder, and the cameras monitored how the shape of the cylinder changed as the cylinder reacted to the extremely gradual increase in pressure. An experiment run could take as much as 90 minutes, since the scientists wanted to monitor what was happening in as much detail as possible. Scientist Stein Sture, with the University of Colorado at Boulder, said, "It's like watching grass grow." But once something interesting happened, the experiment could be stopped, with the sand and cylinder "frozen" in place: without gravity, every grain of sand remains in place and doesn't move.

After the first test, the sand collapsed due to the pressure. What could Sture do? He asked the astronauts to hit the cylinder sharply several times to get rid of the bulge. The crew transmitted a video of Rick's efforts to "spank" MGM to remove the bulge, and Laurel quipped, "This gives new meaning to packing sand in space." Sture said, "You could say the old time-tried spanking the kid was really an attempt [the crew] did. I think we may have overdone it because we ended up with a dramatic kink in the specimen, it went unstable in some way. Later on we found out; leaving the specimen [on its own and letting the grains] drift all on their own, worked much nicer." So, perhaps in this case spanking the experiment wasn't the best solution. This book's website includes a video clip of Rick Husband "spanking" MGM.

MGM had nine runs planned, but ended up getting a bonus run because there was enough time. Two of the runs successfully simulated how solid ground can become similar to a liquid during an earthquake, and a new technique was developed so the experiment could be reset and reused, increasing the number of tests that could be performed during a mission. One of the specimens was recovered in Columbia's debris and was able to return useful science.

The wakeup call for the red shift was "Prabhati," by Ravi Shankar, sent up for K.C., who replied, "Good morning. Linda, and thank you very much. That's a very favorite piece of music of mine, and I think it's titled "East meets West" and it fits this because we're going around the Earth all day long. Thank you very much for the cosmic music, and we're ready to go."

I've been asked many times whether or not all of NASA's microgravity experiments are really justified. Do the experiments really deliver on the promises NASA makes? Certainly many of the claims are hard to believe. At a 1992 press conference on zeolite crystals, NASA noted that a 1 percent improvement in the efficiency of refining gasoline would save billions of dollars each year. A nice figure, but nothing whatsoever to do with the particular experiment: in no way was its goal to increase the efficiency of gasoline production or create a zeolite that could. A status report during a 1995 mission stated, "Some of the zeolites are the type used to crack crude oil into refined petroleum. Increasing their efficiency could result in cheaper gasoline"—ignoring the fact that increasing the efficiency for oil refining zeolites was not one of the mission's goals.

In other well publicized instances, NASA promoted John Glenn's flight on STS-95 as research into aging, and Coca-Cola was flown on the STS-63 and STS-77 flights to test how taste buds change, and it was even mentioned that the "experiment" could lead to improving Coke's taste.

Rick Husband "spanks" the MGM experiment to try to remove a bulge in the sand.

But when the people who make these claims are pushed to show actual results, they tend to become silent or claim it takes time for the researchers on the ground to complete the scientific investigation and any practical benefits will occur some unspecified time in the future.

The STS-107 experiment I really questioned was Astroculture—attempting to create new flavors and fragrances in space for new perfumes. The flowers were a hybrid rose (Jerry-O) and rice flower (Agalia Odoratae). They were grown to see if their essential oils have aromas different from flowers grown on Earth. Professional snifters working for perfume companies would decide if the new fragrance was any good and should be used in new perfumes. If the "noses" like the aroma from the space flowers, their oils would then be chemically analyzed and artificially reproduced. Astroculture was hyped as enabling "research to literally create scents that are out of this world."

At technical shows, Astroculture's promoters encourage attendees to do a smell test. They're presented with two bottles of fragrance oils—one ordinary rose oil and one that's described as "grown on John Glenn's flight," and the subject is asked which one smells better. Not surprisingly, in this highly unscientific test, most people preferred the version created in space better—mainly because they're told in advance which one is which! That's what real scientists mockingly call a "double non-blind experiment"—both the person promoting the test and the subject are aware of the contents of both bottles, aware of the test's objectives, and already have a built-in bias for which is "better." Growing flowers in space to create new smells for perfumes isn't science—it's marketing.

One of my broadcast media clients informed me on flight day 7 about the piece of debris falling off the ET and hitting the wing. But we weren't ready to air a story yet. I was instructed to not ask anything about the incident at the mission status briefing unless the NASA manager giving the briefing brought it up, or some other reporter asked about the incident. The debris was not mentioned at the briefing, so I did not ask any questions about it, but I did make some discreet inquiries afterwards about the availability of the video showing the foam hitting Columbia. I was informed that it was no big secret, and many people were aware of it. But by the next day I was told to "stand down" on the story because it had become a non-issue—the engineers were determining it was not a safety issue.

CHAPTER 26

F L I G H T   D A Y   8 :
S O F B A L L ,   C I B X

Mike Anderson's daughters Kaycee and Sydney chose "Hakuna Matata" from the movie "The Lion King" as the wakeup call for their father.

After the soot experiment was finished, the combustion module was reconfigured for SOFBALL (Structure Of Flame Balls At Low Lewis-number), an experiment that examined how extremely small flames act in space. Flame balls occur only at low Lewis numbers—a technical term which describes a small amount of heat and relatively large mass diffusion. In plain English, flame balls exist only under a rare combination of factors, like those found in microgravity. What makes flame balls unique is they're stable, spherical, tiny, and just float in place—hardly terms that describe ordinary flames.

University of Southern California (USC) scientist Paul Ronney first detected a flame ball in a "drop tower" experiment in 1984. NASA found the flame balls interesting enough to fund a space version, which flew on the Microgravity Science Laboratory mission in 1997. The experimental findings were interesting enough that NASA decided to fly the flame ball experiment again on STS-107.

Ronney noted that flame balls are the most fundamental forms of combustion. They're the simplest, smallest flames, and that makes it easier to predict their behavior and how they'll burn under different conditions. If you look at a flame in a fireplace or campfire, it's extremely random because the wood is not a uniform material—a piece of wood contains a variety of impurities, and is far from consistent in its composition and structure. A candle or gas flame is more consistent and predictable, but flame balls are even more so. By studying flame balls, scientists hope to learn more about how combustion occurs. It's science for the sake of pure science with actual real-world applications.

The SOFBALL team selected a mixture of fuel, oxygen, and inert filler gases for each flame ball run, based on their science goals. How the flame balls actually acted was often different from what the scientists predicted, giving them more to think about.

Ronney pointed out that single flame balls are the best for theoretical research, since their actions are the easiest to predict and analyze. However, multiple flame balls are often the most fascinating, especially how they interact with each other. The astronauts set the combustion module's controls for each test based on what the SOFBALL scientists wanted; 39 tests were performed with a total of 55 flame balls.

SOFBALL set several records, including the weakest flame ever—half a watt, about one hundredth the amount of heat in a birthday cake candle. Another SOFBALL was the leanest flame ever burned—just 8 percent fuel and 92 percent air. In contrast, an automobile engine is about 70 percent gasoline and 30 percent air.

There was a mysterious drift, where flame balls would float off to the sides of the chamber even with the excellent microgravity conditions on the shuttle. Once they hit the walls they were immediately extinguished. Several tests were expected to last for more than an hour, and one was planned for two and a half hours, but none lasted more than 25 minutes because of the drift. Scientists replanned their tests to ones that could be completed before they were cut off by the unexpected drifts.

In the middle of the SOFBALL runs, the astronauts apparently decided to liven things up by giving the flame balls names. Dave announced, "I see one flame ball and its name is Howard." The SOFBALL team wondered, "Why did he give it a name, and who's Howard?" and each of them scribbled down the name in their logs. On Friday January 23 at 11:10 P.M. Dave proudly declared, "This [one's] a girl, and its name is Samantha." Nobody knows why the crew started naming the flame balls, they had never done it during the simulations or the earlier runs on the mission, so it appears to have been a spur-of-the-moment whim on Dave's part. The next day Dave announced, "We had a pair of twin girls—Rachel and Elizabeth," and later Mike named two SOFBALLs "Anne" and "Ed." If for some reason the astronaut didn't assign a name, the POCC would radio up names. At one point they told the crew, "The first has been named Ilan Ramon and the second has been named Mike Anderson." Kalpana Chawla asked, "Which one is the bigger one?" and the reply was, "It's your call." Chawla countered with, "You're avoiding the tough question."

Mike Anderson adjusts the video monitor for a SOFBALL run where three flame balls were created.

Ilan decided to name a pair of flame balls after two important combustion researchers—"Zeldovich," after Russian physicist Yakov Zeldovich, who first predicted the theoretical existence of flame balls in
1944. and "Paul Ronney," after the lead SOFBALL scientist. who discovered them 40 years later. Ronney notes that his flame ball turned out to
be "a weak and wimpy flame ball that lasted only a few minutes, while
Zeldovich lasted far longer."

The very last SOFBALL run was a pleasant surprise. Ronney wanted
to finish the experiment on a high note, with a whole bunch of flame balls,
and to his great pleasure nine appeared. Dave announced, "This is a tough
one, we think we've got a whole control room full of flame balls here. So
we're going to call this the Orbit 2 flame ball crowd." Orbit 2 was the Mission Control shift on duty. Beth Vann replied, "We copy Dave, we can get
all of those flame balls named in the mission summary that we send up."
Dave responded, "Okay, the lead flame ball is named Kelly though" (after
flight director Kelly Beck) and Vann replied, "Obviously." But Mission
Control apparently forgot Dave's instructions to name the main flame ball
Kelly. One of the crew's updates announced "CM-2 SOFBALL
announces the birth of 9-tuplets: Mike, Terri/Terry, Steve, Mark, Jerry,
Frank, Vic, Amy and Hung. Names are compliments of the Orbit 2 team."

Kelly was an unexpected surprise. The scientists predicted that the
test would not last more than 50 minutes and planned it with enough
fuel for a bunch of flame balls. One by one, eight of the flame balls
drifted to the side and were extinguished. Except Kelly, which seemed
immune to drift. Ronney notes, "It was just a beautiful flame ball. It was
just sitting in the middle of the chamber like it was nailed down."

So Kelly had all of the fuel intended for the group and kept burning
and burning. When the POCC asked Mike to shut off the experiment,
over Ronney's objections, Mike exclaimed, "But it's still burning!" wondering why they would want to terminate the unanticipated and fascinating run. After he got verification from the ground that they actually did

The last SOFBALL run with nine
separate flame balls.

want him to turn it off, Mike waited several additional minutes before stopping the experiment. Kelly burned for a record-setting 81 minutes, almost a complete 90-minute orbit around the Earth.

After the accident, Ronney speculated about what was going through Anderson's head. He said, "Mike knew I wanted to fly a flame ball around the world, so he was undoubtedly astonished when he was instructed to shut it off. He apparently let the experiment continue to run for another several minutes until the point where he thought if he let it continue any longer he would jeopardize his next flight assignment...." Ronney says he was looking forward to talking to Mike at the postflight debriefing to verify if his suspicions were correct.

Ronney said, "Kelly almost made it [around the world]. Kelly's experience is a fascinating example of group dynamics among flame balls. She was created in a gaseous mixture of hydrogen, oxygen, and sulfur hexafluoride (the inert filler). All the others began drifting around the chamber, looking for "food," competing with one other, while Kelly remained motionless at the center. Before long, the others were exhausted; they had drifted too close to the walls and winked out. Kelly was left all alone with a chamber full of fuel." NASA's excellent education website put out a story, "A flame ball named Kelly," about the SOF-BALL experiment. The story is on this book's website.

All together, the 39 SOFBALL runs lasted 6 hours 15 minutes. The astronauts and payload controllers named 33 of the 55 flame balls. Ronney sent the astronauts a summary of the preliminary SOFBALL results with his thanks for their hard work. He told them, "The data obtained during the mission will keep combustion scientists busy for many years to come and will help to lead to the development of cleaner, more fuel-efficient engines as well as improved methods for spacecraft fire safety assurance."

Ronney noted that there were already new discoveries. Oscillating flame balls, predicted theoretically about 15 years earlier, were observed. A completely unexpected result Ronney is still trying to explain is why the flame balls drifted. It was frustrating, because all of the flame balls except Kelly would eventually drift, limiting the maximum burn time. Ronney's team was able to adapt, canceling some of the ultra-long tests and replacing them with more medium-length tests. But they still want to determine why the drift occurred.

The flight day 8 red shift wakeup was "Ma ata osheh kesheata kam baboker?" by Arik Einstein, for Ilan Ramon. The Israeli song asks, "What do you do when you wake up in the morning?" For Ilan the answer was, "I'm smiling here, everybody's smiling here. And we're happy for another great day in space for great science."

ITA (Instrumentation Technology Associates) has flown several microgravity experiments on small suborbital rockets, Russia's Mir space station, and the space shuttle including STS-107's CIBX-2 (Commercial

Instrumentation Technology Associates Biomedical Experiments). ITA's shoebox-size DMDA (Dual Materials Dispersion Apparatus) is a miniature laboratory that mixes various chemicals together. An astronaut activates the experiment, and the laboratory automatically mixes ingredients for dozens of different experiments. After several days in operation, the astronaut twists the knob that mixes a fixative into the experiments to deactivate them. It doesn't take much astronaut time, and a single knob can operate many different experiments at once. All together, 480 tubes filled with a variety of samples were flown in the two DMDAs. The experiment also included one Liquids Mixing Apparatus (LMA), where an astronaut manually pushes a bunch of syringes to mix chemicals together. The two DMDAs were found with Columbia's debris, but the LMA was never located.

The CIBX experiment that got the most attention was Planetary Society's Growth of Bacterial Biofilm on Surfaces during Spaceflight (GOBBSS, pronounced "gobs"). The experiment had an interesting twist—two of the team members were Yuval Landau, an Israeli medical student, and Tariq Adwan, a Palestinian from Bethlehem attending college in the United States. Planetary Society head Louis Friedman said, "They had this idea to involve students, a Palestinian student and an Israeli student in a peace experiment—the symbolic value of space exploration as an inspiration for youth to get them involved." The scientific purpose for GOBBSS was to evaluate whether or not bacteria could attach itself to an imitation asteroid in microgravity. If the answer was yes, then it could lend weight to the theory that life on Earth first arrived as a "hitchhiker" on an asteroid. Other scientists involved in the experiment included Dr. Eran Schenker of the Israeli Aerospace Medical Institute, Dr. Johnny Younis of Poria University Hospital in Nazareth, and Dr. Ahmed Tibi, a physician and Arab member of the Israeli Knesset. In the United States, Dr. David Warmflash and Dr. David McKay of NASA's Astrobiology Institute helped design the experiment.

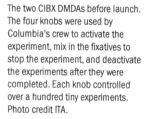

The two CIBX DMDAs before launch. The four knobs were used by Columbia's crew to activate the experiment, mix in the fixatives to stop the experiment, and deactivate the experiments after they were completed. Each knob controlled over a hundred tiny experiments. Photo credit ITA.

Southwest Texas State University flew a different biofilm experiment.

NASA sponsored two of the CIBX experiments. Dr. Dennis Morrison flew urokinase crystals for an anti-cancer drug and microencapsulated anti-tumor drugs and antibiotics. The microencapsulated drugs were the ultimate versions of time-release cold capsules, with a thin soluble coating surrounding a small amount of medicine. In the body the coatings gradually dissolve and the drugs are released. Another urokinase experiment was flown for the Oklahoma Medical Research Foundation.

An experiment from the Israel Aerospace Medicine Institute (IAMI) and Materna, Ltd., was designed to study bacterial reactions in milk and yogurt products. Another Israeli experiment sponsored by the IAMI and the Hadassah-Hebrew University Medical Center studied osteoblasts — bone formation cells. Oakwood College and Biospace Group in Huntsville, Alabama sponsored an experiment to study the regeneration of nerve cell growth factor.

Educational experiments inside CIBX included inorganic crystals for a group of students in Texas, Utah, and New Mexico, tin crystals for a school in Pembroke Pines, Florida, and an experiment for the Milton Academy in Massachusetts to determine how resistant bacteria are to antibiotics.

The astronauts normally leave their video cameras in place. Mission Control calls the astronauts as a courtesy when they're receiving video from the crew cabin. Hopefully astronauts won't get caught unaware and accidentally do something embarrassing on camera. On STS-107, Laurel radioed Mission Control to ask if the camera inside Spacehab was being transmitted to the ground. Capcom Charlie Hobaugh replied, "I don't know—It depends on what you're going to do," which made Laurel laugh. She was sitting on the ARMS ergometer and wanted to explain what it was used for. This book's website has a video clip of Clark's reaction to Hobaugh's off-the-cuff quip.

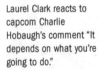
Laurel Clark reacts to capcom Charlie Hobaugh's comment "It depends on what you're going to do."

CHAPTER 27

FLIGHT DAY 9:
S * T * A * R * S , FREESTAR

Dave Brown's wakeup on flight day 9 was "Burning Down the House," by Talking Heads. Capcom Charlie Hobaugh assured Brown, "Your house is safe, we don't want you to think it's not." Dave replied, "We're doing quite a bit of combustion experiments up here. So we hope not to burn down either my house at home or anything here on the shuttle. But we are going to be doing a lot of burning and hope to make a lot of contribution to combustion science today."

The S*T*A*R*S students watched video and downloaded images from the Internet, showing the progress of their experiments. The Australian students noted their spider was active, but they did not see any fruit flies in the video. They put in a request for the astronauts to inspect their experiment and see if they noticed any flies. One of the four motors on the first half of the Israeli chemical garden failed to work properly — it didn't push one of the two blue starter crystals in front of the camera. It wasn't a concern, though, because the experiment had another crystal with the same colors that was working properly, and another set of four crystals was about to be activated.

Ilan opened up the S*T*A*R*S experiment on January 24. The POCC passed on the request from the Australian team to look for fruit flies, in addition to his planned tasks. Ilan was already scheduled to collect some of the spider's webbing for postflight analysis, along with rotat-

A close-up view of the blue colored chemical garden crystals on flight day 8. The left half of the image shows the two ground crystals, the right half has the flight crystals. Note that one of the two rods has failed to extend. Photo credit Spacehab S*T*A*R*S program.

ing the Israeli chemical garden. The chemical garden is about the size of a thick paperback book and has two sets of vials for the crystals. Ramon turned the experiment around so the second side faced the video cameras. Afterwards the Israeli students requested engineers to send the command to activate the second half of their experiment.

Ilan reported, "Access 2 is completed successfully. As far as fruit flies, I couldn't see any of them anywhere. I looked pretty carefully." He noted a pleasant surprise, "On the silkworm from China we noticed one [cocoon] that actually hatched and we have a butterfly in there." He reported that he couldn't see the fish, but that could have been because of his viewing angle and he added, "The bees and the ants and the spider are alive, and the chemical garden is blooming. Thanks for the whole S*T*A*R*S team and all of the students all over the world." A copy of Ilan's remarks is on this book's website.

While Ilan was responsible for most of the S*T*A*R*S activities, his crewmates were also interested in examining the experiment. Later in the flight, Laurel described her excitement at seeing a newly hatched moth. The next day Ilan narrated a video showing the S*T*A*R*S activities.

S*T*A*R*S was allocated times to downlink some of its video, and it attracted a lot of attention. One of the most fascinating to watch was the ants as they crawled/floated through the gel and carved their tunnels. Paceley noted, "I knew the ants experiment were visually going to be the highlights of the flight—they looked cool. Everybody was interested in the ants." Paceley recalls that PhD scientists who had worked on their own experiments for 20 years asked him about how the ants were doing.

I had an invitation from Spacehab to view the backup locker with identical experiments as they were operated on the ground while Ilan performed his activities aboard Columbia. Engineer Mark Rupert shut off the power and opened the hinged top to get access to the six S*T*A*R*S experiments. He inserted a 6-inch rod into the spider chamber and used it to collect some of the webs. Then he took out the spider

The ground control spider experiment.
Photo by Philip Chien.

The ground chemical garden showing the tendrils growing up. Photo by Philip Chien.

chamber and we both examined it closely for fruit flies, but could not see any. Rupert said, "It's up to the student investigators to tell us why the fruit flies aren't there." Whatever problem was preventing the flies from hatching in space may have also caused the same problems in the control unit on the ground.

Rupert then disconnected an electronics cable and removed the Israeli chemical garden. The tiny crystals showed a clear difference from their space-growing counterparts. All of the crystals had a sphere about the size of a pea, but beyond that the space and ground samples were completely different. The ground crystals had fibers growing upwards, away from gravity, while the space crystals grew fibers in a variety of directions.

While it wasn't part of the official activities, Rupert also showed me the ants experiment. On Earth, once the ants were introduced into the chamber they fell to the bottom. They had to climb about 4 inches to reach the opening to their "ant farm." The ground ants had barely succeeded in tunneling one tiny path, about 2 inches long. In contrast, the space ants were incredibly active, making many different tunnels and even moving many small chunks over to the other side of the barrier.

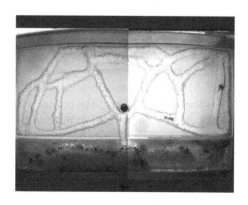

The S*T*A*R*S ants were extremely popular as they made their trails.

Paceley said that during the S*T*A*R*S rehearsal in 2002 the ants did a far better job of climbing up to the entrance of their "farm." For whatever reason, the ground ants during the mission weren't as successful in climbing into their farm. He noted that it might have given a misleading impression during the mission to folks who saw how much better a job the space ants did than their controls on the ground.

The red shift wakeup call wasn't dedicated to any particular astronaut. It was the novelty song "Kung Fu fighting," selected by the STS-107 training team for their crew. If anything could be called a theme song for the crew, that was it. They had a toy hamster that played the song, and they carried it everywhere as their mascot. Rick said, "We really appreciate it. Our training team has been such a fantastic group. We owe a certain debt of gratitude to each and every one of them. We appreciate everything they've taught us, and we know that all that training is paying off as we go through this mission. We want to thank them very much for that. It's great to hear that music as well. It's been one of the enjoyable jokes we had as we've gone through our training. We really appreciate it."

Most of the crew's activities were hands-on experiments within Columbia's cabin or Spacehab. But several experiments didn't need crew interaction or needed to be outside to expose a sensor or antenna to space.

FREESTAR's six experiments were mounted on a truss in the cargo bay behind Spacehab. The primary astronaut activity for four of FREESTAR's experiments was to orient Columbia in a particular attitude and turn it on. In effect, Columbia was a giant pointing platform. SOLSE (the Shuttle Ozone Limb Scattering Experiment) was pointed toward the horizon to measure the amount of ozone in the atmosphere at different altitudes. SOLCON (Solar Constant) needed to be aimed toward the Sun. LPT (Low Power Transceiver) was pointed at a variety of targets—ground stations or a TDRS in a 22,300-mile-high orbit.

MEIDEX was pointed toward Earth while observing dust storms, or toward the horizon while searching for lightning storms. Normally an astronaut would aim MEIDEX's camera toward areas where dust or lightning appeared. In some cases an engineer on the ground would watch the MEIDEX video transmitted from Columbia and use a computer to transmit commands back to MEIDEX to aim the camera toward possible targets.

As Columbia flew over Australia on one pass, Rick Husband reported, "We caught some lightning storms at the beginning of this pass, but we haven't seen any since. But we'll keep an eye open." Mission Control radioed that there might be thunderstorms to the north. Rick replied, "Okay, we copy, full gimbal in the positive direction." In plain English, he said he was going to rotate the camera all the way to one side to point it toward those thunderstorms.

SOLSE had previously flown on STS-87, K.C.'s first spaceflight. Current weather satellites measure the atmosphere's ozone by looking straight down. This gives the total amount of ozone over that location but doesn't give any information about how the ozone is distributed through different altitudes. In effect, it's like trying to examine a multilayered transparent cake by looking down from above—you can tell the thickness of the cake, but not the composition or thickness of its individual layers. The concept for SOLSE is to look sideways, toward the Earth's horizon, at the limb or very edge. The atmosphere in the Earth's limb is an incredibly thin layer of different colors, and it is spectacular at sunrise and sunset—"More colors than you could describe," as more than one astronaut has said. The SOLSE instrument looks at the limb, viewing ultraviolet colors the eye can't see. To use the cake analogy, it's like looking at a slice to see what's in each layer, since SOLSE detects the concentration of ozone at different altitudes.

A complementary instrument was LORE (Limb Ozone Retrieval Experiment). LORE looked at the fine details, while SOLSE looked at the chemical makeup. Scientists flew SOLSE to check out the concept of examining the atmosphere from the side view. The NPOESS (National Polar-orbiting Operational Environmental Satellite System—pronounced "en poes") next-generation weather satellite's Ozone Mapping and Profiling Suite (OMPS) is an operational version that will generate three-dimensional views of the ozone in the stratosphere and display how it changes over time.

Columbia's orbit took it 39 degrees north and south of the equator, never traveling any further north than New York City or further south than Melbourne, Australia. At various times, Columbia rotated to point SOLSE toward the North or South Pole and swept across the polar regions to aim SOLSE toward the arctic ozone holes for measurements. Balloons launched from Colorado, Hawaii, Manitoba, Labrador, and New Zealand during STS-107 provided additional "on the spot" data.

Engineers estimate that only 12 percent of the SOLSE data was transmitted to the ground, but 70 percent of LORE's data was received. Overall, the scientists say most of SOLSE's objectives were met, the key missing factor being SOLSE data at all altitudes.

The Belgian SOLCON experiment studied the sun's energy. Scientists thought the Sun's output stayed at the same level and called it the "solar constant." But more precise instruments showed that there were tiny fluctuations, responsible for climate changes. The Earth's atmosphere absorbs or reflects much of the Sun's energy, only transmitting visible light and certain radio frequencies. It's desirable to have an instrument in space to study all of the Sun's energy. In addition, it's especially important to study the Sun at several different points during its 11-year cycle, including periods of high and low activity. SOLCON has flown on six shuttle missions, plus the European Eureca and SOHO spacecraft, providing its scientists with a large amount of data over a broad period.

The timeline called for SOLCON to get 18 solar observations—two individual observations of the Sun and eight pairs where Columbia pointed SOLCON at the Sun, traveled through the Earth's shadow, and saw the Sun again on the opposite side of its orbit.

All of the data was collected from SOLCON, so the experiment was a 100 percent success. Even more important, SOLCON's many shuttle flights, starting in 1983, spanned two 11-year solar cycles, in which the Sun goes from very little activity to very high activity and back to little activity.

The Critical Viscosity of Xenon (CVX) experiment needed to be maintained just under room temperature, so engineers responsible for Columbia's attitudes had to be careful not to point CVX toward the Sun for extended periods—even though it was sitting right next to SOLCON. CVX's cans were covered with reflective tape to keep the temperature cool inside, and a small heater was used to control ultra-precise temperatures for the experiment.

The critical point for a substance is the temperature and pressure at which the liquid and gas phases become completely merged—and it's impossible to tell the differences between the two. Many properties change radically. Viscosity is best described as how slippery a liquid is. Tar is a viscous liquid, water isn't. The CVX measures the viscosity of Xenon. Xenon was chosen because it's safe, it had been used in similar shuttle experiments, and its critical point at 62° F (16.7° C) is within a practical temperature range for a simple automated experiment.

What's fascinating to the scientists is that near the critical point, the viscosity doesn't behave according to theoretical predictions. This isn't because of anything mysterious—the theoretical models just aren't detailed enough for very specialized cases. By learning more about what actually happens to Xenon near the critical point, better theoretical models can be created.

CVX made five test runs where the chamber was precisely taken through the critical point where the temperature was increased in extremely tiny steps. The ultra-precise temperatures had to be controlled to an accuracy of just 0.005 degrees F at 60 times normal atmospheric pressure.

The Low Power Transceiver (LPT) successfully communicated through TDRS and ground stations. LPT also received GPS signals to determine the shuttle's location, and was tested to determine how it could be used on future rockets. The problems were minor and showed the team how they could make improvements for future tests or use versions of LPT on operational satellites. It was an unqualified success, returning all of its data to the ground during the mission. Even before Columbia's reentry, the experimenters already had all of their data.

One of LPT's most important goals was Mobile IP, where a satellite can transmit its data to a ground station without any concerns about what route the data takes to gets there. Similar to mobile Internet applications on Earth, LPT was able to transmit test data via satellites and ground stations back to the command station. The sample files were drawings made by school children. A file might be sent to the Wallops ground station in Virginia and transmitted to LPT, and then retransmitted through TDRSS and eventually back to the command station. All of the ground communications used NASA's internal Intranet communications network. The key advantage to an approach like this for future spacecraft is the ability to use off-the-shelf software and hardware for communications instead of costly customized software and hardware. Like Mobile IP, the connections are transparent to the user.

Another FREESTAR experiment originally planned for STS-107 was the Prototype Synchrotron Radiation Detector (PSRD). It looked for cosmic rays as a test for the Alpha Magnetic Spectrometer (AMS), an antimatter experiment that will be mounted on ISS's exterior. Because of STS-107's many delays, NASA decided to move PSRD to an earlier flight, the STS-108 mission that flew in December 2001. Scientists from the United States, Germany, Switzerland, Taiwan, and Korea cooperated on PSRD. Scientists from many additional countries are participating on AMS.

Instead of wasting the space which PSRD would have occupied, NASA decided to fly a "Space Experiment Module," or SEM (pronounced "sem"). A SEM is about the size of a 55-gallon barrel filled with up to ten student experiments. These are typically passive samples loaded in vials, about the size of large prescription medicine bottles. Students select the materials they want to fly in space and compare them with control samples kept on the ground. It's not very complicated science, but it helps introduce preschool to high school students to science and how experiments are performed—and it's a good use for space that would otherwise be wasted.

One of the SEM experiments was "Urine in Space" from the Shoshone-Bannack High School in Fort Hall, Idaho. Teacher Ed Galindo explained, "We made a machine which uses zeolite. You can put urine in it and run it through the machine, and it captures the urea part of the urine and you end up with usable water, which we call "space water." The idea is you would use this system to filter urine for water to water plants. If you ran it through the system again and again you could get more pure water." While the water isn't pure enough to drink, it is pure enough to use in a variety of applications, including mixing it with pigments to make paint.

The class called their experiment "More Fun With Urine" (the original "Fun With Urine" flew on STS-108 in 2001), but NASA didn't like the name, so it's officially called "Natural Space Art." Galindo said, "What we decided to do this year, we wanted to paint. We used some traditional dyes which we got from the reservation, and some dyes you could buy in the store." He added, "We got the idea from some of our elders. They were talking about rock painting and messages on rocks." The elders also informed the team about how native artists have used urine as a pigment. Apparently a lot of people offered to donate urine, and Galindo had nightmares he would be stuck with a giant container of urine in his classroom, so he decided to use artificial urine instead. The "Native Art Experiment" became the third experiment on STS-107 involving urine, along with Vapor Compression Distillation and the PhAb4 life science studies. But only the last used actual urine.

The other SEM experiments were similar, passive samples selected by school students from: the Frank D. Whalen Middle School and Peter Tetard Middle School in New York City; City College of New York; the Museum of Science in New York City; Ogdensburg Public School in Ogdensburg, NJ; J.M. Bailey School Kindergarten, Bayonne, NJ; East Norriton Middle School, Norristown, PA; Saint John the Baptist Preschool in New Freedom, PA; Central Park Middle School, Schenectady City School District, and Farnsworth Middle School, Guilderland, NY; Bishop Borgess High School & Academy, Redford, MI; and the Country Centre 4-H Space Science Project.

It's no coincidence many of the SEM experiments came from the New York City area. SEM was added to the mission shortly after 9/11, and the SEM team felt they could help New York heal emotionally by giving school children the opportunity to participate in the space program as an uplifting activity. FREESTAR manager Tom Dixon said, "We felt like we could do some benefit for the Manhattan school district kids. You can still forge ahead and be inspired." Many of the students loaded their samples into the vials in a public ceremony at the Museum of Science on January 29, 2002.

The object for the SEM program—getting students interested in science—was accomplished by the time the students loaded their samples in their vials, but because of the delays, many were disappointed or lost interest. However, many students did follow the mission's preparations through the delays, and some traveled to Florida to see the launch. Certainly those who did keep track of what was happening with "their" mission were ecstatic when Columbia finally did launch with their samples onboard.

As Columbia flew over the Goddard Spaceflight Center on January 24, Rick took the opportunity to thank the FREESTAR team. Husband said, "We just flew over Goddard [and] took some pictures. I'd like to take this chance to say howdy to Tom Dixon and Katie Barthelme, and Tammy Brown and thanks for all of their hard work for putting together a great payload. And also I'd like to send thanks Phil Murray and Alan Birch for the great training we got on SOLSE as well." Capcom Linda Godwin returned the thanks from the Hitchhiker team.

CHAPTER 28

FLIGHT DAY 10:
BDS, BRIC, OSTEO

The flight day 10 blue shift wakeup call for Willie McCool was a change of pace: Instead of music from a professional musician or a class choir, the recording was McCool's son Sean playing the Eagles' "Hotel California" on guitar. Sean's former girlfriend, Josee Julian, sang, and her father, Frank, also played guitar. McCool replied, "It's a real pleasure to hear Sean playing the guitar and Josee singing while we're orbiting the Earth at 160 miles. A real treat, so thank you very much for that."

The Bioreactor Demonstration System (BDS) grew an artificial prostate cancer tumor the size of a golf ball—far larger than anything the scientists had anticipated. The bioreactor hardware looks a bit like a rock polishing kit. A soup-can-size cylinder rotates to gently tumble the samples inside. Scientists describe the bioreactor as a three-dimensional Petri dish. In a normal Petri dish, gravity is a major factor, with the cells sitting on the bottom of the dish. Microgravity, where the cells aren't lying on a flat surface, is a better simulator for cell growth within an organism. Bioreactors on Earth need to rotate at about 25 to 30 RPM to even out gravity as much as possible, but that's a little like trying to pour a bag of sand in a wind storm—everything's always being pushed around, and that limits the size of what can be grown. Space bioreactors rotate slowly, so a much larger cell growth can be achieved.

The question posed by Dr. Leland Chung of Emory University was: What does bone marrow contribute to allow prostate cancer tumors to grow? BDS was loaded with prostate cancer cells and bone cells to see how they would interact and create tissue-like structures. The big surprise was how quickly the prostate tumor grew and how large it became. After only a couple of days the microgravity prostate cancer had already grown as large as a week-old growth on Earth. Five days into the mission, Dr. John Charles said, "BDS is going very well, cells are accumulating and clumping spectacularly. Its growth seems to far exceed growth found on similar cultures on the ground." The artificial prostate cancer tissue had grown to a diameter of about 1.25 inches.

The growth was so large that scientists decided to slow BDS's rotation down from 6 to 3 RPM to avoid damaging the delicate growth. Later in the mission, scientist Tom Goodwin noted that while the ground-based sample was a flurry of small particles, there were no more small pieces in space, "They all began to complex into much larger tumor aggregates…"—they were merging into one large piece of artificial tissue.

Paceley recalled, "They [the scientists] were just ecstatic. They were two-three feet off the ground the whole time—they had just never seen results like they had seen on this flight." The artificial tumor was so large that the engineers discussed how to disassemble the growth chamber after landing to avoid damaging the sample. The BDS sample did not survive the accident, but the scientists did get excellent video showing how large the artificial tumor had grown. They also got some metabolic data, and learned that while the space tumor was large it didn't use any more nutrients than the ground-based controls, indicating that there was a major difference in how the culture grew in space.

The red shift wakeup call was "The Prayer," by Celine Dion and Andrea Bocelli, for Rick Husband.

Biological Research in Canisters (BRIC—pronounced "brick") consisted of boxes filled with Petri dishes with whatever plants or small animals scientists wanted to fly. For STS-107, there were two BRIC experiments, one with moss and one with worms. The worms were a late addition, approved just 16 days before Columbia's launch. It was a totally passive experiment, without any crew interaction. The moss was almost as simple. Several times during the mission, an astronaut would inject a fixative inside one set of canisters to stop the growth.

The fixing was videotaped on flight day 10 and transmitted to the ground. The injection devices looked like miniature caulk guns. As Rick performed this procedure, Ilan described what was happening: "Here's Rick Husband, our commander. The BRIC experiment is out of Ohio State University. It tries to understand the orientation and distribution of the tips of some cells. Rick uses a special gun to fixate the cells after some time."

Rick apparently thought the injectors looked like high-tech weapons a spy might use, so with an injector in each hand, he crossed his arms in a classic James Bond pose, before his face broke out in a broad grin. Ilan continued, "And here are the two guns in the hands of our commander, our great commander Rick Husband."

BRIC scientist Fred Sack watched the video of Husband working on his experiment. Sack said, "Just having that video was extraordinary, to see how much fun he was having, how careful he was being, and just being able to connect. You design an experiment, it takes years and lots of hard work, and then you see your experiment in space and having this wonderful astronaut working on it—it made me feel a connection to how precious his time was. He was working on our experiment just days before the breakup."

Rick Husband works on the
OSTEO experiment.

Canada and Europe flew complementary osteoporosis experiments. OSTEO-2 (Osteoporosis Experiments in Orbit—pronounced "ah stee oh") and ERISTO (European Research in Space and Terrestrial Osteoporosis—pronounced "eh rist oh"). Canada designed and built the hardware and loaned a spare unit to Europe for their experiments.

Astronauts lose bone mass ten times faster than post-menopausal women on Earth. Fortunately for astronauts, the loss reverses when they return to Earth. OSTEO-2 includes a set of automatic syringes and sample containers to mix various chemicals with bone cell cultures. It's hoped that the experiments will result in dual benefits—how to understand and control bone loss in astronauts in space as well as bone loss in people on Earth. This was one of the "commander science" experiments Rick had demonstrated for me a year before the mission.

A typical experiment includes four automatically operated syringes and three modules with bone cells. Two syringes held food for the cells, the third syringe had the chemical to be mixed in with the cells, and the fourth syringe had a fixative to preserve the experiment when it was completed.

The Canadian OSTEO-2 featured three experiments. One examined treating bone cells with estrogen to see if it could combat bone loss. Another used serum from sleep-deprived volunteers to see how it affected bone cultures. On Earth, sleep deprivation can lead to bone loss, and scientists wanted to see if the genes in the immune chemicals affected bone formation. The third experiment examined bone gene regulation patterns. Europe's two ERISTO experiments featured Osteoblasts Mineralization research from France and Osteoclasts in Space research from Italy.

CHAPTER 29
FLIGHT DAY 11: ZCG, BIOPACK

Mike Anderson's wife, Sandy, selected Dionne Warwick's "I Say a Little Prayer" for the blue shift wakeup on flight day 11. Anderson replied, "What a perfect song to wake up to on another perfect day in space. I'd like to thank my wife Sandy for that song and her love and prayers. With that, blue shift is ready to get to work."

The wakeup came around the same time that Columbia flew relatively close to the International Space Station. Capcom Charlie Hobaugh told Rick ISS was off the northern coast of South America, but the lighting conditions weren't right for them to see each other. Husband said, "We appreciate the heads up, and keep 'em coming." Hobaugh acknowledged, "It's always good to know your buddies are out there with you."

NASA has flown Zeolite Crystal Growth (ZCG) on several shuttle flights and on ISS. Zeolites are minerals that act like "chemical sponges." A zeolite will combine chemically with another substance until it's saturated, then it can be "squeezed" with heat to give up what was absorbed. Zeolites can be used as filters to absorb one type of substance and let another substance pass through. Industrially, zeolites are used to refine gasoline. Consumer products with zeolites include water softeners, aquarium filters, and cat litter. The STS-107 zeolites were ones used for refining; 38 tubes filled with zeolites were processed inside a 573-degree furnace.

A rare photo of the STS-107 Zeolite Crystal Growth experiment in progress. The caption on the sign reads "Do not disturb—Experiment in progress."

K.C.'s red shift wakeup call was "Drops of Jupiter" by Train. She talked about seeing an aurora over Australia, and the song reminded her of a very, very dear friend. She announced, "Red shift is up and charged, and we are ready to go for the day."

The European Biopack and Biobox payloads were controlled from the Florida Institute of Technology (FIT, or "Florida Tech") in Melbourne, Florida, 35 miles south of KSC. There were fewer security and access issues for foreign nationals at a university than at a NASA center, and less concern about the various materials the scientists needed for their work. FIT provided laboratory and office space, supplies, and support staff. (Each experiment team sent FIT a "shopping list" of the supplies and laboratory equipment they needed.)

One of the key reasons for using FIT that wasn't publicized was the radioactive materials. BONES included 0.13 microcuries of Calcium-45. FIT was familiar with the proper procedures for handling radioactive materials, and already had licenses from the Atomic Energy Commission and the state of Florida to use radioisotopes in its laboratories.

The January launch meant the mission occurred while school was in session at FIT. Most of the payloads were prepared during the winter break, and by the time the students returned, the only laboratories the scientists needed were those assigned for the control center. FIT had the bonus of exposing its faculty and students to world-class scientists. Life sciences professor Gary Wells said, "A number of students had the opportunity to work with the scientists. It gave our faculty a chance to talk science with international scientists; it gave our students some great experience. Two of our faculty have been invited to Spain to participate in research there." Wells recalled, "[The Biopack scientists] were very willing to give talks to small student groups. Paul Veldhuijzen with BONES was very good. It was obvious he really liked talking to students."

Biopack included miniature centrifuges with space for four small sample containers, plus slots for additional containers. Each container was about the size of a computer mouse, and contained six sets of tubes filled with various chemicals before launch. The astronaut used a screwdriver to turn a rod with openings that let the various chemicals mix. For additional safety, the astronaut performed these activities inside a "glovebox," which has two heavy-duty rubber gloves and a clear top. The astronaut opened up the glovebox, inserted the sample container and screwdriver and closed the glovebox. Then he placed his hands in the gloves and looked through the clear top while he manipulated the contents. Even if the container leaked or there's an unexpected problem, everything's sealed inside and can't contaminate the crew cabin.

After activating the experiment, the astronaut opened up the glovebox and put the sample container into the Biopack centrifuge, or one of the other slots. Biopack included a refrigerator and incubator to process the

sample at whatever gravity level, temperature, and other conditions the scientists wanted. Some of Biopack's eight experiments just sat inside the
incubator for an allocated period; others were spun inside the centrifuge.

After the container was processed in the centrifuge-incubator, the
astronaut put the container back inside the glovebox, and twisted the rod
again to introduce a fixative to deactivate the experiment. The finished
sample container was then put into a Passive Thermal Cooler Unit
(PTCU), basically a super-cold thermos bottle, for the rest of the mission.
The eight Biopack experiments came from Switzerland, Netherlands,
Germany, Belgium, Italy, and the USA.

About three hours after launch McCool started Biopack, and 99
minutes later the first telemetry arrived at FIT, where the ESA scientists
were happy to see that their hardware was working properly.

The first Biopack experiment was Leukin, a collaboration between
Swiss biologist Augusto Cogoli and American scientist Millie Hughes-
Fulford. Hughes-Fulford was intimately familiar with microgravity science; she had flown in space as a payload specialist on the STS-40 mission in 1991. Leukin's purpose on STS-107 was to study immune system
changes in astronauts. During the Apollo missions it was discovered that
the immune system degrades in space and takes a couple of weeks to
return to normal. Leukin didn't study the immune system directly inside
an astronaut's body, but the chemicals involved—how the interleukin-2
receptor affected T-cells in human blood. These isolated experiments
would provide a close-up look at how the immune system is affected by
microgravity. The T-cells came from expired blood from a blood bank.

Willie McCool was the astronaut responsible for operating several of
the Biopack experiments. Hughes-Fulford recalled, "He was just so fast with
it and so complimentary to Biopack and how well everything was working.
We were on page 4 [of the procedures], he was already on page 6—he was so
fast." McCool injected the activator cells into the T-cells. After the experi-

Willie McCool handles the
Leukin experiment inside
a glovebox.

ment was completed, he put the Leukin cartridges into the PTCU. The instructions Willie followed for Leukin are on this book's website.

While McCool was a pilot without a biochemistry education, he was certainly interested in the science. Later in the mission he told me, "I would say the most interesting and probably most challenging thing was an experiment called Leukin. That particular experiment is looking to see how white blood cells are activated to fight disease in microgravity, in an attempt to understand why astronauts are more susceptible to diseases." An audio clip of Willie talking about Leukin is available on this book's website.

Next on the schedule were the BONES and STROMA experiments. BONES studied the role of bone cells and skeletal tissues in microgravity. It was finished relatively quickly, but STROMA needed to stay within the incubator for almost the entire mission. STROMA studied bone marrow stromal cells (connective cells within the bone marrow). REPAIR, which examined how DNA repairs radiation-induced damage was simultaneously performed with the other two experiments.

Then trouble. Laurel reported that after she opened the door of the cooler/freezer, Biopack suddenly shut down. The Biopack team made the decision to stop using the freezer and transfer the samples directly to the PTCU as soon as they were finished. In cases where the experiment needed to be cooled quickly, ESA got permission to use the freezer inside Spacehab to cool the experiment cartridges before they were put in the PTCU.

Two days later, there was a problem with pushing the incubator tray back in, and the door remained open. That resulted in the experiments getting cooled to cabin temperature. By this point the REPAIR experiment was completed, and the next up was CONNECT, two related experiments that studied gene expression in human cells and bacteria.

Two days later, Laurel reported unusual noises from the fan, and an electrical odor coming from the incubator tray when the door was open.

Dr. Enno Brinckmann demonstrates the ground control Biopack hardware to the author. The open drawer shows the small centrifuge and space inside the incubator for additional Biopack cartridges. Photo credit: Philip Chien.

The engineers believed the odor was due to the temperature of the electronics, and not a short circuit.

It was frustrating for Biopack's engineers to have their equipment constantly overheating. Hughes-Fulford believes the Biopack filter was clogged by food particles. She recalled that on her flight, "Anything which escaped our mouths ended up on the refrigerator-freezer screens." The crew attempted to clean a filter when the problems first started. Biopack asked for a more intense cleaning with a vacuum cleaner, and asked to delay the next experiment by a day. McCool made a hole in the air inlet filter and vacuumed the cooling air duct. But that was only partially successful.

On Sunday January 26, the crew tried to troubleshoot the Biopack hardware. Laurel Clark reported, "POCC, I've done the 30 second hold [of the circuit breaker] and I let those switches go back to the middle position." Beth Vann replied, "We copy and appreciate the extra effort." Laurel said, "No problem, we're standing by watching and keeping our fingers and toes crossed," and Vann replied, "We're doing the same down here." Unfortunately full Biopack operations were never restored.

The remaining Biopack scientists were informed about the changes and replanned how they wanted to conduct their experiments. Next on the list was BACTER, which grew *Pseudomonas aeruginosa*, a bacterium typically found in drinking water. The plan was to finish BACTER, and for the final two experiments, BIOKIN and YSTERS, have the incubator switched off with the door open and the centrifuges off—a compromise aimed at getting as much data as possible. BIOKIN studied *Xanthobacter autotrophicus* bacterial cultures and YSTRES examined baker's yeast (*Saccharomyces cerevisiae*) cultures.

On January 29, the final Biopack experiments were completed. The STROMA containers with their bone marrow cell cultures were put in their PTCU, and Biopack was shut down. While the problems with Biopack had been frustrating, the scientists were able to work around many of them and maximize what could be done. Overall, they were very satisfied with what they had accomplished and looked forward to receiving their samples.

Biobox was a similar set of four experiments that didn't require any interaction from the crew. It was completely autonomous and had also flown on automated (unmanned) spacecraft. Biobox starts when a gravity sensor detects that the experiment is in space. Then a computer controls the rest, including running a centrifuge, incubator, injecting fixatives, and a freezer. For the shuttle version, the engineers added the ability to send commands and receive some telemetry. The four Biobox experiments, from Belgium, France, and Italy, studied bone cells— their formation, how they degrade, and the functioning of bone marrow.

All of the Biopack and Biobox experiments were a complete loss, since the scientists needed to get their samples back intact.

CHAPTER 30

FLIGHT DAY 12:
SHIP-TO-SHIP CALL,
FRESH

The flight day 12 blue shift wakeup was "When Day is Done," by Django Reinhardt and Stephane Grappelli, for Dave Brown. Capcom Charlie Hobaugh said, "We're doing great down here, and I feel cultured." Dave replied, "It's about time." Later, Queen's "Love of My Life" was radioed for Ilan. One of the blue shift astronauts videotaped the sleeping bunks showing the red shift as they woke up, and Ilan reaching for the microphone. Ilan said, "Good morning, Ken, and good morning to all. And a special good morning to my wife Rona, the love of my life." Ramon continued in Hebrew saying he loved his wife and kids and he missed them. He finished in English, "So here we come for another wonderful day in space." Capcom Ken Ham replied, "Ilan, we're looking forward to it and we'll get you home in a week or so." and Ramon, in no way anxious to end his incredible experience, replied, "Don't hurry up." The videotape survived the accident and shows the typical wakeup activities on the shuttle.

The entire crew gathered inside Spacehab for a public affairs broadcast. Appropriately, a bright green apple was floating as Rick and Mike talked about the opportunities for teachers to become "educator mission specialists"—former elementary or high school teachers who, as astronauts, would educate students as well as carry out their responsibilities as mission specialists.

Ilan Ramon talks to capcom Ken Ham after "Love Of My Life" was played as his wakeup music.

Later in the day, a ship-to-ship call was made between the shuttle and ISS. Technically it was simple — the control rooms are down the hall from each other at JSC in Houston, so it's simple for the audio technicians to connect the two communications circuits whenever needed. Obviously, ISS's crew had to be awake along with at least one shift on Columbia. Both spaceships had to be within communications range and, of course, the timeliners had to find an opening in Columbia's busy schedule where the astronauts could take a break. Public affairs also had an additional requirement: the shuttle had to have video downlink capability.

The flight planners selected Monday, January 27, at 12:34 P.M. The contact took place as ISS was passing over Russia and Columbia was almost halfway around the world, over South America.

Americans Ken Bowersox and Don Pettit on ISS had the opportunity to talk to the red shift on Columbia while the blue shift slept. Russian crewmate Nikolai Budarin did not participate. Bowersox, Pettit, and Budarin had been in space since November, and were scheduled to come home in March on the next shuttle mission. After greetings were exchanged, Bowersox, recognizing Columbia's many delays, said, "We're so glad to see you guys made it into orbit."

Rick noted that he had seen ISS once, but the other astronauts hadn't. Pettit said, "We haven't been able to see you." He added that he hoped he would get the opportunity to see Columbia before its landing.

Pettit told Columbia, "We're doing just fine up here, it's certainly an amazing place to live and work." Bowersox asked about the rookies, "So what do Ilan and Laurel think of space?" Clark enthusiastically moved both arms with a "thumbs up" happy gesture, and Ilan gave his own thumbs up. That couldn't be seen on ISS, which only had an audio connection. Rick said, "I think they're loving it a lot. I'll let you talk to them."

Ilan, envious that ISS's crew got to stay in space eight times as long, said, "Hey, Don, this is Ilan. I wish I could stay some more time like you guys. But we're shorter." Pettit replied, "Good to hear your voice, Ilan,

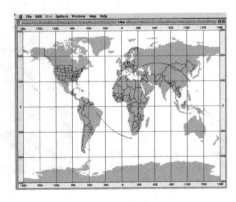

A satellite tracking map shows the paths of Columbia over South America and ISS over Russia as they talked to each other from halfway around the world. Map by Philip Chien.

and good to know you're up on Columbia getting your experiments done." For Bowersox, Ilan said, "Our kids are very good friends and enjoy each other. I saw your oldest not long ago, and he sends kisses from here—from space to space." Bowersox replied, "Hope you'll be able to tell them all hello from me when you get back. We'd love to have you here visiting as well, so you could spend a little more time."

Laurel asked, "Don, I was wondering how your [two-year-old twin boys] are doing and what you might do to help them remember you?" Pettit replied, "They seem to remember me. They know who I am. I'm doing the best job I can to maintain contact with them as they're going through the terrible twos."

Pettit was in the same astronaut class as Laurel, Willie, and Dave. Laurel said, "It's just really cool to be up here in space at the same time. We've got three sardines [astronauts from the 1996 class] up here plus Ilan, so we've got a lot of new folks."

Pettit said, "I guess Willie's on the team that's working the night shift." McCool and Pettit were carrying an unusual correspondence—a long-distance chess match via e-mail—real long distance. Rick said, "He did give us a message that he wanted us to give to you. He wanted to say 'hi,' and he was sorry he's on the shift that's asleep right now. He just wanted to say, 'It's your move,' and he sent E2 to E4 via e-mail a long time ago and it's time for you to respond on that one." Pettit replied, "Copy. E2 to E4 and it's my move. I tell him—*en garde*."

Rick said Willie was taking plenty of pictures over Australia as a tribute to Pettit and his unusual musical instrument, an Australian aboriginal didgeridoo. Pettit said, "We've had a couple of nice passes over Australia, we got some good shots of the Perth-Dongara area. and I'm looking to seeing how those [photos] turn out." Bowersox interrupted, "We're having lots of good didgeridoo serenades too." On Christmas day, Pettit had played "Jingle Bells" on his didgeridoo while Bowersox sang to entertain Mission Control.

Pettit had a reputation for taking literally thousands of photos for environmental scientists. Laurel said, "I heard you were doing such a great job taking pictures that it was taking them days or weeks to download all of the images. It sounds like you're doing a fantastic job for the Earth observations folks." Pettit joked, "What we can't have in quality we make up in quantity." Bowersox added, "Your spaceflight doesn't exist if you don't have pictures of it."

Bowersox also said, "I just wanted to let you guys know how proud we are of you and the great work you're doing. Hope you can keep it up and enjoy the rest of your mission." Rick replied, "We certainly wish the best to all of you and Nikolai and hope that he's doing as well also, and you have a great mission. We look forward to seeing you when you get back." Pettit finished, "Remember to fly safe." and Husband replied, "Okay we'll do it, Don. We'll see you when you get back." All four astronauts on Columbia chorused "Bye bye!" together.

After the event was finished, capcom Charlie "Scorch" Hobaugh announced, "We're going to resume normal communications and put you back to work." The four astronauts chimed, "We copy, Scorch" together, and Hobaugh replied, "Work, work, work, work, work," and added, "Good job by all you guys."

A copy of the ship-to-ship call is on the book's website.

The FRESH (Fundamental Rodent Experiments Supporting Health) consisted of 13 male rats flying for three experiments. All had been approved by NASA's oversight panels, which examined the scientific value and considered whether there was any way the experiments could be done without animals, minimizing the number of animals needed, and ensuring that the animals would be treated well and killed as painlessly as possible.

Eight of the rats were 9 weeks old, with a weight of some 10 ounces (300 grams) each, and five rats were 10 weeks old and weighed 9 ounces (250 grams). Because the scientists needed rats of specific ages and weights, a large number of animals had to be obtained. The scientists selected the animals for flight and the ground controls based on the needs of the experiment and those that appear healthy. Originally NASA destroyed the unneeded rats, but now there's a policy in which excess rats are killed and given to the Florida Audubon society to feed injured wild birds being nursed back to health.

NASA is extremely serious about how experimental rats are treated. If a rat appears unhealthy, the astronauts consult with NASA's chief veterinarian. If the vet decides the rats are suffering, the astronauts are instructed to euthanize them, even though it would affect the science results. On a previous mission, an astronaut had to do that to several rats when the newborns wouldn't eat and were becoming sicker.

Many people are against NASA—or anybody—using animals for scientific research. One person who had mixed feelings about the rats was Kalpana Chawla, a Hindu and life-long vegetarian. Before the flight she told me, "From my point of view, I know the benefits of that research. I personally have thought about experiments like that when we used animals, basically the thought process that goes behind studies like that and the justifications the researchers have. I never try to get too involved and just trust the people who are doing it. Because if you take it really personally, when your own sibling or parent needs the results of research like that, then you are obviously going to say, 'Yes, this is a good idea'. The bottom line ends up making sure the people who are doing the research are putting the right thought process into the procedures and making sure the best comes out of it and nothing is wasted. I really leave it up to them because it's not my area [of experience] at all." K.C. readily admitted she was happy she was not responsible for the rats during the mission. An audio clip of K.C. talking about the rats is on this book's website.

Every couple of days, an astronaut checked on the rats' health and provided them with more water. Their food was a bar of rat food. There were concerns that the rats were becoming overheated because of Space-hab's high temperature. Mike removed the Animal Enclosure Module mufflers to increase the airflow. The rats were also provided with additional water. The astronauts reported the rats appeared okay and did not appear to be heat stressed.

The three FRESH experiments studied the balance organs and cardiovascular system.

Dr. Michael Delp, with Texas A&M University in College Station, Texas, studied "Arterial Remodeling and Functional Adaptations Induced by Microgravity." It studied how the cardiovascular system reacts to microgravity. A long-term goal was to develop treatments or countermeasures to improve the health of astronauts when they return to Earth.

Dr. Jaqueline Gabrion, with the University of Pierre and Marie Curie in Paris, France, studied "Choroidal Regulation Involved in the Cerebral Fluid Response to Altered Gravity." The choroid plexus is a portion of the brain that produces cerebrospinal fluid, and previous flights indicated that less fluid is produced in microgravity. The experiment's purpose was to find information about fluid production in the brain and body and reveal information about cerebral adaptation and fluid balance during spaceflight.

Dr. Gay Holstein, with the Mount Sinai School of Medicine in New York, developed "Anatomical Studies of Central Vestibular Adaptation," which studied the rats' balance mechanism and how it was affected by the return to Earth's gravity. Unlike the other two experiments, in which the rats would be sacrificed shortly after landing, these rats would be kept alive for another day to allow the scientists to study them and see how they readapted to gravity and regained their balance.

Each of the three experiments required specific rat organs, and NASA offered the remaining rat tissues to the scientific community. A total of 14 investigators from the United States, Canada, Europe, and Japan were selected for 34 tissue samples. No portions of the rats would be wasted.

Mike Anderson inspects the three lockers containing the rats for the FRESH experiment.

CHAPTER 31

The flight day 13 blue shift wakeup was "Slow Boat to Rio," by Earl Klugh, for Mike Anderson. "Good morning, Houston, that's a wonderful song. Ready to start off another wonderful day in space. I'd just like to thank my wife and my two kids for all of their support for this flight, and we're ready to get busy on another day in orbit."

The Australian students were disappointed that their spider wasn't weaving webs. They asked for an astronaut to flip the tab to let a backup spider into the chamber. Mike performed that task early on January 28.

The red wakeup call was "Running to the Light," by the Scottish band Runrig, for Laurel. She said, "Sending out love to my husband, who's been a great dad, and to my son Iain—can't wait to see you guys in a few days. And to my entire family, who's been supportive and terrific as a family for me through this big exciting adventure."

Both shifts got their second off-duty periods, a half-day to unwind, look out the windows, answer e-mail, and talk to their families. These private video teleconferences turned out to be the last time the astronauts got to talk to their families.

The first two combustion experiments, Laminar Soot Process and SOFBALL, completed their runs, and the crew reconfigured the Combustion Module for the Water Mist Fire Suppression Experiment. Unlike the other experiments that came from university laboratories, Mist was partially sponsored by a commercial laboratory at the Colorado School of Mines. The concept behind Mist is to use a very fine spray— almost like a fog—to put out fires.

Water used to put out fires often causes as much damage as the fire itself. If the amount of water could be reduced to a small fraction of what's normally used—just enough to put out the fire—there would be far less water damage. The system developed for Mist used one-tenth of the water in a fire sprinkler system, a small enough amount that it would not do much damage and theoretically would not even conduct electricity. Mist's scientists foresee applications in computer rooms, where a normal sprinkler system can damage electronics; spacecraft, aircraft, and ships, where a sprinkler system is far too heavy; and even historic buildings, where it's especially important to minimize water damage.

Ilan and K.C. set up the
Mist experiment.

On the ground, the Mist experiments are hampered by gravity. The interaction between the water particles and flame front can be measured much more accurately in microgravity, and with those measurements, scientists would be better able to crack the puzzle of how to use mists to put out flames.

The crew had difficulties setting up Mist on January 27 because a vacuum pump wouldn't remove all of the air in the flame tube. Engineers thought the problem was just caused by stiff hoses, and the crew was asked to disconnect everything and hook it up again. CIC Lora Keiser told the crew, "Angel (the lead scientist) says to send you all a big hug." But that didn't work, and the team tried some more troubleshooting. It took another shift until Mike believed he found the problem: He sent down video showing a loose connector in the gas chromatograph sampling line that verifies the gas mixture, and held up an O-ring to the camera. But even with the replaced O-ring, the sampling line still leaked.

The Mist team thought about what it would take to seal off the line. One person suggested kinking it, just like a twisted garden hose prevents water from flowing. But would the kink hold enough pressure? Somebody picked up a $14.99 manually operated bicycle pump from a local hardware store. The engineers used the pump to verify the idea would work. The Mist team developed a two-page procedure for the astronauts to remove the line and twist it in two places, tape it to ensure the kinked section would stay in place, and put it all back together. The repair would take several hours to complete.

The repairs would have to come out of Mist's remaining time for its experiment runs. The scientists looked enviously at the off-duty time the crew had on their schedule, and wondered if it could be used. The rules say Mission Control can't ask the crew to give up their off-duty periods, because the astronauts might feel obligated. But the astronauts can offer to give up the time on their own. As soon as K.C.'s off-duty time came, she radioed that she'd like to offer it to help to troubleshoot Mist.

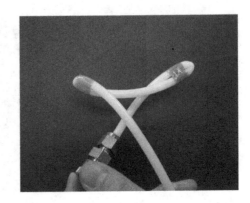

The solution to the Mist experiment malfunction was to isolate the leaking line by twisting it and using Kapton tape to hold the twist in place. Photo courtesy of Dr. Angel Abbud-Madrid.

Lead scientist Angel Abbud-Madrid said, "Out of the blue it came—we weren't expecting it. She knew the experiment so well because of her training. When she volunteered, we uplinked the procedures. She looked at it and started working on it immediately. She just worked through her off-duty—five straight hours." K.C. had other activities scheduled, like exercise and lunch, but she ignored them. Abbud-Madrid said, "She just went through those. She went into the pre-sleep time." Other crewmembers also assisted with the troubleshooting. K.C. had to go to sleep but asked about the progress for the first run with the repaired hose. By that point it was 30 hours after the experiment was supposed to start. All together, the repairs took two days, and the efforts of five of the astronauts to get Mist running. It was a great team effort by the entire crew to ensure that everything else was also accomplished while the Mist repairs were underway.

CIC Brad Korb told the crew, "We greatly appreciate all of the effort that you put today to recover the Mist EMS (Experiment Mounting Structure). We are now going to prepare to do a ground commanded test to verify that everything's working properly." K.C. replied, "That sounds

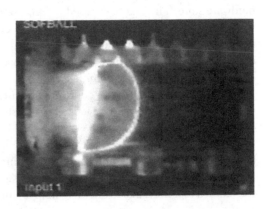

Even though the video display still says "SOF-BALL" it's actually a Mist test with the curved flamefront moving from left to right where it will encounter the fine water mist which will put it out.

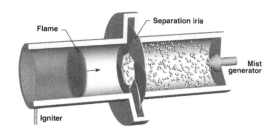

A drawing showing how Mist works. A fine spray of water is injected into a tube, then an iris opens, permitting the flame to enter where it's extinguished by the mist. Drawing courtesy of Dr. Angel Abbud-Madrid.

good, and Brad you are very welcome. We are here to do all of these experiments and help out in any way that we can. Please give us a call when we need to do anything with Mist."

Mist used an ultrasonic vibrator to create an extremely fine fog in a 19-inch transparent tube. Abbud-Madrid described the fog: "Very fine, very small droplets of water—one tenth the diameter of a human hair." A set of lasers verified that the mist had the proper density and distribution. Then an iris opened, similar to a camera lens. An igniter lit the premixed fuel and air in the left side of the tube. The flame front traveled toward the right, where it encountered the mist and was quickly extinguished. Video cameras, lasers, and radiometers documented each test.

Abbud-Madrid said, "The mist has been able to stay suspended for minutes in a very uniform way. We have very exciting results to report to the scientific community and our industrial partners on what may become very important information for developing the next generation of fire-suppression systems."

The Mist team could activate their experiment when the crew wasn't actively involved. At one point, Korb notified the crew that if they were interested in watching, another test was about to start. Afterwards Ilan said, "Let Angel know that we've just looked at harmony in space on the screen."

Scientists originally planed 34 tests using different mixtures of fuel and air. Because of the delays they got 32, but they were very happy with the results. 90 percent of the data was transmitted during the mission. The various tests used propane-air mixtures with a very lean fuel-to-air ratio, a lean ratio, a theoretically optimum ratio, and a rich ratio. The flames were exposed to mists with droplets ranging from 20 to 30 microns in size at different water concentrations.

As each test was performed, the scientists examined the data and replanned the next test based on the previous test's results. They gradually moved toward their target—a mist that would slow the flame front to almost a stop.

Abbud-Madrid explained, "What we're trying to do is solve a complex problem in small pieces. What happens when the droplets hit the flame. Once we know that information, we can put that into our model, which has to include how to get the mist to the fire, how much convection plays a role, the pressures that are used. Knowing that information, you can complete the study [and] you can have a system that will work both on Earth and in space."

One experiment wasn't started until just three days before landing. On Earth, plants use gravity to determine which way to grow—roots always grow down while plants grow up. Other cues include light and magnetic fields. The University of Louisiana at Lafayette's Dr. Karl Hasenstein developed Biotube to examine how strong magnetic fields affect plant growth.

In the experiment, flax seeds were launched dry. During the mission water was added and the experiment ran on its own for 48 hours, including an injector which put in a fixative to stop the growth and preserve the roots. Biotube was intentionally started as late as possible to ensure that the plants would remain in the fixative solution as little time as possible before they were returned to the scientists for analysis.

Video was transmitted to the ground starting 15 hours after the seeds were germinated and for another 24 hours. The scientists were able to determine the roots sprouted at the same rate as on Earth, but grew more slowly in the absence of gravity. Most important, the roots did not grow past the strongest portion of the magnetic field—almost as if there was an invisible wall in the way. That was a surprise to the scientists, who expected the roots to curve away from the magnetic field or get slowed down, not stopped. They called it an "unexpected avoidance reaction."

Tuesday, January 28, marked the 17th anniversary of the Challenger accident. Seven astronauts were killed and a shuttle was lost due to the management decision to launch in too cold conditions despite recommendations not to launch by engineers. (The previous day marked the

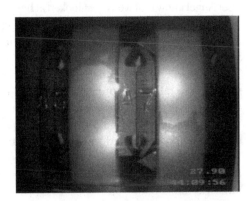

Still video frames show the tiny Biotube shoots as they grow in microgravity in a magnetic field. Courtesy of Dr. Karl Hasenstein.

36th anniversary of the Apollo 1 fire caused by an electrical short, suffocating the three astronauts in a capsule on the launch pad.) It's a NASA
tradition to remember the Challenger astronauts with 73 seconds of
silence — the length of time Challenger flew before it was destroyed.

Columbia's astronauts followed the tradition. Rick Husband said,
"We've got an announcement we'd like to make on behalf of the STS-107
crew. It is today that we remember and honor the crews of Apollo 1 and
Challenger. They made the ultimate sacrifice giving their lives in service
to their country and for all mankind. Their dedication and devotion to
the exploration of space was an inspiration to each of us, and still moti-
vates people around the world to achieve great things in service to others.
As we orbit the Earth we will join the entire NASA family for a moment
of silence in their memory. Our thoughts and prayers go to their families
as well." An audio clip of Rick's remarks is on this book's website.

None of Columbia's crew was with NASA at the time of the Challenger accident. According to their bios, Husband was an instructor pilot
at George AFB in California, McCool was in flight training, Chawla was
working on her doctorate degree at the University of Colorado, Brown
was a flight surgeon on the aircraft carrier USS Carl Vinson, Anderson
was assigned to Randolph AFB, Clark was studying for her medical
degree at the University of Wisconsin-Madison, and Ramon was a student at the University of Tel Aviv.

STS-107 was only the third time NASA has had a shuttle in space on
the Challenger anniversary. The previous flights were another microgravity research mission in 1992 and Mike Anderson's first flight, the
STS-89 mission in 1998.

Mist scientist Angel Abbud-Madrid was in a meeting with a group of
engineers, troubleshooting his experiment, when the anniversary took place.
At 10:38 Central Time all work stopped for the moment of silence. Abbud-
Madrid recalls, "And then we just started up again without losing a beat."

JSC did not have a formal ceremony, but groups of workers planned
their own get-togethers to commemorate the Challenger crew.

I asked STS-107 manager Phil Engelhauf, mission scientist John
Charles, and scientists Arny Ferrando and Scott Smith about their recol-
lections of the Challenger accident. An audio clip with their thoughts is
on this book's website.

Phil Engelhauf, now a lead in Mission Control, was a flight planner
in 1986. He said, "I was in a conference room. It was the third floor of
building 4 [at JSC] with a bunch of my cohorts in a conference room
watching the launch. I'm not sure I can adequately describe publicly the
reaction each of us had. These were our friends that lived down the hall in
our offices. They were heroes every day to us. This was our life, our job,
our career. There was a lot of uncertainty as to what this was going to
mean. Within hours people had adjusted to a state of optimism that we

were going to find out what happened and pick up and carry on. 17 years later here it's kind of nice to see we did that and picked up with that legacy and are here today with a space station occupied 365 days a year and an independent shuttle program servicing the space station and the broader science community and the objectives that all those people stood for."

Dr. John Charles recalled, "I had just come from an off-site meeting and parked my car to go to the conference room in building 37 to watch the launch. So I arrived just in time to see everything unfold. The rest is as Phil said, the uncertainty and then followed very quickly to do whatever we had to do among the scientific community to get the program back going again."

Protein Turnover scientist Arny Ferrando remembered, "I was in active duty officer in the Army at the time, an instructor pilot. We actually watched [the launch] between flight shifts. Back then [shuttle flights] were still very novel and everything came to a halt. This was at Fort Rucker, Alabama, the Army's training site. So it was quite a sobering event as you can imagine."

In contrast, Dr. Scott Smith, the Calcium Kinetics scientist, was a graduate student with no connection to the space program. He said, "I was working in a basement laboratory without any exposure from the outside. So it actually took 30 minutes to an hour before somebody called down to let us know what happened. The reaction's hard to describe. But obviously to be here 17 years later is an incredible testament to the people who come before us and continue to strive to take that next step."

The 51-L Challenger astronauts (L-R) Christa McAuliffe, Greg Jarvis, Judy Resnik, Dick Scobee, Ron McNair, Mike Smith, and Ellison Onizuka pose by Challenger's hatch during their dress rehearsal in January 1986.

CHAPTER 32

FLIGHT DAY 14:
PRESS CONFERENCE

The flight day 14 wakeup calls were pop tunes with lyrics that had a sort of spaceflight connection. "I Get Around," by The Beach Boys, was played for Dave and the blue shift. He replied, "I can tell you from looking out the window that we are getting around pretty well. And certainly up here the kids are hip." The red shift wakeup call was "Up on the Roof," by James Taylor, for Rick. He said, "Red shift is ready to go for another day. It's really great to hear James Taylor; he's one of my favorites. We are way up on the roof up here; you can look at the city lights below from up here, it happens to be dark right now. It's a gorgeous view up here; we're really enjoying our mission. We've got just a couple of days to go before we will be coming back and seeing everybody who's supported us so well on this mission, including our families, and I'd like to take this opportunity to say hi to all of them as well, and thanks for all of their support."

NASA offers several opportunities to selected press to talk to the astronauts during each shuttle mission, but many of the reporters have little knowledge about the mission, and for some reason seem to think the astronauts have an opinions on every conceivable topic. A reporter from KNSD-TV in San Diego had the opportunity to interview the blue shift a couple of days earlier. He didn't care about what they were doing on the mission or their science, but used his time to ask how they felt about preparations for the upcoming war in Iraq, and apparently felt there was a need to ask the astronauts about Super Bowl XXXVIII. He asked, "What's your sense about the game on Sunday—what about the Super Bowl?" McCool said, "I'm going to root for Oakland, I have memories of watching football games with my dad and fond memories of watching the Oakland Raiders play. Tampa Bay didn't have a football team back then, so I don't quite have the same memories as I do of the Oakland Raiders, so I'm going to go for Oakland." The reporter pushed the other astronauts to respond, and Brown said, "I'd have to say that having been launched out of Florida, there's a heck of a lot of people who worked really hard from that state to get us up here. So I'm going to go for Tampa." The reporter apparently wanted to break the tie so he asked, "Mike, you're not off the hook. Colonel, who are you rooting for?" and Anderson joked, "I'm rooting for the Houston Texans. But I guess I'll have to wait for a few years." Mission Control did notify the crew about the game's results and uplinked a short digital movie of the highlights for any of the crew who wanted to watch it during their spare time.

There's also a press conference for the entire crew. Reporters have to be located at one of the NASA centers, normally JSC in Texas or KSC in Florida. For STS-107, reporters in Washington, D.C., were also offered the opportunity. Only four American reporters showed up in Florida, plus one Israeli television reporter in Washington D.C. The reporters in Florida were Chris Kridler from *Florida Today*, Bill Harwood with CBS, Jim Banke with Space.com, and myself. With just four of us—all "space beat" reporters—sharing 35 minutes there were opportunities for each of us to ask several questions, and even share some humor with the astronauts.

I had thought about asking Rick about what he had been told about the foam strike, but chose not to, since I had been assured that the engineers on the ground had determined it was a non-issue.

The astronauts prepared by hanging large American and Israeli flags in Spacehab's front half. The flags were positioned over the vents, and all seven crewmembers were in the module. On the ground, the Spacehab team nervously watched the temperature creep up because of the reduced ventilation and additional heat with all of the crewmembers inside together. Fortunately, the press conference and posing for group photos didn't take long, and after the flags were put away and the crew resumed their normal activities, the temperature went back down.

Cameras had been set up, and Rick Husband kept adjusting a digital camera to get the best shots of his crew. Ilan Ramon held up a scroll with the Israeli Declaration of Independence for the camera, and when he let go the floating scroll quickly rewound itself. He also posed with his Kiddush cup. Nobody saw those scenes until the videotape was recovered after the accident.

Ilan Ramon floats with his Kiddush cup inside Spacehab.

When Ilan Ramon let go of a copy of Israel's Declaration of Independence it quickly rolled itself back up.

The on-orbit press conference with the STS-107 crew in their color-coded shirts.

Astronauts often joke that if you don't have photos of all of the crewmembers on a spaceflight, they weren't actually there. The crew posed for group shots wearing a variety of different shirts—long-sleeve white, blue, and blue and red—color-coded for the two shifts. They remained in the blue and red shirts for the press conference.

On January 29, I started the press conference with my favorite question, the one every astronaut gets asked thousands of times before their first spaceflight—by reporters, by school children, and by VIPs: "What's it like to fly in space?" Asking this question is a bit like asking a teenager about to get a learner's permit how he likes driving a car, or asking an engaged person how he or she likes married life.

I always enjoy becoming the first person to publicly ask ex-rookie astronauts to talk about what it's like to fly in space. I usually get a broad smile on the astronaut's face, since he or she can finally answer the question! Willie said, "I liken it to a smorgasbord of new and invigorating and sometimes fun and sometimes not so fun experiences. All in all it's just been overwhelming." Ilan told me, "It's great to be here in space. I wish I could go again and again." He added that while he missed his kids and wife, "It's even greater to fly over the US and Israel [in daylight]. The world as a whole looks marvelous from up here." Laurel said, "This has been a great experience for me. The first couple of days you don't always feel too well. I feel wonderful now. The first couple of days you adjust to the fluid shifting, how to fly through space without hitting things or anybody else. But then after a couple of days you get in a groove. It's just an incredibly magical place." Dave talked about doing zero-G gymnastics: "I got to do some backflips here in the center of the [Spacehab] module, it's big enough to do that. That was quite a bit of fun."

Mike talked about the science. "A lot of our experiments have exceeded our expectations by 100 percent. We've seen things which we never expected to see onboard this flight. We had a flame ball burning for an hour and a half, which is a new world record, and something we really didn't expect to see. We've got a very large cancer cell that's probably 100 times larger than we could have predicted. It will go a long way in the area of prostate cancer research. So overall this flight has been absolutely fantastic, the science has been spectacular, and we just can't wait to bring it all home so the scientists can really take a close look at what we have done."

K.C. said, "Combustion Module is easily the most fun payload in the Spacehab module to work with. It's very interactive; there's lots of hardware that we integrate into the module, which just makes you feel really good. Sort of like raw work with your hands. Then you get to see the studies that the investigators planned." She mentioned that the other astronauts would be interested in what was happening, asking, "What did you see? What did you see?" K.C. noted after Dave started to name the flame balls, "Everybody got involved in that thing. 'How long did yours last?' 'Was that the leanest?' So it has been a lot of fun working on that payload."

K.C. talked about one of the favorite activities for most astronauts — looking out the window. She said, "The coolest thing for me is the experience of floating and hanging by a window just after sunset and watching the stars in the big black dome of the sky as the Earth moves underneath. I somehow try to find 10 or 15 minutes every day to do that. I continue to postpone my meals so I can do that. I have to watch where my food is [floating] because my eyes are glued on the outside. It's just an absolutely amazing magical wonderful feeling to do that."

Ilan said he didn't have the opportunity to commemorate the Sabbath the previous week. "When you are in space you lose [track] of the days down on Earth. If you ask me what day it is I have no clue. I know that we are on flight day 14. I think I will catch this Friday [the night before landing] and try to do something. The one thing that I do is eat Kosher food up here, which was provided for me by NASA. My crewmates share with me and they love it," and Laurel gave an enthusiastic thumbs-up gesture.

Talking about the incredible views Ilan said, "I think flying over Israel is great — to see your country is safe from up there, so calm, so nice, so neat. It's really great. I wish we will have very soon, the quiet and the peace we see Israel and the Middle East from up here [in space] down there."

Ilan said, "The world looks marvelous from up here, so peaceful, so wonderful, and so fragile. The atmosphere is so thin and fragile. I think everybody, all of us down there, not just in Israel, have to keep it clean and good — it saves our lives. To the people of Israel, I wish we will have a peaceful land to live in very soon."

About the temperature inside Spacehab, Laurel admitted, "A pina-colada would have been really nice. It was warmer; we all got along just fine. We're glad the temperatures are a little cooler [now]. Several of the payloads were starting to get warm. We did a lot of things to help them out — put some fans up, opened lockers, things like that. They all got through it just fine. We're all very comfy right now."

Mike talked about the animals, "They're all doing well—the spiders, the ants, the moths, rats, they're all just doing great, and I think we're getting some great science from it. Most of those experiments are educational experiments, and we're beaming down video to students at schools all around the world. I'm sure they're really excited about what they're seeing, and I'm sure they're excited for us to bring those specimens back so they can take a closer look at them."

Anderson was asked to compare their science with the science on ISS. He said, "A lot of what we're doing now is really in preparation for future flights aboard the space station. A lot of experiments that we have are really just being demonstrated and developed, and once they're fully developed they'll reside on board the space station. The scientists on board the space station will have years to conduct the experiments that we're trying to do here in a relatively short period of time. The science we're doing here is great and it's fantastic, it's leading edge, but I think once we get a seven-member crew on board the space station, you're really going to see some outstanding science in space. This flight has been absolutely fantastic."

Rick noted, "We've had a great time up here. We are very happy with the way things have gone. To be honest, Willie and I are really looking forward to [landing]. We're really excited about it. Probably what we'll be looking at is an on-time landing if the weather's good, and if the weather's bad we may extend a day or so."

The best question came from *Florida Today* reporter Chris Kridler. She asked each of the astronauts to talk about an "oh wow" moment, and they all gave excellent responses, some listing multiple "oh wows."

Dave: "At the time, I had no idea [I had discovered an ELVE—an upward burst of lightning]. It was early in the mission, I [just] ran the video recorders, the cameras in the payload bay. It was the ground that actually discovered this very brief phenomenon. It's not something we've been able to see with our eyes—at least I haven't. So the 'oh wow' time came a couple of days later when we got a message from the ground and a picture showing this thing that we had filmed, and I was very fortunate enough to be the camera operator. So that was the 'oh wow' moment, getting out of bed and going to the printer and seeing this picture that the ground had processed and sent up, and that was one of the real highlights on the flight for me."

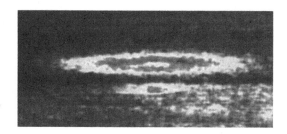

The first ELVE discovered by MEIDEX. Photo courtesy of the MEIDEX team.

Mike Anderson works on the
Combustion Module.

Mike: "I'd have to say with the Mist experiment, after spending two hard days working on it, it was a really a big 'oh wow' when we saw the first Mist flame go down the flame tube. For me that was a really exciting time, the people who had worked on the experiment on the ground had so much time and energy tied up into it, I just could not imagine having a failure with that experiment. After working on it for two days and finally getting it to work, that was a real a big 'oh wow.'"

Willie: "A big part of my time is spent on the flight deck, maneuvering the vehicle to support the experiments that have pointing requirements. So the 'oh wow' for me is oh wow, I've had the opportunity to be on the flight deck more than most of my crewmates to look outside and really soak up the sunrises and sunsets and the moonrises and moonsets, the views of the Himalayas, Australia, all of the continents except for

A spectacular shot of Mt. Fuji really shows its height.

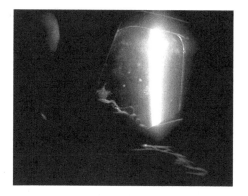

Ilan Ramon (barely visible in the darkened crew cabin) admires the view outside Columbia's window.

[North America]. So the 'oh wow' for me is just as a rookie astronaut being thankful that my job is maneuvering the vehicle so I can be up there and look outside."

Ilan: "Looking out the window is a big 'oh wow.' This morning is also a big wow for me, being able to conduct a press conference from up in space as the first Israeli astronaut. Looking at the U.S. and Israeli flag behind us, I think it's a big 'oh wow' at least for Israel and a big start of scientific international collaboration. As you know, we have international involvement in all of our experiments, including Arab scientists, and that's a big wow for me."

Laurel talked about watching a sunset from space: "The first time I got to see the orbiter as the Sun set or rose. There's a flash—the whole payload bay turns this rosy pink. It only lasts about 15 seconds and then it's gone. It's very ethereal and extremely beautiful and unexpected, I hadn't heard about that before." She added, "Anybody who knows me knows how much I love plants and animals. One day we took the video camera and hooked it up in order to do some sampling. There were roses in there. They had been buds; they had opened up to bloom. It was so magical to have roses

A typical sunset in space. This particular photo was shot with a digital camera two days after launch, as Columbia passed over Australia.

growing in our laboratory in space." Laurel also talked about the silkworm experiment, saying, "One of the silkworm cocoons had just recently hatched. There was a moth in there, and it still had its wings crumpled up and it was just starting to pump its wings up so it would be able to fly. Life continues in lots of places, and life is a magical thing."

K.C. said this: "Early in the mission we were very fortunate, we had the full Moon when we launched. The orbiter moved from one attitude to another; if you were looking at the tail the whole spaceship was glowing in silver light from the Moon. The attitude was changing and you really felt the spaceship was headed somewhere." She added, "We've been able to get some magnificent views of the Milky Way. The very first time I saw the Milky Way almost close to the horizon it looked like a silver dust cloud." She called several of her crewmates and they discussed what it could be, but it took a few days before the Moon dimmed enough for them to realize that what they were looking at was the Milky Way. K.C. said, "It really sort of overpowered the rest of the sky because it's very very dark."

Chawla then gave the most poetic description for what she saw—seeing the Earth reflected in her own eye. She said, "One day I was in the flight deck looking outside. It was starting to get dim so you got to see your own reflection [in the shuttle's windows]. The Earth limb is outside and you can still see the Earth's surface and the dark side. I could see my reflection in the window and in the retina of my eye, the whole Earth and the sky could be seen reflected, so I called all the crew members one by one, and they saw it, and everybody said, 'Oh, wow!'"

Rick concluded: "From my perspective as the commander—my 'oh wow' is watching everybody perform as a team. One of the big aspects of

The moon as photographed by one of the Columbia astronauts.

The Orbit 2 shift of flight controllers posed for a group photo before launch.

that is when we hit Main Engine cutoff, how everybody kicked into gear in orbit. The crew has performed just marvelously, I really love seeing the team come together." Recognizing the contributions by Mission Control and the others who made STS-107 possible, Husband continued, "The bigger team which makes this possible is the team on the ground."

The time remaining after Florida's 35 minutes was given to the Israeli television reporter in Washington, D.C.. He conducted most of his interview in Hebrew with Ilan. Ramon tried to summarize the questions and answers in English for his crewmates, but the reporter interrupted several times and prevented Ramon from continuing with the translations.

Ilan was asked about his private family conferences and he replied, "It's very emotional, especially the first time after six or seven days up in space and it's very emotional for them as well, to see me up in space floating and doing tricks with M&Ms." He also mentioned, "I think floating, especially when you float in slow motion, gives you the feeling you belong to the universe. Although we fly very fast up here, the motions of the universe are very, very slow."

The reporter asked the rest of the crew in English, "With all of the time you've spent with Ilan in space, can you say something to us in Hebrew?" The crew laughed, discussed it for a couple of seconds before they all said "Shalom" together and the reporter replied, "Very, very good. Shalom and peace. Thank you very much for your time and this interview with Israeli television channel 10."

Later that day I received a copy of the flight day 12 Mission Evaluation Report clearing the foam strike, confirming what I had been told six days earlier.

CHAPTER 33

Capcom Linda Godwin radioed John Lennon's "Imagine" for Willie's wakeup song. He replied, "That song makes us think that from our orbital vantage point we observe an Earth without borders, full of peace, beauty, and magnificence. We pray that humanity as a whole can imagine a borderless world as we see it and tries to live as one in peace." Ilan translated McCool's remarks into Hebrew. Willie and Ilan's remarks are on this book's website.

The next day Mission Control sent a message: "Willie, we wanted to let you know that we appreciated your words and perspective of the world following your wakeup song yesterday. Too bad the whole world can't look over the shoulders of you and your crew and see what you see!" They joked, "We're also about out of wakeup music, so you're going to have to come home soon!"

Throughout the mission, the crew was limited to using one of the two intercom channels. Mission Control asked the crew to try to troubleshoot the system. It worked, and Willie admitted the intercom was misconfigured due to "first time flyer syndrome."

The MEIDEX payload's primary objective was to look for dust storms. In addition, the scientists hoped to observe electrical activity in thunderstorms.

Israel Space Agency Director General Aby Har-Even told me that the MEIDEX experiment needs a trained operator on orbit to look for dust storms to determine whether they're actually dust storms or just clouds. While Ilan was the primary astronaut for MEIDEX, many of the other crewmembers were also trained.

Sprites and ELVEs (Emission of Light and Very low frequency perturbations due to Electromagnetic pulse sources, pronounced "elfs") are very rapid electrical bursts and were only discovered nine years before the mission. They're responsible for the 'crackle' in shortwave radio communications. Sprites and ELVEs had been seen by the shuttle's video cameras by chance, but never with a calibrated scientific camera.

The ELVE which Dave Brown mentioned during the press conference was discovered a couple of days into the mission. Israeli scientist Zeb Levin explained, "[Dave] aimed a camera in the South Pacific in the direction we identified as a potential direction for ELVEs. It was above a very big thunderstorm area in the South Pacific. He observed ELVEs in the altitude between 85 and 95 kilometers, so we are very excited. It's the first time it's been observed with a calibrated spectral camera." Later, Dr. Yoav Yair remarked, "It exceeded my optimistic prediction that we could get the sprites and ELVEs with our camera."

Unfortunately, Columbia ended up launching in January, the time of year in which there's the fewest dust storms. Four days after launch, Levin said he was hopeful that dust storms would occur during the flight. He said, "The weather is going to clear in a day or two, so we expect to have a better chance of seeing dust storms than we have had until now." But he admitted that the chances for a dust storm before Columbia had to return home were "Not very big. We were unlucky by the fact that the shuttle was postponed to the winter; originally it was supposed to be in the summer or spring, when the chances are much higher."

Later in the mission Ilan noted, "In the first seven days we didn't have any dust. But it looks like they waited until my shift. We had a big dust storm over the Atlantic, which lasted about two or three days, so I think the scientists are very, very successful with their experiment."

While the scientists were happy to collect the dust storm data over the Atlantic Ocean off the equatorial African coast, since they had waited so long to view one from space, it wasn't everything that they had planned for. A specially equipped plane was based in Greece. The scientists hoped to fly the plane underneath a dust storm while MEIDEX simultaneously viewed the storm from above, but that would only be possible if a storm occurred over the Mediterranean Sea—the airplane couldn't fly to the other locations, like the Atlantic Ocean region, quickly enough.

Fortune smiled on the Israeli scientists when forecasters predicted a dust storm over their primary region. After two weeks of hoping for dust storms over the Mediterranean Sea, the Israeli scientists were ecstatic

Atlantic dust plume, Jan. 26th

MODIS image

Xybion mosiac

A dust cloud off the African coast is observed by the MODIS instrument on NASA's Terra satellite and also by MEIDEX's Xybion camera's six filters. Photo credit MEIDEX team

when one actually appeared. By coincidence, the astronaut who was operating the MEIDEX camera was Ilan Ramon. It took 188 orbits until the dust showed up over the Mediterranean—it was the last orbit before the flight plan called for the experiment to be shut down.

Dr. Joachim Joseph said, "We hoped to carry out [the simultaneous shuttle and aircraft operations] many times during the mission, but it did not happen. At the very last orbit over the Mediterranean, we got a nice dust storm in the area of Israel, which we overflew [while] the plane flew through it. We just lucked out that there was a dust storm. You see me smiling in retrospect now. We were not expecting it. We didn't know if there was going to be any dust, but lo and behold, we got it."

Overall, though, MEIDEX was a mixed success. The hardware worked properly, and the scientists got some dust observations and lots of lightning data. Data was collected on more than 50 orbits, and all together, 392 minutes of video was transmitted to the ground. The scientists produced some preliminary findings, such as the changes caused by burning a rainforest in Brazil. Dr. Joseph said the MEIDEX data verified the theory that smoke dissipates cloud cover in its vicinity and lets more sunlight to enter. He said, "If this kind of thing happens, biomass burning all over the world—and it is happening all the time all over lower latitudes—if the clouds do that, then this is a factor that has to be taken into account when you try to model climate and greenhouse effect on climate more accurately." The failure to collect more dust data over the primary region was because of the January launch, not because of any problems with the hardware or its operations.

Capcom Stephanie Wilson radioed the Indian song, "Yaar ko hamne ja ba ja dekha." by Abita Parveen, for K.C.'s wakeup. In a video recovered after the accident, K.C. smiles as Wilson pronounced the title. It's hard to tell whether she's thinking, "Boy, Stephanie really did a good job pronouncing that name," or, "At least she's trying to pronounce it as well as she can." K.C. radioed, "You did a really, really good job in saying the title of the song. I woke up just in the nick of time and recognized it. It says I've met my friends, they're in front of me and they are behind me and they are all around. I certainly feel that way. I think the red team is up all around and we're ready to go back to work; thank you." Wilson admitted she had some coaching for the proper pronunciation from K.C.'s sister.

Three payloads were mounted on Spacehab's roof: Texas A&M's StarNav, the Air Force's MSTRS, and the European Space Agency's Com2Plex (Combined Two Phase Loop Experiment).

Com2Plex consisted of three heat pipes—passive devices to maintain temperature in orbit. Spacecraft thermal control is a tricky task. The vacuum in space acts like a Thermos bottle, with the Sun generating a lot of heat. The spacecraft has to dissipate the excess heat to maintain a stable temperature. Without gravity, colder liquids don't settle to the bot-

K.C. smiles as she hears capcom
Stephanie Wilson pronounce the
name of an Indian wakeup song.

tom of a container, so some method has to be used to "pump" the heat
out of the spacecraft. Mechanical pumps and radiators can be used, but
that involves a lot of components that can fail over time. A heat pipe has
no moving parts. It's a sealed tube with thin inner ribs, running along its
length, that attracts drops of a liquid like ammonia. Capillary action,
similar to how water crawls up a napkin dipped in a glass of water, sucks
the fluid along the pipe. The colder fluid is pulled through the pipe by
capillary action to the hot end, where it absorbs the waste heat from the
spacecraft's electronics. The liquid boils into a gas and spreads through
the pipe. On the cold end of the pipe the gas cools back into a fluid and
the process repeats. The three Com₂Plex heat pipes came from
ASTRIUM in Toulouse, France, SABCA in Brussels, Belgium, and
TAIS in Moscow, Russia cooperating with OHB in Bremen, Germany.
Com₂Plex radioed back all of its data before Columbia's reentry.

The Air Force's "Miniature Satellite Threat Reporting System"
(MSTRS—pronounced "misters") was a radio receiver designed to mon-
itor the normal radio frequencies and strength at the shuttle's altitude.
The ultimate goal for the research is a detector for future satellites that

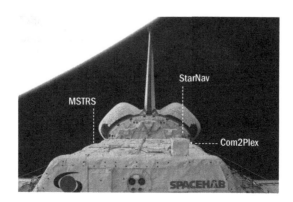

The three external
Spacehab payloads—
StarNav, MSTRS, and
Com2Plex.

can sound a warning if an enemy is trying to overwhelm the satellite with a blast of radio waves so it can't hear its commands. MSTRS was able to radio back most of its data. It was the only military experiment on STS-107, not classified, and flying as a commercial payload. But after the accident, conspiracists wondered whether or not MSTRS played a sinister role in a government coverup.

StarNav was a star tracker, built by students at Texas A&M University. The team's goal was to design an inexpensive star tracker to determine the shuttle's attitude. A CCD camera was mounted in a tube with a single-board computer. The computer converted those images into readable star charts, and its Lost In Space Algorithm (LISA) program calculated the spacecraft's orientation. The software had to isolate debris and other unwanted defects in the photos so accurate star patterns could be determined.

There was a problem the StarNav team had not anticipated: stray light entering the camera. Imagine holding an empty paper towel tube to your eye and using it to look at the Moon when you're standing near a streetlight. The tube prevents the streetlight from entering your eye. The StarNav camera had a set of baffles to limit stray light, but some light still got in because of all of the reflective surfaces in Columbia's cargo bay. The team asked the payload planners to change their viewing opportunities to ones when they were within the Earth's shadow and less likely to encounter any stray light.

Before the mission, the StarNav team made a list of restrictions for when their instrument could be turned on — no viewing the Sun, Moon, brightly lit Earth, or when Columbia was passing through a high radiation region over the South Atlantic Ocean. Unfortunately, six days after launch StarNav was powered while going through the South Atlantic high-radiation region. Its hardware was built from non-radiation-hardened components and, not surprisingly, it stopped working. The StarNav team could not send commands to shut off their unit in time because of other higher priority communications. Later, the crew was sent a maintenance procedure instructing them to send commands from a laptop computer to the payload. StarNav scientist Igor Carron noted, "Eventually we had to stop, as the camera never was able to perform the duties that were expected from it [after the radiation problems]."

So, was StarNav a failure? Actually no. The main purpose was for the students to develop a technique that could be used on future spacecraft sensors. They were able to design their star tracker and build it out of off-the-shelf electronics, test it ahead of time to ensure it would work in outer space, and make it safe enough to fly on the shuttle. All together, they had thirty 25-minute opportunities to use StarNav and they received 32 digital images. The onboard computer was able to use those images to determine where the instrument was pointed.

Rick Husband shows off the freezer filled with the blood, urine, and saliva samples contributed by the crew.

The goal for the experiment was a prototype star tracker, and that was extremely successful. The problems the StarNav team encountered were useful problems because they showed the limitations of their concept and where they needed improvements. Texas A&M has used the knowledge from StarNav to develop the Khalstar Star Tracker, which is commercially available.

The red shift completed their final science operations on Thursday, January 30. Before Rick left Spacehab for the last time, he showed off the 1-cubic-foot EOR/F (Enhanced Orbiter Refrigerator/Freezer), filled with the crew's blood, urine, and saliva samples. Laurel, noting the color of the urine and its value to the investigators, called the samples "liquid gold." The crew had worked long and hard to fill the freezer with those samples.

A rare photo shows Columbia flying over the Altun mountains in western China.

CHAPTER 34

The blue shift wakeup on flight day 16 was "Silver Inches," by Enya, for Dave Brown. He said, "The blue shift is up, that really is a pleasant song to start the day with. We've got a busy day as we finish up our science today. We're also starting some activities to get ready to come home. I think as much as we've enjoyed it up here, we're also starting to looking forward to seeing all of the people back on Earth that we miss and love so much."

Several automated payloads flew without much astronaut involvement. At most, the astronauts were asked to turn on an experiment and clean the some filters.

Protein Crystal Growth (PCG) was the shuttle's frequent flyer payload. It has flown scores of times, carrying thousands of samples. On Earth, protein crystals can grow only so large because of gravity. In space, crystals grow far larger, and larger crystals are easier to analyze. The hope is to find a protein crystal with useful qualities for fighting diseases which can be used in laboratory synthesized drugs that can be replicated on Earth. In effect, the PCG scientists are the equivalent of a hunter using a shotgun filled with buckshot aimed at a flock of birds—the more times you try, the greater your chances for success. Critics say the PCG experiments only show that it's possible to grow crystals at thousands of times the cost of growing protein crystals on Earth. Proponents say the larger crystals created in microgravity are more useful for researchers analyzing the structure of the proteins, and the information gathered from microgravity experiments has led to better techniques for growing crystals on Earth.

STS-107's Advanced Protein Crystallization Facility (APCF) was sponsored by ESA, and 38 experiments were flown for eight European scientists. A separate protein crystal experiment featured 500 proteins selected by commercial pharmaceutical companies. Some of the specific diseases targeted included Alpha interferon (hairy cell leukemia); AIDS-related Kaposi's sarcoma, venereal warts, multiple myeloma, melanoma, and chronic hepatitis B and C. The University of Alabama at Birmingham sponsored another protein crystal growth experiment. Useful protein crystals sell for $1.7 million per gram—so they're incredibly valuable materials. Surprisingly some of the delicate protein crystals survived.

The Orbital Acceleration Research Experiment (OARE) was permanently mounted on the floor of Columbia's cargo bay. It measured long-duration, low-frequency accelerations, like the shuttle's drift and drag. SAMS (Space Acceleration Measurement System, pronounced "sams") measured higher frequency accelerations like pumps. Both OARE and SAMS were sensitive to crew motions. Extremely sensitive microgravity experiments, like the combustion experiments, needed the SAMS and OARE data to verify the quality of the microgravity during their tests. Much of the OARE and SAMS data was downlinked during the mission.

One of the final experiments completed was the European Facility for Adsorption and Surface Tension (FAST, pronounced "fast"). It blew bubbles. Bubbles hold their shape because of surface tension, but they're affected by gravity. If you blow a soap bubble, there's a little more liquid at the bottom. FAST permitted scientists to study surface tension and elasticity in a "pure" environment, without gravity's influence. A pump would apply a precise amount of pressure, and by measuring the size of the bubble the scientists could determine the amount of surface tension. The bubbles were tiny—just 1 mm. (0.04 inches) across. In some cases the pressure was quickly varied to see how the bubble expanded and contracted. A video clip on this book's website shows a typical FAST run.

There was a variety of FAST experiments. Libero Liggieri with the National Research Council in Genoa, Italy, studied "Adsorption Dynamics and Rheology (internal flow properties) of Liquid-Liquid Interfaces." It used water and hexane with a detergent solution. Reinhard Miller with the Max Planck Institute in Berlin, Germany, examined "Fast Surface Relaxation at the Solution-Air Interface." His experiment used detergent in water with rapidly changing pressures to see how the bubble would react. Giusepppe Loglio at the University of Florence in Italy studied the "Slow Surface Relaxation at the Solution-Air Interface" using a soapy water solution. The chemical name for the detergent was n-dode-cyl-dimethyl-phosphine-oxide.

Each of the experiments was repeated many times with different temperatures and concentrations of the chemical solutions to get a wide range of data. About 2,000 FAST tests were completed. At the end, the

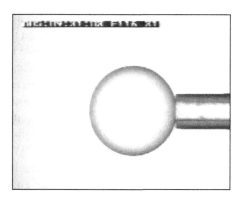

A typical FAST bubble.

team decided to have a little fun with the experiment. It looked like the machine was on overload—blowing bubbles all over the place! A video clip of the FAST 'fireworks' celebration is available on this book's website. All of FAST's data was transmitted during the mission, so the scientists were still able to perform their science.

The red wakeup was "Shalom lach eretz nehederet" by Yehoram Gaon, for Ilan Ramon. The tune is the same as Arlo Guthrie's "City of New Orleans." Ilan said, "Good morning, Stephanie, it was wonderful to hear your voice in Hebrew. It was perfect timing while we flew over Israel to hear this song, which says I've been to the North Pole and to everywhere but there's no place better than Israel."

Most of Friday, January 31, was spent packing up and reconfiguring Columbia for return to Earth. Everything the crew took out during their two weeks in space was packed away, preferably where it could be found, after landing, by the technicians responsible for removing experiments and other critical items. A special bag marked "return to Houston" included the exposed film, tapes, crew notebooks, and anything else that needed to be returned home quickly with the astronauts. There was a request by the MMT to rush back to Houston the film Mike and Dave shot of the External Tank showing the bipod area, so it could be analyzed as soon as possible. Depending on how much foam had fallen off the bipod, it could affect the launch date for the next shuttle flight, scheduled for early March.

From the perspective of collecting scientific data, STS-107 was a major success. Dr. John Charles noted, "We had talked even during the mission about having another STS-107-like mission because it had gone so well." Because of NASA's changing priorities, even before the accident, the chances were low for any additional dedicated shuttle science flights. There were some initial discussions about joint ISS missions, where part of the Spacehab module would be filled with cargo and the rest occupied by microgravity science experiments, but even that possibility disappeared due to higher-priority ISS payloads needing that space.

Overall, STS-107's science throughput was incredible. The only conflict was when a Zeolite Crystal Growth experiment took place at the same time an astronaut was doing a run on the ARMS ergometer. The ZCG team wanted as pristine a microgravity condition as possible without the astronauts creating vibrations on the ARMS exercise device, but the planners on the ground weren't able to reschedule the ZCG run or the ARMS activities, so they had to occur at the same time.

Before launch, STS-107's flight plan called for 468 hours of astronaut time for the experiments. During the mission, the payload team received an additional 23 hours that the crew donated, giving up their free time to ensure all of the experiments were completed.

Dr. John Charles noted that it was "What I believe is one of the most complex multidisciplinary missions which NASA has flown to date." He said, "With the mission coming to an end, the data analysis is only just beginning, and it may take months to complete. The final results from the scientific experiments will be published in the open literature, and we're planning to gather all of the NASA investigators together at a post-flight symposium, maybe 18 months after the mission, so we can hear first-hand what they've learned from this mission."

By this point the payload team members on the ground were packing up their support equipment and shipping it back to their laboratories. In many cases they already had all of the data from their experiments and were anxious to start their research. Payload teams who needed to pick up their experiments as soon as possible after Columbia landed traveled to Florida to be there for the landing.

As a matter of routine, the day before a shuttle lands, the pilots and flight engineer check out the hydraulics, thrusters, and other systems. As Columbia flew over KSC and the backup landing site in California, the UHF radio was checked out. It's basically an off-the-shelf military aircraft radio and serves as a backup to the sophisticated TDRSS communications. One key advantage to the UHF radio is its simplicity—as long as it's above a ground station, the shuttle can communicate with Mission Control. Entry flight director LeRoy Cain said, "The vehicle performed flawlessly today [during the checkouts] as it has the entire mission."

Husband and McCool also used a flight simulator program, very similar to consumer computer games. The space shuttle simulator is an extremely sophisticated piece of software, certified as an astronaut training tool. It includes the same displays the astronauts use during the actual reentry and has a fairly realistic simulation of a shuttle landing. On long shuttle missions, it's important as a "refresher" training tool because of the time since the pilots last flew the STA.

Columbia operated almost perfectly throughout the mission. Manager Phil Engelhauf said, "Columbia in its 28th flight is just nearly flawless. It's extraordinarily clean, for such a long mission to have only the two or three trivial squawks (minor problems) against the hardware that we have is just really great."

The evening before landing was a Friday, and Ilan's last opportunity to perform a Kiddush. This was after the final e-mails to and from Columbia, so Ramon would not have had the capability to inform anyone afterwards. It's possible he didn't have enough time. It's possible he did make the time for the Kiddush, and may have asked his crewmates to participate. But nobody will ever know.

The red shift went to sleep for the final time on Friday at 5:39 A.M. Weather forecasters predicted excellent conditions for the planned landing on February 1: light winds, few clouds, 7 miles of visibility, 56 degrees Fahrenheit, and no chance for rain. Columbia had enough supplies to remain in orbit for an additional four days, but it was desirable to land as soon as possible to give the scientists their payloads back in as good condition as possible.

Amateur astronomer Rick Baldridge
took a wide-angle time exposure of
Columbia's reentry over California.
Photo courtesy of Rick Baldridge.

PART IV

FEBRUARY 1, 2003

Landing day is a set of mixed emotions for most astronauts—satisfaction they've done a good job and accomplished their tasks, and pleasure they're soon going to see their families and friends again. But also a bit of disappointment—the big adventure is coming to a close. Landing itself is a bittersweet moment—exhilaration the mission has been completed successfully and you're safely back on Earth, but a bit of sadness your mission wasn't extended, and realization the six people you've shared the last two weeks with—and much of your lives for the past several years—will soon be separated.

The American astronauts realize they will probably fly again if they want to, but it may be several years. With a large number of astronauts and a fairly low flight rate, it could be 3 to 6 years before an astronaut has a chance to fly on another mission. And Ilan, of course, had to realize no matter how much he wanted to fly again, his spaceflight was almost certainly a once-in-a-lifetime experience. (He had sent an e-mail to managers insisting the entire crew get immediately assigned to another flight, but he had to realize there was no way his suggestion would be taken seriously.) It wasn't likely the U.S. would invite Israel to fly another passenger for free, and no way Israel could afford to become a more active participant in international space programs.

On February 1, 2003, everything looked perfect for Columbia's landing. While the shuttle had enough supplies to stay in orbit for another four days, the mission had accomplished all of its planned science activities.

CHAPTER 35

REENTRY

STS-107's final day in space began for Mike and Dave at 4:39 P.M. January 31. Their wakeup music was "If You've Been Delivered," by Kirk Franklin, for Mike, radioed by capcom Linda Godwin. He said, "Thank you, Houston, that should certainly wake you up. Thank you for waking us up on what is blue shift's last day in space. It's kind of mixed emotions as we get ready to come home, but we have enough memories to last us for a lifetime. We'd like to thank our families, our wives and our kids for providing us with all of the wonderful wakeup music each day, getting us off to a great start. Linda, as you know this is a very busy day today: we have a lot to do to prepare the orbiter to come home. We're going to get to work and get all those things done and hopefully we'll see everyone soon." Godwin replied, "We've all enjoyed the mission down here, I think that music woke up the control center too." Willie was given an extra two hours of sleep to ensure he would be well rested and in peak condition for the landing.

Anderson and Brown spent most of their final shift packing everything for reentry. Dave's last science activity was to check on the rats and refill their water bottles. He was the "deorbit guru," responsible for making sure everything that was taken out was put into the appropriate locker or storage location. The seats were installed and the seven orange launch and entry suits were prepared.

The red shift was awakened at 1:39 A.M. by "Scotland the Brave," by The Black Watch and the band of the 51st Highland Brigade. Laurel replied, "Good morning, Houston. We're getting ready for a big day up here, had a great time on orbit and [are] really excited to come back home. Hearing that song reminds me of all the different places down on Earth and all the friends and family I have all over the world. Thanks, and it's been great working with you and all the other folks."

Entry flight director LeRoy Cain and his flight controllers came on duty in Mission Control at 2:30 A.M. EST. They were the ones responsible for bringing Columbia home.

The only concern was Columbia's weight. On a flight like STS-107, without any satellite deployments, everything launched returns home with the shuttle, except for the hydrogen and oxygen used to create power and water. Excess water from the fuel cells is dumped overboard periodically. Columbia was running slightly more energy-efficient than anticipated, so less of the supplies were used than planned. As a result, it was just over its planned reentry weight. Engineers calculated Columbia would weigh 234,011 pounds at reentry—11 pounds over the limit for a planned mission.

The extremely slight excess was not a big concern. A 1997 Columbia flight had a much heavier landing because its 16-day mission was cut short due to a technical problem. Still, one manager suggested—seriously—turning on some electronics not normally used on orbit. The extra power would consume more hydrogen and oxygen, creating water which could easily be dumped: A simple way to get rid of a few pounds, but it wasn't needed.

The Spacehab hatch was sealed around 4 A.M. Even if Columbia had to remain in space because of bad weather or a technical problem, there was no need to go back inside: the rats and other animals had enough food and water for several days on their own.

The updated forecast called for some clouds at 3,500 feet, 7 miles visibility, winds from 6 to 9 knots, a temperature of 58 degrees Fahrenheit with a relative humidity of 63 percent—excellent landing weather.

Chief astronaut Kent Rominger was back in Florida. His responsibility was to fly the Shuttle Training Aircraft and give a pilot's perspective on the weather. There was an early-morning fog over the runway, but it burned off as the Sun warmed the ground.

One of the more bizarre reentry activities is "fluid loading." The blood tends to concentrate liquids in the torso and head in microgravity. Astronauts drink salty liquids a couple of hours before landing in an effort to push more fluids into the arms and legs, to help the body readapt to gravity. The amount of liquid depends on the astronaut's weight; from 24 ounces for K.C. up to 48 ounces for Husband and Brown. Each astronaut chooses water with salt tablets, chicken consommé, or salt tablets mixed with a drink. Rick chose chicken consommé, Dave Orangeade, Laurel imitation lemonade, and the rest selected plain water.

If there's a wave-off, then they drink more of the salty liquid. If the decision is made to stay in space, there's a long line to use the shuttle's toilet! Mission Control is very aware of this situation, and tells the crew when to start their fluid loading, or to hold off if the weather is iffy.

The Mission Control team considered holding off for one orbit to let the fog dissipate and see if the winds subsided, but that wasn't necessary. Once Cain was satisfied with the weather forecast, he gave the go-ahead. Capcom Charlie Hobaugh radioed, "I guess you've been wondering, but you are 'go' for the deorbit burn."

Rick, Willie, K.C., and Laurel took their seats on Columbia's flight deck. Laurel used a camcorder to videotape their activities during the reentry. Mike, Dave, and Ilan were in the middeck. They got to look at the lockers in front of them but could watch the video from Laurel's camcorder on portable video monitors.

The deorbit burn began at 8:15 A.M. over the Indian Ocean. It was the 255th orbit. Columbia was flying "backwards," with the engines facing the direction it was flying. Husband and McCool fired the two Orbital Maneuvering System engines (OMS), which hadn't been used since Columbia arrived on orbit. From inside the crew cabin, they sounded like cannons going off as they slowed Columbia out of orbit. Then the pilots flipped Columbia "right side up" to expose the black tiles on the underside to the atmosphere.

All of the launch energy has to be removed during the reentry, and that's a lot of heat—heat that's reflected away or absorbed by the thermal protection system. As the atmosphere slowed Columbia, the astronauts started to feel gravity for the first time in 16 days, a little at first, then more and more. It isn't as drastic a change as during launch, but it feels like more because their bodies are now used to being weightless.

Five days after the accident a videotape was found near Palestine, Texas. It was the video Clark shot during reentry and showed the crew performing their ordinary tasks and includes nothing unusual. The video lasts about 13 minutes and shows flight deck activity beginning about 8:35 A.M. EST.

The tape shows the astronauts laughing and enjoying themselves—clearly pleased their mission had gone so well. Laurel said, "There's some good stuff outside. I'm filming overhead right now" and Willie replied, "It's kind of dull." Husband assured his crew "Oh, it'll be obvious when the time comes."

McCool mentioned, "It's going pretty good, now. It's really neat, just a bright orange yellow out over the nose, all around the nose," and Husband told him, "Wait until you start seeing the swirl patterns out your left and right windows." Willie replied, "Wow," and Rick verified, "Looks like a blast furnace." Husband dropped a checklist and said, "Let's see here

Rick Husband (right background)
and Willie McCool (left foreground)
during Columbia's reentry.

. . . look at that"—for the first time in over two weeks it fell down instead of floating in place.

Willie said, "This is amazing, it's really getting fairly bright out there," and Rick replied, "Yep. Yeah, you definitely don't want to be outside now." K.C. found that rather amusing and quipped, "What, like we did before?" and Rick acknowledged, "Good point."

The videotape breaks up at 8:48 A.M., when Columbia was over the eastern Pacific Ocean, southwest of the San Francisco Bay area. The reentry heat destroyed the rest of the tape after Columbia broke up. The destroyed portions would have shown more activities as Columbia entered its heaviest heating. Normally Clark would have continued filming through the landing, but at the first indication of anything wrong she would have shut off the camcorder, put it away, and taken out an emergency checklist. A transcript of the astronauts' comments and video is on this book's website.

A month later, CAIB head Hal Gehman noted the videotape only showed the first portion of the reentry, not the period when the accident occurred, so it wasn't useful to the investigation. In one way, the videotape was extremely valuable. It showed the crew was happy, enjoying themselves and joking with each other. They had had an incredibly productive 16 days of science, and were looking forward to the landing and reuniting with their families. Laurel's husband Jon said, "It was kind of like a message or gift. A special blessing, their final moments joking about how hot it was out there and you could just see them in high spirits. To me it was a very joyous thing." Dave Brown's brother Doug noted the video showed the families without a shadow of a doubt the crew was happy and had no knowledge about any impending danger, contrary to some of the speculation and rumors the astronauts knew they were doomed.

Around 8:49 A.M. Columbia started to slow its descent. It was at an altitude of 40 miles and traveling 24.6 times the speed of sound. The intensity of the heating increased as Columbia traveled through thicker atmosphere. Unnoticed, a hot blast furnace started to enter Columbia's left wing through the area damaged by the foam strike. In Mission Control, telemetry was received showing an extremely gradual rise in the left main gear temperature. It was too subtle for anybody to notice or cause any concern at the time.

In stark contrast to all of the extra security for launch, there was just ordinary security for landing. The KSC visitor center was open and put out a press release saying it was the closest location the public had access to, and if Columbia approached from the south, it would be possible to view it as it went overhead. The lack of security also applied to the families. While they spent the launch at a military base with heavy security and police escorts wherever they traveled, for landing they were basically on their own, staying in condos in Cocoa Beach.

The escort astronauts picked up the families from their condos and drove them in for what promised to be a picture-perfect Saturday morning landing.

CNN was the only U.S. television network to provide updates about Columbia's reentry. Throughout the morning, reporter Miles O'Brien informed viewers about the progress of Columbia's reentry, describing how it would be visible in many Western states. At 9 A.M. O'Brien said, "Good morning, Texas. If you hear a boom-boom, that's the space shuttle. Take a look outside, you should see what looks like a streaking meteor. It's actually the space shuttle Columbia coming back."

CNN asked Dallas affiliate WFAA to videotape Columbia as it traveled over Texas both because it's an incredibly exciting visual, and just in case something bad occurred, which would be more newsworthy.

As usual there were only about 20 reporters who decided to attend the landing—the local television stations, newspapers, wire photographers, and a couple of members of the media who are space fans. A television station and a couple of print reporters represented Israel.

To the north of the press area were about 100 VIPs. They included NASA managers, scientists anxious to get their experiments back, VIPs who'd attended the launch and decided to come back for the landing, and ordinary workers who came in on a Saturday morning to see Columbia's landing. Further north of the VIPs was the roped off area reserved for the astronaut families and their guests.

Far less people are interested in landing than launch. Launches attract more crowds because of the noise, vibrations, and sheer excitement when people ride a 500,000 gallon fireball of explosive propellants into space. In contrast a shuttle landing is extremely quiet—almost a work of poetry in its subtlety. Without engines the shuttle lands almost silently—just a whooshing sound as it pushes the air aside. There's more noise from the vehicles in the landing convoy and the STA.

Many VIPs feel it isn't worth the extra cost and time to travel back for the landing, especially since landings can be put on hold due to bad weather, or the shuttle can be diverted to California if the weather is bad in Florida. For example Laurel's family present for the landing consisted of her husband, son, and sister. Dave's family for landing was just his brother Doug.

CNN doesn't normally bring a team to Florida for shuttle landings. It receives NASA's satellite feed at its headquarters in Atlanta, Georgia. I was CNN's person on the spot for the landing, just in case something happened. I was connected to the CNN producer in Atlanta by cell phone and ready to go on the air if there was a something that justified a major news story. I had never expected to actually be used for that purpose, but was always conscious it might become necessary some day.

CHAPTER 36

T H E   A C C I D E N T

As far as anybody waiting at KSC knew, everything was going perfectly with the scheduled 9:15:50 A.M. landing. The NASA broadcast consisted of the audio of the astronauts, capcom Charlie Hobaugh, and public affairs announcer James Hartsfield. (The same audio is played on speakers at the landing site.) Those inside Mission Control could also listen to the flight director's channel and other internal communications loops.

At 8:58:44 A.M. Rick radioed, "And, uh Hou. . ." but got cut off in mid-syllable. Hobaugh replied, "And Columbia, Houston we see your tire pressure messages and did not copy your last." Translated to English this meant there was a warning in Columbia's cockpit that a tire pressure sensor had gone bad. Husband or McCool had pressed a button to acknowledge they saw the warning indicator and Hobaugh was telling Rick he was cut off and should repeat whatever he had previously said. Husband replied, "Roger . . . bu– " and was cut off again. While out of the ordinary, there wasn't any indication of any problems other than the loss of communications.

Earlier spacecraft had communications blackouts during reentry. As a spacecraft enters the Earth's atmosphere, the intense heat ionizes the air, making it impossible for radio signals to get through to the ground. However, the ionization does not prevent antennas from transmitting "up," toward space. Since TDRS (Tracking Data and Relay Satellite) became operational in 1989, the shuttle has been able to stay in near-continuous communications, including during reentry. There are some small communications dropouts, but nothing like the three-minute blackouts experienced before TDRS. The loss of communications during Columbia's reentry was unusual, but there were many possible explanations.

CNN producer Dave Santucci asked me, "Hey, Phil, what's up with the loss of comm?" I told him, "Don't worry, Dave, they're smart boys and girls—they know how to fly the shuttle on their own." There was absolutely no concern in my mind—loss of communications happens for a variety of reasons for a couple of seconds, and it isn't a big deal if the shuttle is out of communications for a longer period. Even if communications isn't restored, there are dozens of possible reasons. There could be a problem with a TDRS in orbit, the TDRS ground station, the link to Mission Control in Houston, or a technical problem at Mission Control. Something as simple as a noisy phone line could cause the problem, so I wasn't worried.

Communications was not restored. Hobaugh called, "Columbia, comm check," without any response. The giant displays in Mission Control were frozen with the last data received from Columbia. Those monitoring NASA's broadcast and at the landing site heard only Hobaugh's "comm check" call and Hartsfield announcing, "Flight controllers are continuing to stand by to regain communications with the spacecraft."

As the lack of communications stretched out to over three minutes, I wondered what could be the problem that was taking so long to correct. Then, at 9:03:35 A.M., Hobaugh said the words I will remember for the rest of my life: "Columbia, Houston, UHF comm check," he radioed, and got no response. The UHF is a completely separate radio from the sophisticated S-Band radio used by TDRS. Columbia should have just passed over the horizon at KSC, and if it was a relay satellite problem, then the Merritt Island tracking station should still have been able to contact Columbia directly on UHF. For the life of me, I could not think of anything that would cause two separate, independent radios with different communications paths to fail. I had a bad feeling in my gut—realizing there had to be a major problem, probably catastrophic. I told my producer, "Dave, something really wrong has happened."

A minute later an Internet space website reported, "We're getting reports from Texas of debris behind the shuttle's plasma trail during reentry."

About this time, Rick's wife Evelyn and their children Laura and Matthew posed for a photo in front of the runway. They were all smiles and looking forward to seeing their husband and father again. Dave's

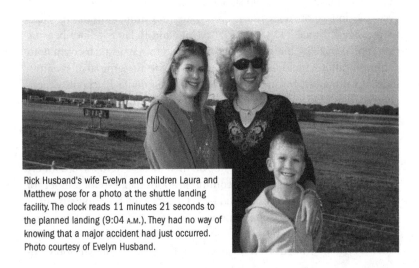

Rick Husband's wife Evelyn and children Laura and Matthew pose for a photo at the shuttle landing facility. The clock reads 11 minutes 21 seconds to the planned landing (9:04 A.M.). They had no way of knowing that a major accident had just occurred.
Photo courtesy of Evelyn Husband.

brother Doug was busy setting up a camera so he could get his own photos of his brother's space shuttle as it landed.

Air Force radars were having as little luck finding Columbia as the
UHF radio. The tracking dishes send out powerful microwave C-Band
signals in a tight beam. The computers aim the dishes at the precise location the shuttle is supposed to appear above the horizon and follow it
through landing. The radar tracking is incredibly precise, and part of the
data Mission Control uses to determine the shuttle's flight path and
whether or not the commander needs to make any adjustments during
the final approach to the runway.

At 9:06:52 A.M. Hartsfield reported, "Flight controllers are still standing by for C-band tracking data from the Merritt Island tracking station
of Columbia and UHF communications." At that point there was no
doubt in my mind something had gone drastically wrong. It might be
possible for two separate radios to go bad, but how could Columbia hide
from radar? I practically shouted into my phone, "Dave, let's get on the
air now. Something really bad has happened." In my brain I realized that
there was a drastic accident and the crew was dead, but I couldn't get
myself to say it out loud. What happened? There was no indication of
anything wrong—just the lack of communications. That could indicate
the time of the accident—the communications were lost when whatever
happened to Columbia happened. But what? There were no warnings or
indications anything had gone wrong. I had a copy of the Flight Day 12
Mission Evaluation Report, which said the engineers had completed
their analysis, and the foam strike would not be a problem. It was considered a closed issue, so that wasn't even in my thoughts.

While this was going on, other reporters tried to talk to me, but I waved
them away and wouldn't talk to anybody. Nobody in my area would have
access to any more information than what was being broadcast on the
speakers, and I needed to concentrate and try to comprehend what was happening. My brain told me a major accident had happened. But in my heart
I still hoped I was somehow incorrect and I had reached the wrong conclusion. These weren't just astronauts. they were people—people I knew, people I shared jokes with, people I had talked to just three days earlier.

A pair of sonic booms precedes a shuttle landing. It sounds like a
rock drummer hitting the bass drum twice in a row while you're standing
next to the loudspeaker—something that really gets your attention, you
feel it in your bones. The sonic booms follow the shuttle all the way
through entry. Early in the reentry, the shuttle's altitude is so great the
sonic booms are barely audible—like listening to a jet aircraft when it's
far above you. But close to the landing site, the shuttle's altitude is much
less, so the booms are far louder—like listening to a jet flying overhead as
it's coming in for landing. By coincidence, the sonic booms are heard at
KSC at roughly the same time the shuttle passes from supersonic to sub-

sonic, so public affairs commentators and reporters often incorrectly claim the shuttle passing through the sound barrier causes the booms.

The booms are heard at KSC about 2.5 minutes before landing. The large digital clocks that show the time until landing are hooked into the same timing circuits Mission Control uses, so they're extremely precise. I watched as they counted down through the time the sonic booms should have been heard. I had little doubt we would not be hearing them, but I hoped I was wrong. More than anything else, I wanted to hear those two welcome booms but they didn't come. Many of the viewers started to realize something was wrong when they didn't hear the booms.

It wasn't until 09:14:23 A.M., two minutes before the planned landing time, Hartsfield announced, "This is Mission Control, Houston. Flight Controllers continue to seek tracking or communications with Columbia through [the] Merritt Island tracking station. Last communications with Columbia was at 8 A.M. Central Standard Time, approximately above Texas as it approached the Kennedy Space Center for its landing. Flight director LeRoy Cain is now instructing controllers to get out their contingency procedures and begin to follow those." But Hartsfield didn't spell it out and say an accident had occurred.

At the viewing site, the countdown clocks continued toward zero, and more people realized something was really wrong. But many weren't aware anything was wrong until the planned landing time of 9:16 A.M. came and went and the clocks started to count up. At least one person speculated the clocks might not be correct.

Some media broadcast Columbia was late for its landing—not realizing the significance of that simple statement. The shuttle *can't* be late for landing: once it fires its engines for reentry, it's basically falling to Earth in a controlled manner. It's possible for the shuttle to divert to another airport if there's a major problem at the shuttle runway, but that clearly hadn't happened. There's no way for the shuttle to wave off and try a second time or take an alternate route. The landing time can only change by a couple of seconds unless something really drastic happens. Something really drastic had happened.

At 9:20 A.M., CNN started to broadcast the video from Texas television station WFAA of Columbia traveling over Dallas. It should have shown a single bright object moving across the sky but instead showed several objects, including pieces coming off and an occasional bright flash.

CNN got me on the air at 9:21 A.M., and Miles O'Brien asked me what information we had at the landing site. I said, "We don't know much more than what you know. About 9 A.M. Eastern we lost communications with the shuttle. The crew's very well trained if they lose communications with the ground. They can proceed on their own. However, when the shuttle [should have] come over the horizon, the capcom in Houston, Charlie Hobaugh, kept on calling the shuttle on the UHF. It's a

CNN's video from WFAA in Dallas showed Columbia as it came apart over Texas. Photo from the author's collection.

totally separate radio from the other [communications] systems on the shuttle. He could not get any contact. Even if that radio was out, it would have still been tracked by the ground-based radars, which it was not. And in addition we would have heard the sonic booms, which we did not. So obviously something happened." I still couldn't get myself to say out loud the astronauts had died, even though I had no doubt the "something" I mentioned was a major accident.

From Atlanta O'Brien asked me, "Can you tell us where that occurred, where in the world that communications loss occurred?" and I replied, "Roughly 9 A.M. the shuttle would have been traveling over Texas, around the same time you saw the video from Dallas."

By this point, CNN had gotten NASA spokesperson Kyle Herring on the phone. Herring was in Mission Control, assisting commentator James Hartsfield. O'Brien said, "Stand by Phil, we've got Kyle Herring on the line." O'Brien told CNN's audience, "He's the public affairs officer from Houston, NASA. Kyle what do we know?" Herring said, "At this point we know we lost communications around 9 o'clock Eastern time. That would have placed . . . the shuttle . . . over the northeast portions of Texas, tracking right on its ground track toward the Florida landing."

O'Brien asked him, "As part of the contingency procedures, what's going on right now in Mission Control?" Herring said, "It's a very detailed plan, obviously. The first thing to do is pull out a contingency plan and start following that. The flight dynamics team obviously continues to try to make contact with its counterparts to see if they had seen anything. Obviously, it's a very methodical approach, a very business-like approach, even in a contingency situation."

O'Brien asked, "We're told this communications loss occurred somewhere over Texas. As part of the contingency plan are they calling local authorities for any search and rescue?" Herring said, "I do not know if that's part of the contingency plan but I would assume that something like that would be taking place."

So far Herring hadn't said anything but the obvious. But then he added, almost as an afterthought, "The only additional information I have is around that time some of the flight controllers did report that some of the sensor data was lost on some of the hydraulic systems on the shuttle. That was the only additional information that we had prior to the loss of communications."

Both O'Brien and myself instantly realized this last statement was an important piece of new information. No sensor losses had been mentioned in Hartsfield's commentary. Finding out about the lost sensors hit me like a sack of bricks. This was significant. It also implied whatever happened to Columbia didn't happen all at once, clearly things were failing over time before whatever happened that caused the accident.

O'Brien told CNN's viewers, "The key thing here is the shuttle operates on hydraulics as it comes in. The key thing to look at here is the Auxiliary Power Units which power the hydraulics." He asked, "Do we have any indication, Kyle, of any sort of failure or problem with those Auxiliary Power Units as they fired up for descent?" Herring said, "We didn't have any indication of anything like that. The only indications we had is we lost some sensor data on some of the hydraulics systems that would indicate tire pressure—those kind of things. When we lost the sensor data, shortly after that we lost the communications link. So at that point we didn't have any data to show any other APU data . . . those kinds of things. At that point we had a total loss of communications."

As CNN aired the WFAA video from Dallas, O'Brien said, "I don't know if there's anything you or I could discern from looking at that shot. But I see multiple trails, multiple trails. It reminds me a little bit of when Mir went down," referring to the intentional controlled reentry of the Russian space station a year earlier, as it burned up and split into several pieces. O'Brien continued, "What does that tell you?" Herring was cautious not to commit to anything, saying, "Well, obviously it would be sheer speculation until we had confirmation about that. Obviously multiple targets is not something that you want to see. I'd be foolish to tell you that that wasn't something that looked pretty bad. We always hope for the best and we're just following the contingency plan as we can at this point, and try to make any contact with folks that can help us with this." O'Brien interrupted, "Kyle, where does it go from here?" and Herring replied, "We're just going to continue to try to contact the folks that can help us find out what happened and following any additional procedures necessary under a contingency."

O'Brien came back to me, asking, "Phil Chien, I don't know if you're near a CNN monitor. This is a very telling piece of videotape. Can you see the video where you are?" Unfortunately I wasn't near any television set. I desperately wanted to see what he was describing—multiple trails certainly were not good news or what I wanted to hear. O'Brien described to his audience, "Multiple trails, indication of multiple targets as the shuttle streaked over Dallas, Texas. It doesn't take a lot of elaboration to tell you what multiple trails and multiple targets would indicate. Phil—anything to add on that?" I stammered out, "There's no doubt we've had a bad day. As you said before—everything's like clockwork with landing. Even if all communications was lost, it should have come up on radar once it came over the horizon."

O'Brien asked me, "What are the other areas that would be high on the list to look at as the shuttle's coming in?" I replied, "There's three Auxiliary Power Units. They operate on launch. They use one of them to check out the shuttle the day before for landing, they turned on all three during landing today. Those are used to operate the wings and elevons, same as on an airplane. Obviously if there was a major disaster on one of those it would be a bad day." In my mind I was going through all of the critical systems on the shuttle and what could have happened, but the foam strike was still not in my thoughts. CNN rebroadcast the WFAA video as I continued, "The shuttle's an electrically operated vehicle; it uses three fuel cells for power. It's not likely all three could have failed, but that would certainly be a possibility."

O'Brien asked me about the significance of the multiple pieces, but my reply was lost as my cell phone signal broke up. O'Brien explained what happens during the reentry. Herring, watching the CNN-WFAA video of pieces coming off Columbia, acknowledged, "Oh, definitely. It's definitely something that occurred. Obviously at this point we don't know what that is that caused this."

At 9:34 A.M. O'Brien reported CNN had received a phone call from a Palestine, Texas, viewer who reported "A loud impact."

O'Brien summarized everything known at that point: "What we're seeing here is very ominous indeed. These are pictures which tell the story. That is clearly the shuttle breaking up as it passes south of Dallas, Texas, near Palestine as it's coming in. Communications was lost approximately 15 minutes prior to its anticipated landing at the Kennedy Space Center at 9:16 A.M. Eastern Time."

But while Miles O'Brien was talking about the accident, none of us—NASA public affairs officers James Hartsfield and Kyle Herring, O'Brien, or myself—could get ourselves to say the words straight out, that the astronauts had died—at least not yet. It was just too overwhelming a concept to comprehend at that moment.

In its 15-day 22-hour 36-minute mission, Columbia traveled 2.1 million miles around the Earth. It missed getting home by just 900 miles and 15 minutes.

No announcement was made that an accident had occurred. The VIPs and press were just asked to get on their busses. The astronauts' family members were asked to go to the vans by their escorts. By this point, many had come to the conclusion something had gone wrong—but nobody wanted to say anything out loud. They rode back to the crew quarters in silence.

As the media got on the busses to go back to the press site, I saw a row of mini-vans and realized the families of the astronauts were inside. Driving the first van was family escort Steve Lindsey. He had tears running down his face and a cell phone in one ear. He was reacting that a very close friend had just died. I realized that whatever accident had occurred changed his title from "family escort" to "casualty assistance call officer."

The astronauts' guests were taken to an auditorium and asked to wait until official information became available. The escorts were just as confused as the guests. Many of the VIPs didn't realize anything was wrong, or at most thought there was a change they weren't aware of. Some believed Columbia was landing at the alternate site in California, or the landing was going to be delayed until the next orbit or next day.

Some folks found out from experienced observers who realized things were wrong, others only found out through a news broadcast or from a phone call. In many cases friends were watching the landing on television, and called friends who were at the landing site, hoping they might have more information. Ironically, they provided information to the people at the landing site.

As the bus traveled back to the press site my mind was reeling. Was it possible the CNN video was just a daytime meteor, or out of focus, so it only looked like there were multiple objects? What had happened? Was there any chance Rick could have flown the crippled shuttle down to low enough an altitude and speed where the crew could bailout and parachute to safety? While I knew the odds were miniscule, my mind wanted to grasp any possibility—no matter how slim.

Astonished people in East Texas heard loud sonic booms followed by pieces falling out of the sky. CNN put its first eyewitness on the air at 9:45 A.M. Bob Muller of Palestine said at 8:05 A.M. Central Time he heard a noise, went outside and saw the trail in the sky.

Miles O'Brien told his viewers at 10:03 A.M. about the debris which had come off during the launch, and engineers said it wasn't a safety concern.

At one point CNN producer Dave Santucci asked me two questions in order—Was I wearing a clean shirt? and Was there a video camera I could get in front of to do a live report from Florida? He knew me well enough to first ask whether or not I was wearing a clean shirt! I assured him I was. The only television stations present were the Israeli and local Florida stations. They were busy transmitting their own stories and would not have extra camera and satellite time they could lease to another news organization, so there was no way to make arrangements to get me on the air.

By this point much of my brain was on overload, and my emotions had shut down. Seven people I knew had just died, and the space shuttle's future was in question.

CHAPTER 37

MISSION CONTROL ON LANDING DAY

The astronauts get the most attention—and they should. They get the glamorous job of flying in space and are certainly the most visible portion of any shuttle mission. But just as important is Mission Control, the team of engineers who operate as the crew's extended hands and eyes—informing them of what they need to know and acting as their backups on the ground.

Anybody who's watched the movie *Apollo 13* realizes how complicated activities get in Mission Control. Most space fans are aware of the "Flight Control Room," which insiders call the "front room." That's where the flight director oversees a team of a couple of dozen specialized engineers. In effect, the flight director is the conductor of an orchestra, and the flight controllers play the individual instruments. The flight director is the most prestigious and stressful position in Mission Control—it carries full responsibility and control over the entire Mission Control team. The NASA flight rules read:

> THE NASA FLIGHT DIRECTOR (FD) IS THE FINAL
> AUTHORITY IN CONTINUING THE SPACE SHUTTLE
> INTERFACE TO THE PAYLOAD. THE FLIGHT DIREC-
> TOR, AFTER ANALYSIS OF THE FLIGHT CONDITION,
> MAY CHOOSE TO TAKE ANY NECESSARY ACTION
> REQUIRED FOR THE SUCCESSFUL COMPLETION OF
> THE FLIGHT CONSISTENT WITH CREW SAFETY.

In other words, if the flight director believes the shuttle can't make it back to the runway successfully, he can order the crew to abandon the shuttle, even though that would guarantee the loss of a $2 Billion vehicle—and that's a lot of responsibility. While there are higher-level managers in the room, the flight director has the ultimate authority in Mission Control. The STS-107 entry flight director was LeRoy Cain, a 15-year NASA veteran who had overseen many previous launches and entries.

The "back rooms" at Mission Control are staffed with more specialized engineers, who support the front room. The term "back room" shouldn't be taken literally, though; some are in the same building, others are in separate locations.

Like the astronauts, the Mission Control team does countless simulations before each mission: normal ones that simulate what they expect during the mission, and simulations in which the training team creates a variety of problems. These simulations can range from minor problems which can be put aside, to life-threatening situations where the crew has to be instructed to abandon the shuttle and parachute to safety. The flight controllers have to recognize which problems can be set aside and which ones are warnings that a problem is serious.

The training is so intense that the controllers' actions become instinctive, so they can react immediately, without having to think about a particular situation. When there are actual problems in flight, the Mission Control team relies on this training, operating on their instincts.

Each flight controller has his or her own communications "loops" (channels) where they can talk to their back rooms and other support personnel. In addition, there's a flight director's loop, in which the flight director keeps in contact with everybody in the front room and others as required, plus the air-to-ground circuit the capcom uses to talk to the shuttle. As a rule, most controllers monitor the flight director's loop, their own loops with their engineers, and the air-to-ground transmissions. Normally the public hears only the air-to-ground communications plus commentary from NASA's public affairs staff. But because of the high interest in the Columbia accident, NASA released the audio from the flight director's loop ten days after the shuttle broke up. The recording covers about an hour, from 8:34 A.M. EST, when Columbia was traveling over the Pacific Ocean to the west of Hawaii, until after the accident. Throughout the entire recording, the flight controllers are extremely professional—at no point was there any noticeable stress in the engineers' voices on the flight director's loop. The book's website has an exclusive annotated transcript of the flight director and air-to-ground communications loops from reentry through the accident, and a time-compressed audio recording.

Maintenance, Mechanical and Crew Systems (MMACS, nicknamed "Max") is the engineer responsible for monitoring the structural and mechanical systems. MMACS was the first flight controller who saw problems on Columbia. Jeff Kling was MMACS that day. His team reported to him that some of the sensors inside the left wing had gone bad. The readings didn't go up or down—they just stopped. Shortly after the accident, manager Ron Dittemore said, "An easy way to think about that is the measurement was no longer reading. It was not giving an indication. It's as if someone just cut the wire." At the time, of course, Dittemore didn't know how accurate his analogy was. The wire was indeed cut—by the super-hot reentry heat that was starting to burn up Columbia's left wing.

Some of the STS-107 crew members posed with LeRoy Cain's ascent/entry team for a group photo in Mission Control.

Kling told Cain, "I just lost four separate temperature transducers on the left side of the vehicle, the hydraulic return temperatures." Kling and Cain both wondered how separate sensors would go bad in rapid sequence, and Cain asked if the sensors had any common electrical connections. It's similar to having lights go out in different rooms in your house and then wondering, quite logically, if those lights were on the same circuit breaker. Kling replied there were no common components. It was far too early for anybody to recognize the cabling for the failed sensors all went past a common point—the left landing gear. A couple of bad sensors isn't anything to really worry about, just something unusual. The priority was to keep proceeding with the landing. Other sensors went bad, and then the tire pressure sensors for the two left landing gear tires went bad quickly, one just after the other. Controllers wondered what was causing multiple failures, but kept their emotions in check and concentrated on their data.

A couple of months later, flight director and manager Phil Engelhauf recalled, "When Jeff Kling indicated that he had lost the four transducers, I made a mental connection there to the impact during the ascent [from the bipod foam]. But at that point, I wasn't even close to thinking about a catastrophic outcome. I thought maybe a connector had been jarred loose by the impact, and even though at first blush, Jeff couldn't find any commonality, kind of deep down inside I was expecting him to come up with, 'Oh, I found it now, and now I know where the commonality is. They all go through this connector,' or some similar sort of thing. When we lost some more transducers, part of it was this was a difficult puzzle we were trying to figure out, and I was waiting to get some more data, so we could figure this puzzle out."

A view inside the Flight Control Room after communications was lost with Columbia.

Columbia's onboard displays only showed the tire pressure indicator going offline. There are far more eyes and brains in Mission Control, monitoring far more information than what's available to the astronauts. The astronauts are alerted only when it's something extremely important that needs their attention—like the tire pressure. But a bad tire pressure sensor only indicates a lost sensor. If the tire pressure had gone up or down then it would have concerned the crew because an over-pressurized tire bursting or a flat tire could result in a disastrous landing. Rick and Willie noticed the warning, and one of them responded by pushing a button to acknowledge the warning. That's the only indication the crew had before the loss of communications of anything wrong.

Then communications was lost. The INCO—Instrumentation Navigation and Communications Officer (pronounced "In Coe")—told Cain the telemetry from Columbia was getting really poor—"ratty" is the lingo NASA uses—and the communications was worse than predicted. When the audio was lost, all of the telemetry was also lost. The big display in Mission Control with the shuttle's telemetry and location froze with the final good data it received. The icon showing Columbia's location remained over Dallas, Texas.

While many in Mission Control certainly realized it was possible Columbia was destroyed, it was best to assume the bad sensors and lost communications were for some other reason, and Columbia was flying on its own.

Engelhauf said, "When we didn't get comm back, you sort of get that sick feeling in your stomach that this is not good. But in my heart, I was still holding on to hope that, yeah, comm will come back in a minute and then we will know why we were out of comm and we will have some more data, we can solve this puzzle."

Contrary to an Internet myth, MMACS Jeff Kling was not discussing with his engineers the possibility of recommending to Cain when they regained comm that the crew should bail out due to the heat in the wheel well during reentry. Kling, Cain, and others did discuss the possibility of having the crew bail out if there was a problem with the landing gear because of the damage from the foam strike, but that was a couple of days earlier in "what-if" e-mails idly speculating with other engineers.

Engelhauf noted, "When we got to the point where we should have had radar data and we didn't, it was, in my mind, pretty clear we were having a real bad day and we probably had lost the vehicle. It wasn't 100 percent certain, but I was pretty sure."

Somebody notified managers about television broadcasts showing Columbia coming apart. Deputy chief of flight operations Ellen Ochoa was sitting next to Engelhauf at the manager's console. She had flown with Husband on his first spaceflight and reacted in grief when she realized that Columbia and its crew were lost. She quickly pulled herself together and got out the contingency procedures manual.

Engelhauf said, "At that point, it took me a minute. A couple of the crew members were people that I knew fairly well, and, you know, I reflected on that for a minute or two. But then your training does take over, and everybody is going to want to know what happened during the accident, and our piece of helping that happen is to secure the data in the control center and make sure we don't lose anything."

Cain accepted the inevitable conclusion—a major accident had occurred. He recalled, "That gave me great pause." He quickly put the preplanned emergency plans into action, giving the order to lock the doors to prevent any outsider from entering. Communications was restricted to the official recorded "loops" in case there was anything said

Ellen Ochoa reacts to the news that Columbia was destroyed.

Only after completing his critical tasks did LeRoy Cain give himself a moment to react to what had happened.

which might be of use in determining what happened. All flight controllers were told to save all of the data on their computers and write down any notes that could be of use.

Only after Cain gave his instructions and had a moment to himself did he take a deep breath and let his emotions show.

Then Cain had what was certainly the most difficult conversation in his career. He asked Landing Support Officer Marty Linde, a support person in a separate room, if he had contacted the Department of Defense to send any search and rescue personnel they had in the Dallas area to find whatever remained from Columbia.

All of the personnel in Mission Control had to finish storing their logs and computer archives before they were permitted to leave. It wasn't a concern about somebody intentionally erasing damning information, it was to ensure nothing was accidentally erased or forgotten. Several hours later, the communications restrictions were lifted so the personnel could have the opportunity to tell their families when they might be coming home.

LeRoy Cain held a press conference on February 14, two weeks after the accident. It was clear, even before the audio of the flight director's loop was released, there was nothing the reentry team could have done which would have made any difference. Cain said, "I never had any doubt and I still don't doubt what [Mission Control] did or didn't do. The kind of problem we've suffered on this day—there isn't anything in my estimation the flight control team could have done differently or should have done differently."

Cain added, "Part of the message I want to give you today, and the main message I want to give you, is I was very proud of the way the team performed on Saturday [February 1] in the face of the tragic events. I remain proud of them in these days in the aftermath, in the work they're doing and the commitment that they show. They're a very professional group of individuals. They remain very dignified and showed a lot of integrity in the face of adversity. And I'm very proud of them for that."

CHAPTER 38

WHAT SHOULD HAVE HAPPENED

Imagine if the bipod foam hit the black tiles on the lower wing, where the engineers believed it hit—and it wasn't a fatal blow to Columbia's heat shield. Commander Rick Husband would have landed Columbia on KSC's three-mile runway. It would have gotten about a minute of live coverage on the news in the U.S., a couple of additional minutes in Israel. Only in the hometowns of a couple of the astronauts would it be front-page news, most newspapers wouldn't have even bothered to mention the 112th safe shuttle landing, or at most would have run a small wire story on an inside page.

The astronauts would shut down Columbia's systems. Technicians wearing airtight suits would approach the shuttle with sensitive electronic sniffers to ensure that none of the poisonous fluids were leaking. After an "all clear," a set of stairs would roll up to Columbia and technicians would open the hatch. The first whiffs of fresh Florida air would enter the crew cabin. The flight surgeon would enter and make sure all of the astronauts were healthy. Since Rick, Willie, and K.C. didn't participate in the on-orbit medical experiments, they could exit Columbia by themselves if they felt up to it, or with assistance from the technicians and flight surgeons. Anderson, Brown, and Ramon would remain seated until technicians were ready to load them on stretchers. This wasn't because of any health concern, but to minimize the changes as their bodies adapted back to Earth gravity. Since Laurel was sitting upstairs on the flight deck, she would have to climb down the ladder to the middeck. Before the mission she told me, "I'll have to climb down with some assistance from the astronaut support personnel. They'll come in and help us get unstrapped from our seats and go through the landing checklist. We have to walk/crawl out of the hatch," and then she'd lie down on a stretcher too. Once in the Crew Transport Vehicle, a battery of tests would be performed on the four science crewmembers and normal medical checks on Husband, McCool, and Chawla. Willie would be given a balance test for a separate experiment.

While the crew was helped out of their seats, other technicians would hook up Columbia to the support vehicles, including air conditioning trucks and power vehicles.

Most of the time the commander decides to walk around the shuttle just to admire the vehicle and see how well it came through reentry. It's an opportunity to reflect on the mission and its successful completion. Willie and Kalpana may have decided to join Rick for the walk around, but the science crew would remain lying down for their experiments. Laurel said before the mission, "Instead of walking off, we're going to go straight to crew quarters." High-level managers would have been waiting to greet the crew and chat with them about their flight.

Everybody would have been astonished to see the gouge caused by the foam. No doubt the crew would think, "Boy we're lucky there wasn't more damage." The shuttle is inspected and videotaped after landing, with the video transmitted to the engineering teams at other NASA centers.

Even though NASA was under internal pressure to keep ISS assembly on schedule they would have to put launches on hold until changes were made to prevent bipod foam from falling again. There would have been an internal engineering analysis to prevent a recurrence: clearly, a large gouge could not be ignored after so much damage on two out of the last three missions. Like the flowliners problem, settling this issue would have resulted in a couple of months delay, at a minimum.

The Crew Transport Vehicle would drive the astronauts back to the Operations and Checkout Building, the same place they stayed before launch. There they would get reunited with their families. The youngest kids are just happy to be back with their parents. The pre-teens to early teens are usually excited about their parent's spaceflight and what it was like. The older teens—well who can figure out what teens are thinking?

Doug Brown says he was looking forward to seeing Dave stumbling and dizzy as he readapted to gravity and teasing his brother about learning how to walk again.

The next items in priority are a freshly cooked meal and a long shower. More medical checks are performed with more sophisticated equipment. Laurel told me before the flight she and the other science crew would have, "A full day of testing after we land."

Technicians would remove many of Columbia's time-critical payloads. These included rats, protein crystals, moss, and payloads which required power, like the freezer filled with Laurel's "liquid gold." All together, 1,500 to 2,000 pounds of payloads were planned for removal on the runway so they could be given back to the scientists as soon as possible.

Eight of the rats would be killed as quickly as possible to minimize readapting back to gravity. The other five rats would be kept alive another day so the scientists could study their behavior as they readapted to gravity; then they, too, would be sacrificed.

One of the reasons the worm experiment was able to join the flight as a last-minute addition was the lack of crew activities, not even having the astronauts inject a fixative to stop the worms' growth. As soon as the scientists got their canisters back after the landing they would inject fixatives.

They would count the worms and inspect how well the new food worked. In contrast, the moss experiment should have been well preserved by the astronauts, and scientist Fred Sack was anxious to see if once again they grew in clockwise spirals. Dr. Karl Hasenstein had asked for Biotube to be started as late as possible so his flax seeds would be in their fixative solution for as little time as possible before landing. Because Columbia was landing without any delays, he would get access to his plants just one day after the experiment was completed.

The Biopack and Biobox experiments would be taken to FIT in Melbourne, where dozens of scientists were anxiously waiting. Each group had a laboratory assigned. In some cases the experiments would be packed in dry ice and FedExed back to the experimenters' laboratories, in other cases the scientists would remain at FIT for the next several days to study their samples closely before returning home.

The student S*T*A*R*S payload would get delivered to Astrotech in Titusville, where Japanese student Maki Niihori was waiting to disassemble her experiment and see how the medaka fish had grown in comparison with the control experiment. None of the other S*T*A*R*S teams needed to get anything quickly. In some cases their experiments were completed once S*T*A*R*S was shut down in orbit—there never were any plans to return the ant farm or silkworms to their students. At most, the gel and ants would be scooped out and put into a bowl and mailed back to Fowler High School. The crystals from the chemical garden, spider webs, and balsa block from the bees experiment would get shipped back to their student teams. I had made arrangements with Spacehab to take photos of all of the experiments a couple of hours after Columbia's landing. I was especially curious whether or not the delicate "chemical garden" crystals would survive through the landing.

One of the early items to come off any mission is the "Return to Houston" bag. The astronauts fill this bag with any important items they want to return to JSC as quickly as possible, including all of the film they shot in space. The Mission Management Team was anxious to examine the photos of the ET to determine how much foam had fallen from the bipod this time.

Several hours after landing Columbia would get towed back to the Orbiter Processing Facility, with many of the support vehicles still attached. The hazardous materials like excess propellants get drained first. Then lower priority items are removed. A couple of weeks after landing, the Spacehab Research Double Module, FREESTAR bridge, and Extended Duration Orbiter (EDO) pallet, which held the extra hydrogen and oxygen tanks Columbia needed for its two weeks-plus in space, would get removed and transported to another building. Within a couple of weeks the Spacehab module would get transported to Spacehab's facility in Cape Canaveral, where technicians would remove the non-time-critical items to return to the scientists. The FREESTAR pallet would be taken to a clean room where it would get disassembled. Scientists were anxious to recalibrate the SOLSE experiment, since they made some late changes and did not have the opportunity to calibrate SOLSE before launch. Eventually, all of the payloads would get returned to their scientists. Some would be refurbished and reflown on future space missions, others would be disassembled and their parts would be used for other purposes, and still others would be put on display in museums or universities, or even sold as surplus. The EDO pallet would be put in long-term storage because NASA had no future plans for shuttle missions longer than 13 days.

In the crew quarters the blue team members might have decided to take a nap after their initial medical tests. After all, they had been up since 4:30 P.M. the previous day, almost 24 hours by now. About six hours after landing, the seven astronauts were scheduled to meet the press. I was looking forward to seeing all of them again, and talking to them about their experiences. Only a handful of reporters were expected to show up, including of course the Israeli reporters who returned to Florida. In some cases the press conferences are rushed, with the astronauts leaving quickly; in other cases they've been more relaxed, and there's even time to chat informally afterwards. Certainly the first question from an Israeli reporter would be whether or not Ilan commemorated the sabbath the evening before the landing.

And then something rather unusual would have taken place—the crew would be given some free time. It's an opportunity to spend time with their families, especially the young children they haven't hugged since they entered quarantine over three weeks earlier. If they're not completely exhausted, the astronaut and family may go out for dinner. If the kids are along, it may be fast food, but if it's just the spouse, then it's probably at a nice restaurant. They're pretty much incognito, and can go to dinner anywhere without being recognized.

The next day the crew and their families fly back to Houston. Many of the astronauts' children are extremely excited when they get invited to sit in the cockpit with the pilots for the trip home. There's a major celebration at the NASA hangar, with people who worked on the mission, neighbors, friends, and the public coming out to greet the returning astronauts. There's a podium, and each of the crew is asked to say a few words.

Dave's neighbor Cindy Swindells was looking forward to seeing him at the reunion—but it would be a bittersweet moment—the second he saw her Dave would certainly think about his dog Duggins.

After shaking hands, signing autographs, and posing for photos, the crew finally gets to leave, but it's under escort, to make sure they don't get into an accident on the way home. One astronaut commented his kids weren't too excited about his mission but they really enjoyed riding home with a police escort and waving to their friends as they came up their block! Quite often there's also a block party when they get home.

Brown's neighbor Al Saylor had a big celebration planned for Dave's return. He said, "We just couldn't wait for him to get back and tell us about his great adventures. We knew we were going to get to see every-thing that happened—who did what and how it went. Our anticipation of his return and sharing his experience with us was huge."

Within a day after landing, all of the onboard film would be developed. A handful of photos including each of the astronauts, a group shot, and the mission's activities would get released publicly through NASA's website. Later, the crew would choose additional photos for their own presentations.

The crew would go through an intense series of debriefings on every-thing—Columbia, their activities, and their experiments. SOFBALL sci-entist Paul Ronney was anxious to ask them how they came up with the wonderful idea of naming the flame balls. Who was "Howard" named after? Did Mike read in between the lines while "Kelly" was burning for a record length of time and try to delay shutting it off, to keep it burning as long as he could, and if so why did he stop just nine minutes short of an entire orbit? On a more professional note, Ronney wanted to collect any thoughts the crew had on his experiment, anything unusual which they might have noticed, and any recommendations for future improvements.

The payload crew would have extensive medical tests for the next several days, and then less often over the next two months for the post-flight data collection necessary for some of their experiments.

Within a couple of weeks the crew would put together a movie, not the fancy professional-quality documentary Dave was planning, but a 15- to 20-minute highlight reel. They would show their movie and photos at a presentation for the JSC workers. Each of the astronauts would receive a NASA Spaceflight medal. The U.S. rookie astronauts would also receive gold astronaut logo lapel pins that had flown in space. As a pay-load specialist, Ilan would receive the medal but not an astronaut pin.

Willie had volunteered for an unusual experiment to measure how an astronaut's balance mechanism can be confused by mixed signals even several weeks after returning from space. A month after the landing Willie would lie down sideways in a spinning chair. The chair would spin him around while various images were displayed to confuse his ori-entation. The object behind this test was to see if the spinning chair and visual patterns could confuse the body's inner ear balance sensors, simi-lar to how you feel dizzy after spinning around for a long time.

NASA planned to carry the Space Experiment Module back to New York, where it would be opened up in front of the students who had placed their samples inside over a year before. The vials with the mold, bubble gum, seeds, and other materials selected by the students would be returned so they could examine them and see how they compared with the control samples which had not flown in space.

A high priority for the entire crew would be a VIP trip to Israel, where Ilan would be honored as a national hero. K.C. had told me before the flight that she hoped the entire crew could visit India with her. The crew would also visit the various NASA centers responsible for their mission and its payloads, and possibly the European and Canadian teams, too. Some of the astronauts could have been invited to grand hometown celebrations like Rick's Amarillo celebration after his first flight, but that's pretty rare. During all of these tours, they would show their movie and photos and answer questions from their audiences. Now the ex-rookies could honestly answer the question, "What's it like to fly in space?" in great detail.

And then, several months after the end of the mission and almost three years after they were put together as a team, the crew would get assigned to other tasks. As Dave noted in an e-mail to his family and friends, the landing would mark the start of the end for the STS-107 "family." They would get broken up as the postflight activities were completed. In some cases, former shuttle crews remain fairly tight with each other, in other cases they barely talk to each other—like any other extended family. The STS-107 crew was certainly a crew that got along and enjoyed sharing their personal lives. Before the mission Willie mentioned his favorite Israeli food was baba ganouj, an eggplant dip. Willie said, "It's my favorite. [Ilan makes it with] toasted pita bread. He does it on the grill with the eggplants and everything, and takes the skin off and mushes it all up and adds all this garlic and it's tremendous."

After all of the postflight activities settled down, Ilan and his family would move back to Israel, where presumably he would get promoted to a high-ranking position in the Israeli Air Force. Many folks have speculated that Ilan might decide to go into politics. As the son of one of Israel's founders and a Holocaust survivor, Ilan was already perceived as somebody special, and his military experience defending his country only added to his status as a national celebrity. When you add becoming your country's first space traveler on top of that, you come up with an individual of incredibly high visibility and potential mass appeal.

Closer to home, though, there would be no more automatic promotions for the NASA astronauts—that went away a while ago. They would have had to go through the normal military system to get their promotions. Navy Commander Laurel Clark was a "captain selectee," and would get her formal promotion that summer.

Dave planned to take the hundreds of hours of video he shot during the STS-107 training and mission and edit it on his Macintosh computer into a professional-quality interactive movie about what drove the crew. He expected to make about 1,500 copies to distribute to colleagues and friends and hoped that it could be distributed commercially.

The NASA astronauts would get technical assignments supporting shuttle missions, ISS, future projects, or other support functions. Eventually, they hoped, they would get assigned to another spaceflight. In some cases, astronauts who have flown together in the past fly together again, but it's more the exception than the rule. It's also possible one or more astronauts could decide to retire—either to a high-paying civilian job with an aerospace contractor, a university teaching position, or some other new career. Some could have stayed with NASA and the space program, perhaps in a management position, others could have left to search for new goals to conquer. Dave had told friends that he wanted to become a capcom in Mission Control and hoped to eventually do a spacewalk on a future flight. On a more personal note Dave was serious about his girlfriend Janneke, and they had even discussed baby names. Presumably they would have gotten married and started a family. Willie had a long-range plan. Whenever he decided to retire from NASA and the Navy he planned to move back to the house in Anacortes, Washington, his family called home. The American STS-107 astronauts would become just another six relatively anonymous astronauts instead of national heroes.

The next mission after STS-107 was the STS-114 ISS flight. Atlantis was scheduled to roll out to its launch pad the day after Columbia's landing and launch in the predawn hours in early March. Its primary objective was to exchange long-duration crews—the Expedition 7 crew of Yuri Malenchenko, Alexandr Kaleri, and Ed Lu would ride up to ISS, and the Expedition 6 crew of Ken Bowersox, Don Pettit, and Nikolai Budarin would take their places for the return journey. Three spacewalks were planned to make repairs and maintenance. In addition, the shuttle was carrying tons of supplies, including three brand-new scientific racks which would increase ISS's science capabilities. One of the racks, the Human Research Facility 2 (HRF-2), was a more advanced version of STS-107's ARMS experiment.

STS-114 would have been put on hold after the damage to Columbia. Depending on the length of the delay mangers could have made the decision to exchange the Expedition 6 and 7 crews via the Russian Soyuz instead of via the shuttle or stretch the length of time in space for the Expedition 6 crew until the bipod foam issue was fixed.

Columbia's next mission was STS-118, scheduled for November 2003. It would be Columbia's first flight to ISS.

A year after STS-107, the scientists would get together for a technical conference where they would give presentations on their experiments, how well they performed in space, and whatever results they wanted to share with their colleagues. Scientists would submit papers for peer review to technical journals and present them at scientific conferences in their own fields. Some of the experiments, like MPFE and Biopack, would describe the problems they had with their hardware and what they did to maximize whatever science they could collect. Others would be simply ecstatic about how everything went and how they got more science than originally planned.

CHAPTER 39

THE MEDIA'S
COVERAGE
OF STS-107

NASA accredited 340 press for STS-107's launch. That included about a dozen reporters who cover the space program on a daily basis, the local television stations, Israeli media who traveled to the United States for the first launch of an Israeli astronaut, assorted space fans who were able to wangle press badges to get a closeup view of a shuttle launch, and Florida newspapers who send reporters to launches "just in case" an accident occurs.

Columbia's launch made page 3 of the *Washington Post* and page 14 of the *New York Times*. It was front-page news in only a few newspapers, primarily in Israel the hometowns of the astronauts. In contrast, the accident led on the front page of newspapers around the globe.

Within minutes of its occurrence, the Columbia accident became *the* news story. Everything else, even planning for the upcoming war in Iraq, became a non-issue. News directors and editors scrambled to get any video clip, photograph, or other bit of information they could about Columbia's crew. Small television stations and newspapers in the middle of Texas suddenly found themselves at the center of attention—anything they could transmit of the debris was quickly broadcast to the world. Even though it was Saturday, the news quickly spread in Israel that Ilan Ramon had died.

The public was intensely interested in what was happening, even though only a small amount of information was initially available. The CNN.com website reported nine times as many "hits" as on a typical Saturday morning.

At least one Israeli newspaper got caught jumping the gun. *Yediot Aharonot*, Israel's largest daily, reported on its website Ynet.co.il that Ramon had landed safely. In Hebrew, the site declared, "Ilan Ramon Returned from Space—The Shuttle Columbia descended [touched down] successfully in Florida. The Space Shuttle Columbia descended just a short while ago with success in Cape Canaveral, Florida. The Astronauts, amongst them 'Aluf Mishne' [Ilan's title in the Air Force] Ilan Ramon, will undergo a series of physical examinations, and nine hours after touchdown they will meet with the local media." The newspaper quickly pulled the premature announcement.

A Hebrew newspaper in Israel used its website to announce the safe landing of Ilan Ramon. It quickly pulled the incorrect story. Photo from the author's collection.

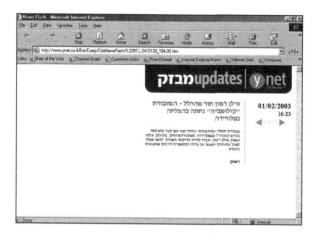

Afterwards, an Israeli reporter asked me how I knew before anybody else something was wrong. "Just experience" was my reply.

Many reporters were prepared for the possibility of a shuttle landing accident, but thought it would happen just before the landing, or during the landing—not earlier, during the reentry. Many television stations, networks, and reporters were caught unprepared and had to do what they could—which included stealing their better-informed competitors' stories. CNN producer Dave Santucci was sitting in the control room fielding calls, informing managers that a major story was breaking, and giving information to on-air anchor Miles O'Brien. At the same time, Santucci was monitoring all of the major networks, when suddenly he noticed something really strange—the Fox News Channel broadcast was identical to CNN's, except the network "bug" identifying CNN was fuzzed out. Santucci immediately suspected Fox was pirating the CNN feed. He asked his technician to put up CNN in small letters in the upper left corner at 10 A.M. Sure enough, "CNN" popped up on the Fox feed. Santucci realized he had caught Fox, and decided to make sure everybody else knew it too. He had his technician switch from the replayed video of Columbia over Dallas to the studio camera on Miles O'Brien. CNN viewers were confused why CNN was broadcasting O'Brien's back to the camera for several seconds. Fox News Channel viewers were astonished when for a couple of seconds they saw CNN anchor Miles O'Brien apparently also working for Fox! Fox technicians quickly realized they were caught and switched to the NASA video showing Mission Control.

The justification for such grand larceny is what the television networks call "fair use"—in other words, anything goes when it's a major news story. Everybody can steal whatever they want from whatever sources they can grab, and afterwards pretend they did nothing wrong. In

CNN viewers were surprised to see anchor Miles O'Brien's back. Fox's viewers were even more astonished to see O'Brien on Fox. Photo from the author's collection.

some cases networks did ask for permission to use another network's feed –after they had already used it!

Immediately after the accident NASA's public affairs office at JSC was flooded with requests for over 1,000 additional media badges, above and beyond the nearly 350 who were already badged. An additional 559 media requests were received at KSC. Not surprisingly, most of the media who showed up that day knew nothing about STS-107 before February 1: all they knew was a disaster happened, so they needed to cover it.

A television station in Jacksonville, Florida, immediately sent a crew to KSC, three hours to the south. They had not bothered to cover NASA activities for the past decade, but they wanted a crew there after Columbia went down—not because they thought they could find something their affiliates already there couldn't find, but just so they could broadcast from KSC and say they were there in person. That wasn't an isolated case; 36 satellite uplink trucks showed up at KSC.

NASA administrator Sean O'Keefe and Associate Administrator Bill Readdy came over to the press site at 1:20 P.M. on February 1 and read prepared statements announcing what everybody already knew—a major

After the accident, 36 satellite trucks showed up at the Kennedy Space Center.

NASA Administrator Sean O'Keefe and Associate Administrator for Spaceflight William Readdy make the formal announcement that Columbia was lost.

accident had occurred, resulting in the loss of the Columbia and its crew. O'Keefe said, "This is indeed a tragic day for the NASA family, for the families of the astronauts who flew on STS-107, and likewise is tragic for the nation. Immediately upon indication of a loss of communications from STS-107, at a little after 9:00 A.M. this morning, we began our contingency plan to preserve all the information relative to the flight activities." Readdy gave more details: "This is a truly difficult day for all of us. Many of us were standing alongside the runway waiting to celebrate their triumphant return after a 16-day science mission." He then emphasized, "Those people that may find debris, do not touch it, do not move it. Contact your local authorities. Have them impound it and secure the area." A copy of their statements is on this book's website.

The concern about the debris was serious. NASA could not have conceived of a shuttle breaking up over land, scattering its pieces over a wide area. The shuttle uses many hazardous liquids, like super-pure ammonia for cooling, supercold hydrogen and oxygen to generate power, poisonous nitrogen tetroxide and monomethyl hydrazine propellants, and even formaldehyde for some of the experiments. Some of the liquids like hydrogen and oxygen would quickly evaporate, but others would remain hazards for quite a while. Items potentially dangerous for longer periods included batteries, explosives, and other chemicals used for the scientific experiments. Obviously, anything with a sharp edge will always be a hazard. There was also hazardous medical waste—the used needles from all of the blood draws and injections. The tiles could shed particles that could irritate the skin or be hazardous if they were breathed in. The emphasis was to inform the public to avoid touching any debris, and inform local authorities about how to handle the debris.

Shuttle program director Ron Dittemore and chief flight director Milt Heflin held a press conference at 3:32 P.M. They provided what little information they had: Sensors on the left side of the vehicle had gone bad before the loss of communications. Columbia was hit by debris during the launch, and the engineering analysis indicated it wasn't a safety concern. The NASA press auditoriums at the Johnson Space Center in Houston, Texas, the Kennedy Space Center in Florida, the Marshall Spaceflight Center in Huntsville, Alabama, and NASA headquarters in Washington D.C. were packed with reporters with questions about what had happened, and whether or not there was somebody to blame. One reporter asked, "Should something have been done differently by Mission Control during the reentry?"—apparently accusing Mission Control of not reacting quickly enough or making some other mistake which doomed the astronauts. Others asked questions which showed their lack of knowledge about the space shuttle. Many observers noted Dittemore and Heflin appeared to have a difficult time controlling themselves when they were asked these inane questions. Only a few of the reporters were ones who knew the shuttle program and could ask intelligent questions which Dittemore and Heflin could answer. A corrected transcript of the press conference is on this book's website.

Some reporters have generic shuttle disaster stories written ahead of time—just in case. The stories start "It's too soon to say why the space shuttle Columbia exploded, but the early signs suggest ..." and then speculate about likely causes being cutbacks in NASA safety and quality control, or cutbacks in the budget, or labor issues, or whatever else the reporter wants to speculate about. At most the reporter will insert a couple of lines with the specific details about what actually happened before filing the story. It's certainly the fastest way to put out an accident story, and who cares if the facts are completely wrong?

Media who wanted somebody to say something negative about the space shuttle just had to go to their Rolodexes and look up the same people who always said negative things about the shuttle or space program. Self-proclaimed space experts who were critical of the space program were quick to contact media outlets to offer editorials about why the shuttle program should be stopped. They found a ready audience of media who were willing to hear them repeat everything they had been constantly saying, some of it justified, some of it not, a few pieces of information applicable to the Columbia accident but much of it unrelated.

Many media cited NASA's safety panels, who were concerned about shuttle risks, but not the underlying reason why the panels were concerned and whether or not any of that was pertinent to the Columbia accident.

Commentators who had never met or interviewed the astronauts talked about what wonderful people they were. People who didn't know them and didn't care who they were a day earlier eulogized them as heroes, saints, and supermen. The Indian media praised Chawla as a national hero and tried to condemn anybody who brought up the STS-87 Spartan incident. Any mentions of Spartan were called "shoddy reporting" or attempts to denigrate a woman who had just died and couldn't defend herself.

Not surprisingly, media who weren't interested in the mission quickly changed their minds."Little India" magazine had told me, "We cannot use another piece on Chawla" before the mission even though I had sold them stories for Kalpana Chawla's first flight in 1997. But after the accident, they were immediately interested in a feature about her.

The media certainly has the right to cover any story they consider newsworthy, and the loss of a $2 billion space shuttle and seven lives is most definitely major news. Other news stories involve more deaths or more earth-shattering changes; but only a few events touch the human spirit so much that television networks air 24/7 news, and even make the decision not to air commercials during the most intense period. In contrast, a train accident with over 40 people killed and tensions with Iraq were minor news items on February 1. Major media outlets who don't cover the space program on a regular basis, like the *New York Times* and *Knight-Ridder* newspapers, sent their investigative reporters—top-notch journalists who could dig out information, but without very much technical experience. NASA's public affairs officials gave them priority during the press conferences and access to information not given to reporters who cover the shuttle on a daily basis.

Unfortunately their lack of knowledge resulted in mistakes or attempts to be the first ones to discover "new" pieces of information, which would have been comical had it not been for the fact that they were covering a tragedy. For example, a couple of days after the accident, NBC space reporter Jay Barbaree hyped the fact that he had a copy of the Flight Day 12 Mission Evaluation Room report as an exclusive, and he showed it on the air—the same report I received three days before the accident! And I was not the only reporter who had access to that report; anybody who actually covered the mission and asked for certain materials NASA gives out to reporters only upon request could have received a copy.

There were minor mistakes by all of the media, but that was to be expected in such a rushed period, where there was as much of a need to get things out quickly as to get the facts right, especially when a reporter was on deadline. At one point, CNN's caption declared Columbia was traveling 18 times the speed of light! (It should have said the speed of sound.)

Many space enthusiasts noted one benefit to the Columbia tragedy—NASA released a lot of information it normally claims is unavailable or only for internal use. It seems ironic: When things go well, NASA is often unwilling to go to the effort to release technical information, but when there's an accident and higher media interest, the agency's public affairs and Freedom of Information Act personnel go in high gear to produce whatever is requested to avoid any perception they are trying to hide something. Here's a list of some items NASA released after the Columbia accident that are not provided on a regular basis:

- a copy of the space shuttle flight rules, with the specific rules which applied to STS-107
- a copy of the STS-107 entry checklist
- flight readiness review presentations
- the flight director's audio loop
- a video of the activities in the Mission Control flight control room during entry
- technical documentation for the External Tank and its insulation
- notes from the Mission Evaluation Room team for each day of the mission
- minutes, transcripts and recorded audio of all of the Mission Management Team meetings during STS-107
- video inside the shuttle's crew cabin showing the astronauts during the reentry
- transcripts of press briefings
- minutes of meetings of NASA managers discussing the accident afterwards
- a copy of the NASA contingency procedures

A similar case applies to the Challenger accident. Out of 113 shuttle launches, the only publicly available transcript of the cockpit recorder, with the conversations between the astronauts during a shuttle's ascent, comes from the Challenger accident, where astronauts Dick Scobee, Mike Smith, and Judy Resnik are reacting the same way any astronauts act during a shuttle's climb to orbit.

NASA's official policy for technical documentation is that it only has the resources to provide the information to its personnel, but in many cases, it's actually more effort to put information behind a firewall to limit its distribution and determine who should get access and who shouldn't. In some cases, computer-readable versions or barely readable faxed copies of faxes are also made available to the news media. But the rest of the documents are "for official use only," and unavailable to the public. It seems ironic that the only exception is when there's a tragedy.

CHAPTER 40

THE FIRST HOURS
AFTER
THE ACCIDENT

NASA has a 124-page contingency plan that was refined after the Challenger accident. It covers everything from "close calls" to major emergencies. The biggest emergency is a "Type A Mishap"—greater than $1 million in damages and/or a death. The contingency plan outlines what needs to be done and when, including everything from the public release of information to emergency doctors to communications procedures. A copy of the contingency plan is on this book's website. Shortly after Sean O'Keefe became the NASA administrator, in 2002, he asked the agency to do a simulated shuttle contingency, "confident it wouldn't be needed, but more confident the procedure was a good one in case it would ever be needed," as he put it.

The families were taken to the crew quarters for their privacy. Head of flight crew operations Bob Cabana and chief astronaut Kent Rominger told the families what they already knew, but had yet to really accept—the astronauts had died in an accident. All of the emotions the family members had bottled inside for the last hour came out. Dave Brown's brother Doug said, "It was like somebody snapped their fingers and said 'cry now' and the room just completely fell apart. That was a moment you'll never forget. We were in silence from the time we got in the vans at the runway viewing site until that moment. That moment was truthfully earth shaking. Everybody kind of knew, but nobody wanted to say a word, who would want to say that? You want to believe that something like they landed in the swamp, or any other thing could have happened, but you knew."

A day later Cabana told reporters, "Yesterday was probably the hardest day in my life, to have to sit down with the families of close friends and tell them that their husbands and wives and moms and dads aren't going to be coming home. And if you've never had to do that, I hope you never have to."

In the larger world outside, everybody wanted to tell somebody. When many people found out, whether it was in person, on the news, or from a friend, they quickly passed on the information to others. O'Keefe notified the White House and other government heads. President George W. Bush called Israeli Prime Minister Ariel Sharon to formally inform him about the accident. Sharon, of course, already knew by watching Israeli news broadcasts. Later, Bush talked to the relatives of the astronauts to offer his regrets.

The MMT held their first post-accident meeting an hour after the accident. Managers at the various NASA centers met on a conference phone call to outline what was known and to determine the course of action.

While the issue of the foam striking the wing had been cleared several days earlier, it was still fresh in everybody's mind. Ron Dittemore asked all centers to immediately impound all mission data and to follow their contingency action plans, and for the ET ascent imagery, which showed the bipod foam hitting the wing, to be secured.

Even at this early point it was clear Columbia's left side, where sensor failures were seen in the telemetry, was important to the investigation. It would be hard not to assume the foam strike on the left wing might be related. Nevertheless, managers warned about jumping to a hasty conclusion.

Ralph Roe of the Space Shuttle Vehicle Engineering Office noted the foam debris was being reanalyzed. As Dittemore said later that day in the first press briefing, it was only natural to reinvestigate whatever was out of the ordinary to see if the first analysis made the correct assumptions and came to the correct conclusion.

And another important issue was discussed: Determining where the debris fell, and in particular where the crew compartment had landed, was a critical step toward the top priority—recovery of the crew's bodies.

Rules for information preservation also had to be observed. Any hardware or data which could have had anything to do with the accident was impounded, and the impounding was draconian. Scientists couldn't get access to laboratories until everything was cataloged and transferred to bonded storage — even lab coats were impounded. While impounding data is a standard procedure, it did extend far further than needed. For several days, students working on the S*T*A*R*S experiments couldn't access their own website. (Many noted, perfectly logically, that anything on a public website could easily have been duplicated exactly, with an identical copy made for the investigators while still leaving the public site available. And of course conspiracists immediately speculated whether something was removed from websites so it could be destroyed or altered to fit the cover story.)

The Orbital Information Group (OIG), which handles the public distribution of satellite tracking data, disabled access to STS-107's tracking information until it was allowed to re-release the data a couple of months later. Fortunately, many amateur and professional satellite trackers had their own archives, which were collected during the mission. This book's website has all of the STS-107 tracking data in the "Two Line Element" format.

Payload team members were far more inconvenienced when they were informed their equipment was put into bonded storage until NASA could determine the equipment couldn't have had anything to do with the accident.

Fortunately, one set of data was not impounded—anything released by NASA's Public Affairs Office. The press kits, flight plan updates, photographs, and mission status reports were never removed from the World-wide Web, and have always remained available. Naturally, there was an intense amount of interest after the accident, and it was extremely difficult for many to get access to NASA's web sites, but that wasn't NASA's fault. The difficulties did lead some conspiracists to believe NASA had removed photos after the accident to cover up what had actually happened.

Like many poorly managed websites, many of NASA's early web addresses with information on the accident are no longer valid because items were moved without updating all of the links, not because of nefarious reasons to hide anything.

At 2:04 p.m. George W. Bush spoke to the nation. Ironically it was the very first public statement he made about civilian space activities since he became President. His only previous remarks about the space program were while campaigning during the 2000 elections, in areas populated with large numbers of aerospace workers. The Israeli government also issued a formal statement. Copies of those statements are on this book's website.

The next day, NASA released the names of the accident investigators. As a precaution, and with the hope that their services will never be necessary, suitable candidates for an accident board are named before major events like shuttle missions, rocket launches, and spacecraft arriving at other planets. The logic is that it's better to be prepared and be glad the accident board wasn't needed than to be caught unprepared, and suddenly to have to put together the necessary resources in an extremely tense and emotional situation. In some cases, the candidates were not available for a long investigation and alternates were selected.

The chair was Retired U.S. Navy Admiral Harold "Hal" Gehman, Jr., who co-chaired the independent commission which investigated the attack on the USS Cole in Aden, Yemen, on October 12, 2000. Gehman had also once served as the commander-in-chief of U.S. Joint Forces Command. He had no aerospace experience, but was highly regarded for his work on the Cole investigation. O'Keefe called him the afternoon of the accident to ask him if he would chair the investigation board.

The other initial members of the investigation board were:

- Rear Admiral Stephen Turcotte, Commander, U.S. Naval Safety Center, Norfolk, Virginia.
- Major General John Barry, Director, Plans and Programs, Headquarters Air Force Materiel Command, Wright-Patterson AFB, Ohio

- Major General Kenneth Hess, Commander, U.S. Air Force Chief of Safety, Kirtland AFB, New Mexico.
- Dr. James Hallock, Aviation Safety Division Chief, U.S. Department of Transportation, Cambridge, Massachusetts.
- Steven Wallace, Director of Accident Investigation, Federal Aviation Administration, Washington
- Brigadier General Duane Deal, Commander 21st Space Wing, Peterson AFB, Colorado.
- Scott Hubbard, Director, NASA Ames Research Center, Moffett Field, California.

Hubbard was the only voting member with a NASA connection. Many of the others had no previous contact with NASA.

Supporting the board in non-voting roles were Bryan O'Connor, NASA Associate Administrator Office of Safety and Mission Assurance, and Theron Bradley, Jr., NASA Chief Engineer.

The board had their first meeting at Barksdale AFB in Shreveport, Louisiana the next day. Barksdale was the closest military installation to where Columbia broke up and would become the staging location where the debris would be taken.

By then, Gehman had asked that the awkward name "International Space Station and Space Shuttle Mishap Interagency Investigation Board" be changed to "The Columbia Accident Investigation Board" (CAIB), and NASA agreed.

Later, the CAIB decided to expand its membership by five — adding Dr. Roger Tetrault, Dr. John Lodgson, Dr. Sally Ride, Dr. Sheila Widnall, and Dr. Douglas Osheroff. While the early board members came from the military and from civilian government agencies, the new members, came mostly from academia and some had aerospace experience, most notably ex-astronaut Sally Ride. The CAIB report notes that their 13 members coincidentally made their board the same size as the Rodgers Commission, which investigated the Challenger accident.

The board split into four separate groups to examine (1) NASA management, (2) the inflight performance of the Mission Control team and astronauts, (3) the engineering and technical analysis of the accident and debris, and (4) NASA's history, budget and culture.

NASA restructured its internal investigation teams to match the CAIB's four-part organizational structure. NASA and contractor personnel would do most of the legwork and research to determine what had happened, with the CAIB looking over their shoulders to "keep them honest" and give them instructions if the CAIB felt they weren't asking the correct questions. The CAIB also hired its own engineers, consultants, and support staff.

The thirteen members of the Columbia Accident Investigation Board. From left to right seated are board members G. Scott Hubbard, Dr. James N. Hallock, Dr. Sally Ride, Board Chairman Admiral (retired) Hal Gehman, Steven Wallace, Dr. John Logsdon, and Dr. Sheila Widnall. Standing from left to right are Dr. Douglas D. Osheroff, Maj. General John Barry, Rear Admiral Stephen Turcotte, Brig. General Duane Deal, Maj. General Kenneth W. Hess and Roger E. Tetrault. Photo by the CAIB.

Some parts of the CAIB's testimony were presented in public hearings; other parts were presented behind closed doors. Early in the investigation Gehman said witnesses had nothing to fear from him because the intent was to determine what had happened—not to point fingers or blame anybody who might have been at fault. He joked, "But if you've got a potted plant [where a bug could be placed] or a secretary who talks a lot—all bets are off." The CAIB investigators noted many of the witnesses told them things in confidential testimony that would have gotten them fired if their bosses ever found out. The CAIB was extremely careful with protecting the privileged testimonies, even controlling Congress's access.

The CAIB set up a public website that received over 40 million hits. It featured a feedback section where the public could send messages; it got 3,000 submissions, of which 750 resulted in some action by the CAIB.

The CAIB wanted to ensure nobody would claim there was any conflict of interest, even though NASA was responsible for setting up the CAIB's initial structure. Besides asking for several changes to its charter, the CAIB dismissed Bryan O'Connor and Theron Bradley and requested they return to their normal jobs. The CAIB also asked NASA to remove from its investigation team several mid-level managers who were directly involved with STS-107, and the subject of the CAIB's scrutiny.

CHAPTER 41

VIEWING
COLUMBIA'S
REENTRY

Shuttle reentries are a spectacular sight. When the orbiter reenters the atmosphere, it's a bright fireball as it travels across the sky—bright enough to be seen in daylight. But as it cools down, the plasma trail gradually disappears and it becomes just a speck in the sky.

Most shuttle reentries occur over the Pacific Ocean or Gulf of Mexico, and pass over very few populated areas. STS-107 was one of the rare reentries to travel across a large portion of the continental U.S. Its reentry path would fly over Northern California, Nevada, Arizona, New Mexico, Texas, Louisiana, Mississippi, Alabama, and Florida. As an added bonus, in the westernmost states it would appear in the sky before sunrise, for a truly magnificent pre-dawn sight. East of Texas, it wouldn't be as spectacular because it would no longer have a bright glowing trail.

The general route the shuttle takes is easy to determine in advance. By the time Columbia launched, satellite trackers could estimate the time and path Columbia would take diagonally across the United States from San Francisco to Florida. The precise reentry path is calculated during the mission by Mission Control's flight dynamics officers, based on the shuttle's precise orbit. A set of maps with the reentry path were produced and made available on NASA's Internet site on January 30.

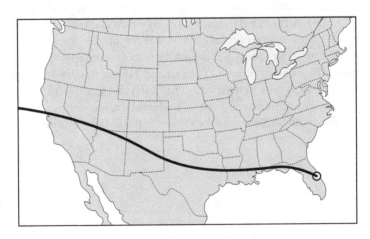

The predicted path Columbia would take across the Southern United States on the way to its Florida landing.

The day before entry. LeRoy Cain said, "This will be a very good visual sighting for folks, in particular on the West Coast and the Arizona—New Mexico area. 5:55 A.M. Pacific Standard Time, Columbia will make landfall in the San Francisco Bay area. It should be a pretty spectacular event. For folks who have never seen a shuttle sighting, in particular at night, it's a sight to see. This should be a very good one."

Amateur astronomers in California and the southwest were enthusiastic. Many JSC workers who weren't on duty planned to look for Columbia as it passed over Texas, even though it would be barely above the horizon in Houston. Special credit must be given to Brian Webb and Rick Baldridge, who sent out e-mails to Internet amateur astronomy and satellite observer lists to inform others about the times and where to look.

Webb publishes an electronic newsletter "Launch Alert," primarily about rocket launches from Vandenberg AFB in California. His instructions for where and when to look for Columbia, with details provided by Baldridge, went out to about 1,000 readers. Many of the people who videotaped or photographed Columbia's reentry did so based on Webb's mailing.

In some cases the observers were experienced astronomers who had seen previous shuttle reentries, while others were just casual observers who heard there was a rare spectacular sight which was worth waking up early to see. Some were equipped with camcorders, binoculars, and cameras; others just viewed it with their naked eyes.

Their reactions and experiences varied widely. Those with good weather in the places closest to the flight path saw a spectacular meteor going across the sky, others further away with cloudy conditions only saw a bright moving object—but they were all excited to see astronauts returning from space. Some of the more experienced observers didn't notice anything out of the ordinary, while others viewing their first shuttle reentries immediately thought something was wrong when they saw bright flashes and what seemed to be pieces breaking off. In some cases, observers didn't find out anything was wrong for another hour, when they heard about it on the news or from somebody else. In a few cases, the observers honestly thought they could be the only ones to have videotaped Columbia's reentry. In some cases the videographers gave away copies to television stations freely, others were paid a token amount of money, some were hoping for more. In many cases NASA purchased their cameras and camcorders so they could analyze them in a professional optical laboratory to ensure the cameras didn't have any defects that could cause some visual effects or obscure others.

Rick Baldridge, an experienced amateur astronomer and satellite observer, noted how spectacular it was the first time he saw a shuttle reentry, in 1994, and how whenever there have been reentries visible from San Francisco, he's tried to see them. STS-107 was his fifth. Baldridge tried to stir up interest in the local media about observing shut-

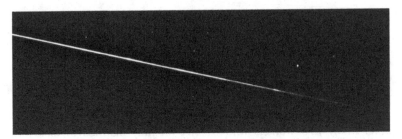

Rick Baldridge's wide-angle lens captured this time exposure of Columbia's reentry over California. Photo courtesy of Rick Baldridge.

tle reentries, but without much success. He was more successful encouraging local astronomers. For STS-107, Baldridge and friend Kevin Sato had a video camera and two still cameras on Mount Hamilton. Baldridge recalled, "It didn't look really that different—it was a little lighter in color, a little brighter. It didn't immediately flag as something unusual. We thought we had seen a nice reentry. I didn't know anything bad had happened until I got home."

Baldridge didn't think his video would have anything unusual, but then he viewed it. He said, "As I played back the tape . . . there were a few brightenings. . . ." Then Baldridge noticed pieces coming from the shuttle and trailing behind it. "I knew it was going to be obviously important at that point. We proceeded to make copies." He went to a place where he could get the film from his still camera developed, and those photos also showed various brightening events.

Baldridge's observations were especially important because he had seen and videotaped previous reentries, so he could give a comparison for how a normal reentry looked. Eventually, NASA asked Baldridge to send his original videotapes and film, plus the videotapes he had shot of the previous shuttle reentries. NASA flight directors Paul Hill and Cathy Koerner talked to him about his observations.

Amateur astronomer Pete Goldie was nearby on Bernal Hill. He set up a digital camera on a tripod and took several photos. Each shot was a time exposure that lasted eight seconds. When he got home, he was astonished to see a corkscrew-like trail intersecting the shuttle's path in the third image. He said, "There was clearly something in the image that did not make sense—I thought that there was something wrong with the camera." Goldie wondered what it could be, and wanted to make sure the images were sent to the engineers who could analyze them properly. He wanted to protect his own privacy and wasn't interested in any fame. Goldie explained, "I did not understand what the photo meant—there were a number of explanations. A natural phenomenon had caused the accident, one which had never been anticipated in the space program, that was one possibility. Another possibility is it meant nothing. There was a big gap between no meaning and important meaning, and it was such a big gap that I felt that I had to proceed very cautiously." The *San Francisco Chronicle* published stories describing his photos but agreed not to use his name, and Goldie did not give them permission to publish the photos.

It turned out that the camera moved as the photo was taken causing the corkscrew shape. After Goldie removed his finger from the shutter there was a tiny bit of motion in the camera, either from the lifting of his finger or a wind gust. The camera rocked, then quickly settled down. Investigators determined that a movement of as little as a twentieth of an inch would have accounted for the squiggle. The bright shuttle registered as an irregular line due to the camera motion, just like a wobbly camera creates a blurry photo. Once the camera settled down, it properly exposed the rest of the shuttle's path as it streaked across the sky, the residual plasma trail from where the shuttle had been, and even some stars in the background. It was a very logical explanation for why the corkscrew appeared in the photo. Nevertheless there were conspiracists who pointed to the photo's existence as "proof" that a lightning bolt had destroyed Columbia — even though Goldie had not released the photograph. One conspiracist even claimed that Goldie had e-mailed his video (sic) of the entry for his "expert analysis."

Also in the Bay area was Dr. Jason Hatton, an experienced amateur satellite observer and a scientist on the Leukin experiment that was coming home inside Columbia. He said, "Columbia appeared in the west as brilliant fluorescent pink ball of light with a bright green trail, again giving a strong impression of raw energy. As it reached its highest elevation,

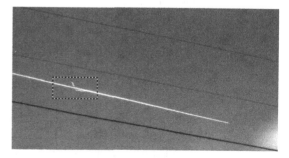

The infamous Goldie photo, showing what at first glance looks something like a lightning bolt hitting Columbia as it traveled from left to right in this time-lapse image. Note the dark electrical power lines (two above and one below) the bright path of the shuttle. In the magnified closeup below, the contrast has been increased to show more clearly the corkscrew shape of the "squiggle," with the residual still-glowing plasma trail to the left, and on the right the properly exposed trail for the rest of Columbia's path. © Pete Goldie, PhD / Contact Press Images, courtesy of the photographer.

due north, a small point of light detached from the main fireball, which faded in a few seconds, followed a few seconds later by another. I stayed outside for another five minutes until I heard the sonic boom." Hatton knew that the pieces he saw coming off were abnormal, and was the very first person to realize something was wrong—even before Mission Control started to see sensors go bad or lose communications.

Brian Webb and Jay McKee went to Mountclef, a high ridge in Thousand Oaks in Southern California. Columbia would be extremely low on the northern horizon. Webb said, "We're looking, and Jay points toward the north-northwest and asks, 'What's that?' We saw it through the thin cirrus clouds. It was a spot of light, like a star moving. Toward the northeast it disappeared because the clouds got thicker. He and I were just both thrilled we saw a manned spacecraft reentry, I had never ever seen anything like that before." A couple of days later, Webb compiled a list of reentry observations that he passed on to NASA. Those observations are on this book's website.

Jay Lawson, in Reno, Nevada, traveled the shortest distance to view Columbia's reentry—just 20 feet, to the end of his driveway. He recalls he barely got dressed. "I wasn't going to have to be working, so I decided to get up and watch one. I had never seen a reentry before. It was very obvious when it appeared in the sky, it was the brightest thing. I knew it was going to pass right in front of Venus, so I zoomed in. That's when you see that bright flash on my videotape. The bright flash startled me, I thought "Whoa, holy crap, that was weird." All of a sudden there's this bright flash, to me it looked like there had been an explosion. I was just calling it like I saw it. I knew almost immediately that there was something wrong or out of the ordinary."

Lawson went to the local planetarium and reviewed the tape with their personnel. Once the TV stations realized there was a major story, they contacted the planetarium for comments and found out about Lawson's videotape.

Chris Valentine had seen Columbia reenter on the STS-109 mission eleven months earlier: "It was spectacular, I couldn't believe it. It was stunning, it was unbelievable." Realizing that reentries that were visible from Arizona were rare, Valentine searched for whatever information he could find on the reentry path, and what time it would be passing over a location he could get to.

Valentine admits "It was a little disappointing for me, it wasn't as bright as I expected it would be. We had high clouds, and the sun was coming up. But following it from horizon-to-horizon, it appeared brighter and dimmer at certain places in the sky. We clearly saw things coming off of it. All it was, was a reaction of surprise. When we saw the bright pieces falling behind, we thought it was just part of the [reentry], but it wasn't. After it was done, it disappeared into a beautiful sunrise. We didn't think anything had gone wrong."

When Valentine listened to the last transmissions from Columbia he thought Rick Husband's last transmission at 8:58 A.M. sounded like, "Feelin' that heat," not "Roger, uh, bu...." It's something Valentine is passionate about, although he does acknowledge he could be wrong.

A group of off-duty workers at Kirtland AFB in New Mexico decided to set up a computer-controlled telescope to track Columbia as it flew overhead. Like dozens of astronomers, they wanted to see if they could capture the rare sight of a shuttle reentry with whatever they could put together. Rick Cleis said, "I just wanted to do it." In effect, they were professional engineers and astronomers doing an amateur activity. They got together equipment readily available in their laboratory.

They were able to get tracking information from Mission Control through a contact, but it came in a format that was incompatible with their equipment. Imagine getting a set of travel instructions as map coordinates instead of which direction you should travel and when you should turn at each landmark. Cleis wrote a program to convert the NASA data into a set of pointing instructions that could be used by their hardware. He said, "I was up for 26 straight hours. Whatever format they gave, it was not what we typically used, it was nothing we've ever used before. We went through a chain of networking heroics to get it onto the little computer we use because it isn't hooked to the network." That computer was a 1992 vintage Macintosh Quadra 950—a high-end computer when it came out, but now a fairly old, underpowered system.

The Macintosh sent commands to motors that rotated mirrors to follow Columbia's path across the sky. The mirrors reflected their images into a 20-year-old 3.5-inch Questar telescope. The one "modern" component was a laboratory CCD camera attached to the telescope's eyepiece and a laptop computer that captured and saved the images.

Cleis said, "It was pretty trying because as the sun was coming up we had to focus the camera again, and I was finishing the [software program]. The last lines of code were written 15 minutes before it was needed—it was that close."

The team wasn't completely confident that everything was set up properly, or that the NASA data was accurate. Cleis said, "We were searching around [the sky] just in case the numbers weren't accurate." They shouldn't have tried to change anything. Cleis said, "It was exactly where they said it would be. We were searching for it when we shouldn't have been because we didn't know [the data] was going to be so accurate."

Cleis's partner Major Robert Johnson operated the more up-to-date computer that was attached to the camera. He just snapped pictures and saved them quickly as they came through the camera into the computer. He said, "We were taking a lot of pictures; as it turned out we only captured a few that had anything in them at all because the field was kind of narrow.

The Kirtland setup with a laboratory CCD camera pointed at a 3.5-inch telescope and a computer-controlled mirror. Photo courtesy of Kirtland AFB.

The Kirtland image of Columbia's reentry, which was publicly released a week after the accident.

It was only after it went over that I looked back at them and realized there were one or two good pictures. . . . We didn't think anything of the picture. We didn't notice anything unusual. We saw the 'feature' [a bump] on the left wing, but we didn't know if it was out of the ordinary or not. We didn't think anything of it—we were just happy we got something."

A few minutes later, the wife of one of the team members called and informed them about the accident. Johnson said, "We were just stunned and said, 'We better back up our data,' because somebody's going to want to see it." Instead of going through the government red-tape, the team used a simple approach—they sent their images to the Internet website NASA had set up for the public to submit any images or video they had of Columbia's reentry.

NASA released one of the still images about a week later. It was not made clear that the images came from a consumer 3.5-inch telescope, not the giant 11.5-foot adaptive optics state-of-the-art Starfire Optical Range also located at Kirtland. The team was instructed not to talk to the media about the images, and couldn't explain why the images that came from a state-of-the-art world-class astronomical facility were of such low quality. NASA public affairs officials would only say the image was what they got from Kirtland AFB, but would not provide any details, adding to the confusion. Naturally the conspiracists had a field day. Eventually NASA and the military permitted the group to talk about how they took their images.

These were just some of the scores of people who submitted their observations of Columbia's reentry to help the investigation. All together, NASA received 150 videos and 1500 still images from the public, and 23 videos from the public and 2 from government sources showed abnormal events, like pieces coming off. Without those videos, NASA would not have been able to determine when those pieces broke off Columbia during its reentry.

CHAPTER 42   WHERE WERE YOU...?

There are five key events in recent history—incidents so stunning you remember exactly where you were when you learned of them, how you found out, and what your reaction was—if you're old enough to remember. In reverse chronological order, they are the Columbia accident, the 9/11 terrorist attacks in 2001, the Challenger accident on January 28, 1986, the moon landing on July 20, 1969, and the Kennedy assassination on November 22, 1963. It's interesting to note that three of those events are about the space program and in the entire list the only positive one is the Apollo 11 moon landing. All of the others involve tragedies.

Other events may be more historically important, but how you found out them isn't as memorable as how, when, and where you found out about these five events. If you were old enough you'll always remember these milestones.

For most people somebody they knew—a friend, neighbor, or relative—called and said, "Quick, turn on the news—it's bad,"—often without specifying the nature of what had happened. But others found out in a variety of ways.

Much of the confusion for the astronauts' family members, VIPs, and most of the media at KSC arose because of a lack of announcements that an accident had occurred or even that there was a problem. They were just asked to go back to their buses, and many who weren't familiar with the shuttle program had no idea anything was wrong.

Here's how some folks with an intimate relationship with the space program or another connection to STS-107 found out about the Columbia accident:

Willie's wife Lani was standing in the bleachers at the Shuttle Landing Facility in an area reserved for the family and their escorts. She said, "Willie had told me to listen for the all of the air to ground communications, there would be a lot of talking." She was concerned when they lost communications, and really worried when they couldn't track Columbia.

Dave's brother Doug was in the same area. He was busy setting up his camera and wondered why the countdown clock to the landing had reached zero and had started to climb up. He assumed the landing had been delayed for whatever reason or the clock was wrong. The escorts asked the families to get in their vehicles. Doug noted, "It took about the time to walk from in front of the bleachers to the side of the bleachers to realize what had happened—about 50 yards of walking." Nobody wanted to talk about the situation, although most had realized there was an accident of some kind and their family members had just died.

Crew secretary Roz Hobgood was nearby with the families' guests. She was busy handing out photos and patches and not really paying attention. She recalled, "I was at the top of the bleachers with some of the families. We weren't sure; we were wondering what was going on. Everybody was so excited about the crew coming home." Another support person, Beth Turner, came up to her and said, "Roz, get ready to get the families together. They've lost communications with the shuttle." Hobgood said, "I just kind of just stood there and looked at her and she kind of looked at me. Within seconds she came back and looked at me and said, "It's time to go." I walked down the bleachers and saw the families being escorted to the vans where they were taking them off." Hobgood wanted to go with the families, but her responsibility was to stay with their guests. She recalled, "I turned around and went to the bleachers full of families and friends. Everybody was wondering. 'Roz what's going on? . . . What's happening?' I had no idea what was going on. We're just standing there holding hands and looking at an empty sky— you want to see something coming through the sky. We watched the time go down [to zero]; it looked like they reset the time, and the time went back up, and finally somebody just got on the [speakers] and asked everybody to go to the buses. The buses took us out to the auditorium where we all sat and waited until they gave us the final word. It was very intense. My first call was from my family, they were in Houston and saw the news. My daughter said, "Mom, I'm so sorry about the crew," and it just hit me and I broke down. It was the most critical thing that could ever happen in a lifetime."

Dave Brown's neighbor Jeff Kling was in Mission Control. Kling was the first flight controller to see problems when sensors started to fail on Columbia's left side. He was allowed to leave around seven hours after the accident. Many of Dave's neighbors had asked Kling about the mission's status, and when Dave would be returning. Kling said, "To have to come back to the neighborhood to everybody that was waiting for him— it broke my heart. Professionally it was a hard thing for me to know we had lost the [astronauts and the] vehicle. Coming back to the neighborhood was probably one of the hardest things I've ever had to do."

Chief Astronaut Kent Rominger flew with Rick on Husband's first spaceflight in 1999. Rominger was in Florida flying the Shuttle Training Aircraft, providing a pilot's perspective on the weather conditions. Rominger said, "I knew they didn't have communications, that put up a flag in the back of my head. They didn't have tracking. On the order of 7 to 8 minutes out from landing, when they still didn't have tracking, that little flag became a pretty big concern."

Rominger kept looking at the location Columbia was supposed to appear in the sky. He said, "I'm looking across and didn't see an orbiter. I was feeling really bad and I'm looking at the glide path. At the time it should have touched down, within a minute after that, Mission Control called me [and gave me instructions to land]. At that point there was no doubt in my mind." After landing, Rominger drove to the crew quarters, where he had the unenviable responsibility of meeting with the astronauts' families and telling them Columbia was lost, with no chance of survivors.

John Herrington, an astronaut on the flight before Columbia, was at home in Houston with his children. Herrington noted, "For some reason they weren't watching Saturday morning cartoons, so I had the NASA channel on, and within a minute or two we knew something big had happened." Herrington thought something was wrong because nothing was heard on air-to-ground. "The amount of silence to me spoke volumes. That's what got my attention—the quiet," he noted.

In contrast, Scott Altman, the commander of Columbia's previous mission, thought it was just a communications problem at first. Astronauts are trained for a wide variety of problems, including loss of communications with Mission Control, and Altman was enthusiastic Rick Husband would get the opportunity to use of all of his training to land Columbia on his own, without Mission Control providing additional information. At the same time, Mission Control would have the opportunity to use their training to solve whatever the communications problem was.

Astronaut Paul Lockhart was at home in Houston with his six- and eleven-year-old daughters. Lockhart said, "I had a call from John Herrington [Lockhart's flight engineer on STS-113]. He said they've lost contact with the orbiter. My first thought was, that happens sometimes when they lose tracking. But I ran it through my mind again; if John is showing a lot of concern, this must be something pretty dramatic. I floated through the different news channels and found it—how they were talking about the loss of audio contact. Then they showed a tape of the breakup over Dallas. As soon as I saw that, I knew things were obviously taking much more [of a] turn for the worse than what I had originally thought." Lockhart's brother Jay was driving from Dallas to Amarillo, and was an eyewitness to Columbia's breakup. Lockhart said, "[My brother] saw what looked like the shuttle breakup, and he was listening to the news on the radio." Lockhart realized, "All of a sudden there's this emotional conclusion that reached inside of me that said, 'Hey, we've just had a major event happen. Seven people I know very closely have just had their lives extinguished.'"

Lockhart said, "It was a gut-wrenching moment when I talked to my brother at the same time I'm watching it on TV, but my first thought was my daughters." They had seen enough of the news to hear the names of people they knew. Lockhart said, "I just tried to tell them that they couldn't find the orbiter at that time. I just didn't want them to understand what had happened [at that point]."

The Lockharts were close to Columbia's crew. Paul Lockhart and Rick Husband knew each other from the time they were in high school. Rick and Evelyn Husband had taken care of the Lockhart daughters just a few weeks earlier, when Paul was busy with STS-113 and Laurel Clark's husband Jon was the flight surgeon on Lockhart's first spaceflight. Lockhart said, "Jon's also basically been there for my family at any time I've needed it. Even before I was assigned to STS-111, Jon put several sutures and stitches into my youngest daughter. He's always been there to help Mary. I've always felt that I've never been able to repay Jon, as much as he's helped me."

By the end of the day, Lockhart was asked to coordinate and lead the flyover of four T-38 jets in the "missing man" formation for the memorial ceremony that would follow a tragedy like a shuttle accident.

Lockhart's wife Mary is an Air Force reserve Colonel, working on tactical weather planning. She was returning from a European trip preparing for the upcoming war with Iraq and had just landed in Dallas-Fort Worth. She found out in an oblique manner. Mary Lockhart said, "I turned on my cell phone and had about 20 messages. They went, 'I'm so sorry,' 'I'm so sorry,' so I'd just skip and go on to the next one, I didn't know what was going on!" Lockhart says she had talked to another passenger on her flight about the shuttle and mentioned her husband was an astronaut who had flown on the previous flight. They were in line at customs and, as Mary described it, "I asked him, 'Did something bad just happen?' and he said, 'Columbia just disintegrated over the skies of Dallas this morning.' My legs went numb, and I can't even explain how I felt. The gentleman was so kind to me; he grabbed my computer bag and said 'Let me help you.' I can still feel it—it was such grief, such pain. I was thinking about the families. I immediately called Paul and got a hold of him. I got on the next flight available from Dallas to Houston. I was just so relieved to be with Paul and the girls. My thoughts were with all of the families; Evelyn [Husband] is a very close friend. I could just really feel for the whole crew."

Astronaut Leroy Chiao was training in Russia. Chiao said, "I was cooking a Chinese New Years dinner for the Americans here. [Fellow astronaut] Bill McArthur called me in my cottage. He said, 'Turn on CNN.' I asked, 'What's up?' and he said, 'It's just bad, turn it on.' It was the biggest shock I've had in my NASA career, and one of the biggest shocks of my life. Dave Brown was a neighbor. I was more close to Dave [than the other members of the crew]. It was very difficult, it's still very difficult."

The three-person crew aboard ISS consisted of astronauts Ken Bowersox and Don Pettit, and Russian cosmonaut Nikolai Budarin. Instead of their normal daily planning meeting, JSC director Jefferson Howell got on the radio and personally told them the news. Several days later Bowersox said, "My first reaction was pure shock. I was numb, and it was hard to believe that what we were experiencing was really happening. And then as reality wore on, we were able to feel some sadness." Pettit said, "When I first heard, at that point, it was not known what the condition of the crew was. We were hoping that there were going to be survivors, and then as it unwound we learned that there were no survivors, and that's when the magnitude of the event really hit me."

Later that day the space station astronauts made their normal weekend phone calls to their spouses. Pettit's wife Micki said Don was still kind of in shock, noting, "They don't see it everywhere [on the news] up there. It just hasn't sunk in yet, it isn't business like usual down here. I don't think he's quite grasped that yet." She added, "He's keeping busy and focused [on his work], and I'm sure that's helping him get through this."

Almost as isolated as the ISS crew was professional space scientist and amateur astronomer Rob Matson, who was by himself in the Mojave Desert watching Columbia's reentry and hunting for meteorites. He saw Columbia but didn't notice anything out of the ordinary. He said, "I heard [on the radio] CBS's Christopher Glenn at 6:17 A.M. Pacific (one minute after the landing time) say they still hadn't heard the signature sonic booms at the Cape. The significance of that simple statement seemed lost on the reporter—it wasn't like the shuttle could show up a minute or two late, like a commercial airliner behind schedule. For those few moments, I felt like I was the only one who knew that the crew and shuttle were lost. What a horrible feeling."

Deputy Chief Astronaut Andy Thomas was assigned to JSC's contingency operations center. NASA has such a team in the emergency center for every landing as a precaution. Thomas noted that since it was a Saturday, he hadn't even bothered to shave before going in, and expected to be back home within an hour. But because of the accident he ended up working a 16-hour day.

Scott Thurston, NASA's vehicle manager for Columbia, was in the VIP area next to the shuttle runway with the high-level NASA managers. He said, "Having been a comm[unications] engineer, I realized when they lost comm and lost tracking something wasn't right. Pretty quick."

Standing in the same area as Thurston was BRIC scientist Fred Sack, who was anxious to get his experiment back so he could see how his moss had grown. Sack had seen Columbia land in 1997, on the previous flight with his experiment. He recalled, "It was a remarkably beautiful sight. It was really humbling—you'd see this little speck in the sky and this perfect landing right in front of you—this huge machine, and literally hours later they'd be handing our experiment back to us. Our little experiment was one little part of a huge machine. I was very expectant for the STS-107 landing, we were just excited about getting our results, but mostly to see that amazing machine emerge out of the sky on a beautiful day." But the landing time came and went. Sack said, "They just asked everybody to get back on the bus, without any real announcement other than that communications had been lost." At that point there was confusion, but Sack, like most of the other visitors, didn't know that something had gone wrong. Sack said, "We were on the bus for a while, and knew obviously that something was wrong, but it wasn't until somebody picked up an AM radio that made us understand how serious it was. It was quite gripping, there was a bus filled with many people—many who knew the astronauts. People just started crying one by one. A lot of hugging and disbelief, just being unable to understand. It was crushing, obviously. The hardest part was seeing the astronaut families there, just seeing the kids and the families there, so heartbreaking—my heart goes out to them."

Florida Tech scientist Gary Wells was also at the VIP site. He said, "We were standing around waiting and waiting. It finally dawned on us that something must have happened. It was just taking too long for Columbia to come into view. I don't actually think they ever even announced that there had been an accident while we were there. It was obvious something had gone wrong. It was quite a shock to everybody." The VIPs were bussed back to their cars but didn't receive any additional information from NASA. Wells said, "On the way back from the Cape I got a call from [my European counterpart] Pasquale Di Palermo. He said, 'We lost the shuttle.' I said, 'What do you mean we've lost the shuttle?' He said, 'It broke up, it's all in pieces, everything's lost.' At that point I pulled over to the side of the road; we just couldn't believe it. It had been a perfect launch, a perfect mission, one of the most fantastic science missions we've ever had. And all gone."

Leukin scientist Millie Hughes-Fulford was in Florida, but not at KSC. She had flown on Columbia in 1991 as a payload specialist. She decided to stay with her European colleagues, who weren't allowed to go to KSC for the landing because of security reasons. Instead of going in by herself, Hughes-Fulford was driving to the FIT control center. Spacehab would deliver the Biopack payload and the scientists' precious samples to them a couple of hours after the landing. Hughes-Fulford was talking on a cell phone to her husband, who was listening to the news. She said her husband told her, "Oh, there's a problem; they're a little bit late getting in." She realized the impact of that simple statement and told him, "If they're late getting in, something's happened." She recalled, "I pulled over to the side of the road, because I was not feeling very well about it, I didn't think I should be driving."

Mission scientist John Charles was at home in Houston. He said, "I was going to step outside and see if I could see the ionized trail over our neighborhood. I was watching NASA TV, looking at the ground track and trying to guess how far above the horizon the shuttle was going to be." But the indicator for the shuttle's location was frozen in one location. Charles said, "I realized they weren't talking anymore, and the ground track hadn't changed. I was watching that and watching that, guessing I didn't understand [the reentry events] as well as I thought I did." Charles had expected the tracking indicator would be moving as Columbia traveled across Texas and wondered what was happening. He didn't want to change the channel in case he might miss something. His wife went to another room and turn on another television set to CNN, where she found out what was happening.

Charles said, "As soon as it became clear what the story was, I just went on into work to Mission Control to see if there was anything I could do to help. I ended up spending the rest of the day running errands for people who were trying to get information together—just generally being a 'gofer' for anybody I could help."

Mist scientist Angel Abbud-Madrid's activities were completed a couple of days earlier, but he decided to watch the landing in the POCC. He said, "This was my first spaceflight experience, so I wanted to participate in all of it. You have the big screen and you can see the different graphics—the flight path, another screen just focusing on the runway, and you can see Mission Control. I hooked up my headphones and selected some of the loops—the flight director, public affairs, and some of the other lines. I was listening to the conversations. I remember Rick talking the last time and getting cut off very quickly. I thought this wasn't unusual—noise on the line or something. Then they started talking about some temperature going up, some pressure readings—I thought this was normal, just monitoring everything. I thought it was normal, but then it was becoming too long. And then you look at the clock—we're getting so close, it was like 8:12 [Central Time, four minutes before the landing time], you look at the runway and nothing shows up. It gets very confusing and you get this strange sensation—some very deep pain in the stomach. And then I heard the flight director say, 'Lock the doors, we're going into contingency procedures.' That's when it starts hitting you. You don't know what's going on. It was quite confusing. I just stayed there. There was a phone call for the other person in the room, and she just started crying. It was a very surreal scene. After that I went up to the third floor and saw the images on CNN. I stayed in the building the whole day until I left in the evening. That's when you realize—when you leave the gates and see all the people outside and the media—the magnitude of an accident like this." He added, "The moment it didn't touch down I felt this sinking feeling, some uneasiness inside. Of course the first thing you think is, 'This can't happen'—you think it's just a communications problem. It's a very sad, sad moment. You think of the people you talked to the last four days who were helping you with your experiment. The last thing you think is, your experiment is lost right in front of you. It's just a very confusing time. It's a very painful moment."

The MEIDEX and FREESTAR-Hitchhiker teams were at their control center at the Goddard Spaceflight Center, in Greenbelt, Maryland. They watched the NASA TV transmission showing the views in Mission Control. Dr. Yoav Yair recalled, "Being a veteran air-traffic controller in the Israeli Air Force, I sensed immediately something was very wrong when the shuttle track stopped moving. Then, in a minute or two, our Hitchhiker friends acknowledged that the shuttle was lost. We were utterly shocked, pained beyond belief. Professor Joseph almost collapsed. We could not believe our eyes. After several minutes, there were clear guidelines of what to do, and we did as told by our NASA colleagues. On the morning of February 2, after a sleepless night, we flew home to Israel and arrived to a pained and sad nation on February 3."

FREESTAR manager Tom Dixon had planned to watch the landing from the Hitchhiker control center with his team, even though there was no requirement for him to be there. But his wife was working late, so Dixon stayed at home to watch their kids. He noted he couldn't find anything about the landing on TV, and that made him feel uneasy. He called the control center and found out Columbia was overdue, and realized the significance of that statement. Dixon recalled, "My heart sank, of course." His wife got home a little bit later, and after telling her what had happened, he went to the control center to be with his team.

Laurel Clark's high school psychology teacher Tom Lewis found out from television—but not by watching it. A local station in Milwaukee called him to ask for an interview, assuming he already knew about the accident. Lewis had followed the mission but didn't get up early for the landing. He assumed it had gone normally, and the station just wanted to talk to him about how excited he was that one of his students had flown in space, and he was looking forward to seeing her come back to town for the high school's upcoming anniversary. It wasn't until Lewis was on camera that he found out what happened. Lewis said, "The first question was, 'What do you think about the accident?' I turned white—that was the first I heard of it. I said, 'This is the first I heard of it, can you turn off the cameras.'" Lewis took a couple of minutes to compose himself before restarting the interview. The reporter assured him they would not air his initial reactions when he found out, but did broadcast their interview with him talking about the sorrow at the loss of one of his former students.

NASA scientist Neil Pellis was on vacation and happened to be watching the news. When a reporter announced the shuttle was late for its landing Pellis, realizing the significance of that simple statement, told his wife, "All hell's just broken loose."

Retired Admiral Hal Gehman Jr. was in Virginia visiting relatives. He found out about the accident the same way millions of others did—a friend called and told him to turn on the television. By that afternoon he got a phone call asking him to head the accident investigation.

CHAPTER 43

THE PUBLIC'S REACTION

Almost immediately after the accident, the public started to express their grief for the astronauts and compassion for their families, whether or not they knew anything about the mission a day earlier.

The STS-107 crew consisted of six Americans and one Israeli who had trained together for over two years to do an extremely complicated set of tasks, with risks they readily accepted. But in their deaths they suddenly entered history to join the names of the seven who died in the Challenger accident and the three who died in the Apollo 1 fire. There's undeniably something special about the human conscience that has an emotional connection to the space program and the people involved—especially if they die.

On the same day as the Columbia accident, a crowded passenger train collided with a freight train carrying flammable liquid in northwestern Zimbabwe, killing at least 40 people and injuring hundreds more. Almost six times as many deaths, but the train wreck barely made the news. The dead were mourned only by their family and friends, not the entire world.

A spontaneous memorial appeared almost immediately outside JSC's main gate in Houston. When the Mission Control team was permitted to leave, they saw the flowers, flags (American, Israeli, and Indian), and other items left by the public. It helped the NASA employees realize how much the accident had impacted their community and the entire world. At KSC,

The public made an impromptu memorial at the main gate to NASA's Johnson Space Center the day of the accident.

A plain photo of the crew in a frame at the Kennedy Space Center.

somebody put a plain crew photo in a frame with six American and one Israeli flags in an empty lot, where the workers could view it as they drove by.

The most unexpected support came from the public directly affected—the communities in East Texas and Louisiana where the debris fell. They had no connection to the space program. These people could have been angry NASA was "littering" their property and endangering their lives with poisonous materials, but they felt as much grief for the Columbia crew as anybody else. They opened up their hearts, providing whatever was needed. If one good thing can be said about a disaster. it is that it brings out the goodness in people. Trays of sandwiches would suddenly show up for the recovery teams. Spare bedrooms were made available. The most common questions residents in the area asked the search teams were, "What do you need?" and "What can I do to help?"

Unfortunately there were also those who took joy in the deaths of the astronauts, or used their deaths to serve their own ends. In Pakistan, there were claims Indian-born Kalpana Chawla had made a mistake, dooming Columbia. In Iraq, some took pleasure an Israeli had died. Some said it was "God's Retribution against the Jews and Ramon." Some chose to use the accident to push political views with messages like, "In light of the tragic loss of the Columbia, the United States should rethink its position with regards to [insert complaint here]."

A notable exception in the Arab world was Palestinian leader Yasser Arafat, who sent a letter expressing his condolences to the White House. Senior Palestinian negotiator Saeb Erakat said, "President Arafat expressed his condolences to the six American families and to the Israeli family for the loss of their loved ones in the explosion of the space shuttle Columbia." Arafat wrote, "We were very sad to hear the news about the dreadful air disaster which hit the American space shuttle on its return from a mission in space, taking the lives of the Israeli astronaut Ilan Ramon and his colleagues. In the name of the Palestinian people and personally, we extend our sincere and heartfelt condolences to you and to our neighbors, the Israeli people, and to the honorable family of the deceased and his friends." One wonders if Jewish officials in Israel would have offered similar condolences if a Palestinian had died.

While Russia was not involved with STS-107, Star City, where the cosmonauts train, reacted almost as if one of their own crews had died. Astronaut Bill McArthur was training in Russia and recalled, "One of the things that was particularly comforting at that time was the outpouring of sympathy and support from the Russians. We had a steady stream of visitors; our Russian colleagues came by and all expressed their condolences." Many of Columbia's crew had spent time training in Russia. Cosmonaut Valeri Tokarev had flown with Rick on STS-96, and cosmonaut Salizhan Sharipov had flown with Mike on STS-89—and both had a close attachment to the Columbia astronauts.

What to do with the three-person crew on ISS was obvious. While they were scheduled to land on Atlantis in March, they were in no immediate danger. Their Soyuz "lifeboat" spacecraft was scheduled to return to Earth in early May, and the crew was fully qualified to fly it back to Earth. Instead of returning to Florida on the shuttle in March, they would land in Kazakhstan on the Soyuz in May. There were jokes about whether or not they would have problems with Customs because they didn't have their passports with them, but those arrangements are spelled out in the international agreements.

The Russian space program would have to bear the burden of keeping ISS supplied until the shuttle was flying again, and it stepped up to the challenge. Initially, Russia echoed what it had said in the past: If the Western countries were willing to pay hard cash, Russia would be willing to build more spacecraft, in particular the Progress cargo ships which carried the food, oxygen, and other supplies needed to keep the ISS crews alive. But that money would not be coming due to international politics. Russia quickly acknowledged the U.S. did more than its share before the Columbia accident, and now it was their responsibility to keep things running and do whatever was needed until the shuttle resumed flying, whenever that would happen.

Scholastic, Inc., publishes many magazines for students. One of the March cover stories was titled "My Mom is an Astronaut," written by Iain Clark with his mother. The magazine was mailed out to schools just before the accident took place. Scholastic considered contacting teachers and asking them to return the magazines or destroy them because it might be morbid to have an article talking about the wonders of space travel when the featured person had just died. But Jonathan Clark told the publishers that Laurel would have still wanted school kids to read those stories. Scholastic sent out his letter with a cover letter. A copy of that letter is on this book's website.

One of Scholastic, Inc.'s, March
2003 magazines featured an article
by 9-year-old Iain Clark. Photo from
the author's collection.

Whenever there's a tragedy, there are always people who behave
well, even nobly, and there are always those willing to exploit it for personal gain. Within hours of the accident, people started selling souvenirs
and other items related to the mission at highly inflated prices. Anything
with the mission's logo was a hot item. and it was a seller's market for as
long as emotions ran high.

The popular eBay Internet auction site was flooded with sellers offering patches, photos, pins, and other memorabilia at heavily inflated
prices. By 5:30 P.M. there were 731 Columbia-related items for sale, compared to just 41 items at 9:30 A.M. In just eight hours, 690 STS-107 items
were put up for auction, and some of the auctions were started as early as
an hour after the accident. Cloth patches with the STS-107 logo which
normally retail for $3 to $5 were sold for $300.

Even more incredible, tasteless, and heartless were the auctions for
items which anybody with Internet access can obtain for free—things
like press kits, crew photos, and other official NASA handouts anybody
can download. One person put up an eBay auction with a minimum bid
of $10,000, guaranteeing a genuine piece of Columbia debris. Ebay offi-
cials promptly took that auction down and issued a press release—"The
handling of any debris from the Space Shuttle Columbia is potentially
dangerous and against Federal law. Any listing of shuttle debris on eBay,
now or in the future, will be immediately removed from the site. In addition, eBay will cooperate fully with law enforcement agencies requesting
information about users attempting to list illegal items."

Not surprisingly, both Internet and actual stores with STS-107-related items quickly sold out. Manufacturers were quickly swamped with orders for far more patches, pins, stickers, and other items than they had made before the mission. Many people offered for sale what they didn't have, expecting to be able to purchase them from wholesalers in time to fill their orders, only to find out the rush on Columbia-related products resulted in wholesalers running out. Manufacturers of STS-107 merchandise, from mom and pop stores which printed their own T-shirts to major suppliers quickly started producing additional batches of merchandise.

There were also some folks with true hearts who fought those who would profiteer from the Columbia tragedy. They put up one-cent auctions where they provided the websites where anybody can obtain the press kits and pictures and told those who read their "auctions" they didn't even have to pay the one-cent fee. The people who put up those auctions paid eBay's posting fees out of their own pockets because of their desire to prevent others from getting scammed by the overpriced auctions.

NASA's spaceflight awareness website includes copies of the STS-107 logo and a special Israeli patch. But the website misspells Ilan Ramon's name — calling it "Colonel Ellon's Mission"!

Within a week of the accident, there were memorial magazines consisting of material pasted together from news articles and NASA's generic publicity materials. Family members and neighbors of the Columbia astronauts were hounded and lied to by the editors of some of these publications for information or photographs. Within a couple of months after the accident, a book came out with similar information compiled by somebody who didn't know the astronauts but was quick to react and take advantage of the situation. The "quickie" publications are filled with mistakes. Evelyn Husband noted she's glad they don't have accurate information about her family because it's one less intrusion into their private lives.

There was one very visible change within NASA after the accident. NASA is a very large organization, with many different locations. Only the Johnson Space Center, in Houston, Texas, Kennedy Space Center in Florida, and Marshall Spaceflight Center, in Huntsville, Alabama, are primarily dedicated to the shuttle. Other centers have a smaller role, like the Goddard Spaceflight Center, in Greenbelt, Maryland, which was responsible for FREESTAR and the relay satellites which the shuttle uses to communicate with Mission Control. However, after the accident, thousands of NASA workers, not just those on the shuttle workforce, wore STS-107 lapel pins, recognizing that the loss of the seven astronauts and Columbia affected all of NASA.

It's sad it takes a great tragedy to get the public's attention. But certainly once a tragedy does happen, there are those who want to help in whatever way they can.

The only usable 70-mm. photo to survive the Columbia accident—a group shot of the STS-107 astronauts inside Spacehab. Credit—STS-107 crew.

PART V

AFTERMATH

With a well-thought-out contingency plan, NASA was able to react in an orderly fashion after the Columbia accident. This was unlike the chaos that followed the Challenger accident. Critical people understood their responsibilities and knew what they had to do. Within a couple of hours, the first recovery teams reached the areas where debris was found. It was the start of the largest organized search in history.

The debris and other evidence determined what happened to Columbia without any shadow of a doubt—all of the evidence pointed in the same direction. A 1.67-pound piece of foam fell off of the bipod region of the ET 81 seconds after launch and slammed into and damaged the leading edge of the left wing. The reentry heat that entered through that breach lead directly to Columbia's destruction. But just as important as the physical cause of the accident were the organizational issues within NASA—what happened to permit the foam to fall off, and why it wasn't identified as an urgent concern by the safety system or even recognized as a safety concern.

Unfortunately, a large number of myths have been spread about Columbia, and many are still repeated today. They range from sick jokes and people who don't understand how NASA works to misinterpreted statements to people who just don't understand the laws of science.

Also here are descriptions of the memorials—literally hundreds of memorials to the Columbia crew on Earth and in outer space, as well as the "memorial" which counts the most—the science which can still be done, even after Columbia's loss.

CHAPTER 44

THE SEARCH FOR DEBRIS

If anything can be said to be fortunate about Columbia's loss, it's where the accident took place. Most of the debris landed in sparsely populated areas in Texas and Louisiana. If bad weather forced Columbia to land in California all of the debris would have ended up in the Pacific Ocean.

NASA's first priority was to recover the astronauts' bodies, if they made it through the reentry and could be found. Within minutes after the accident, there were reports of debris in Texas and later reports of body parts near Hemphill in East Texas. Head of flight crew operations Bob Cabana said, "We're treating those remains with the ultimate respect and care that they deserve."

Recovering the bodies was the most emotionally difficult part of the recovery. When the National Transportation Safety Board investigates an airplane accident, they find the bodies of relatively anonymous people. But in military aviation and test flying, the people collecting the bodies are the friends and neighbors of the ones who died.

Spacehab manager Pete Paceley said he got calls through the night from astronauts who were searching for the bodies. At one point Paceley was asked, "Did you have a golf ball inside your module?" The astronauts never explained why they asked those questions, but it was pretty clear they had found a golf ball and were trying to determine if it was from Columbia, possibly a personal item for an astronaut or one of the payload people. Paceley verified there was no golf ball onboard. He noted he got those kinds of questions for about four days, with the calls stopping when all of the bodies were recovered.

On February 5, the bodies were flown to the Charles C. Carson Center for Mortuary Affairs at Dover AFB in Delaware. That particular morgue was necessary because it's equipped to handle bodies that may have been exposed to hazardous chemicals. After the bodies were cleared they were returned to the families. Ilan's was returned to Israel for a hero's funeral; Rick's was buried in Amarillo, Texas, and Willie's in Anacortes, Washington; KC's cremated remains were scattered in Zion National Park, one of her favorite places. Mike, Laurel, and Dave were buried at Arlington National Cemetery near the Columbia memorial. The memorial is next to the memorial for the space shuttle Challenger.

That crew consisted of commander Dick Scobee, pilot Mike Smith, mission specialists Ellison Onizuka, Judy Resnik, and Ron McNair, and payload specialists Christa McAuliffe and Greg Jarvis. Scobee, a retired Air Force officer, and Smith, a Navy aviator, are buried nearby.

Kathie Scobee Fulgham, Scobee's daughter, wrote a letter to the children of the Columbia astronauts, telling them what she had experienced 17 years earlier. A copy of that letter is on this book's website.

Dave Brown's neighbor Cindy Swindells has the ashes of Dave's dog Duggins and says when she can travel, she plans to scatter them over Dave's grave so they can be together again.

After the recovery of the crew the main focus for the search could move to other priorities. The next priority was electronics boxes with classified components. The boxes could theoretically be reverse-engineered to determine how to decode classified data from military satellites.

Third in priority was anything which could help investigators determine what happened. A burn pattern on a piece of debris could be an important clue for how a piece came off Columbia. Even better would be an avionics box with readable memory. If any magnetic tapes survived, they might include useful data. Since Columbia was the first shuttle, it had performance-monitoring instrumentation that wasn't installed on the other shuttles. In particular, the OEX (Orbiter Experiment) telemetry recorder might have useful data if it was found and its magnetic tape had survived.

The film from engineering cameras that took still photos and movies of the tank as it dropped away from Columbia was extremely high on the priority list. Less useful to the investigation but certainly of interest would be any film, crew tape recorders, videotapes, or laptop computers used for the crew's experiments.

Of course with computer hard drives there was an additional concern—could they survive a 38 mile fall? That depended on how well the drive was insulated and how hard it hit. Certainly an unprotected hard drive hitting a sidewalk would have far less chance of surviving than a hard drive within an exterior box landing in a bush.

Paceley noted, "For the first day or two [there was] real chaos, different agencies determining their roles, staking their territory. After about two days it calmed down, and each agency fell into its respective role. It just all kind of fell in sync; things started to happen very systematically after that. I was very impressed."

NASA repeatedly warned the public to stay away from Columbia's debris because they could be hazardous. The many announcements would also help deter nuisance lawsuits—it would be difficult for somebody to claim they didn't hear unusual objects in the recovery area were dangerous.

Another reason was to discourage souvenir hunters. Certainly a piece of shuttle debris would be a very valuable item, if you could sell it without getting caught or even if you just wanted to keep it for yourself.

Unauthorized possession of debris from either Columbia or Challenger is a federal crime, and the FBI aggressively presses charges against anybody caught with shuttle debris. It's not just a matter of the government keeping pieces out of the public's hands. Each piece of debris, and where it landed, could help solve the puzzle of how Columbia came apart.

Columbia broke up over Dallas, Texas, spreading most of its debris in a strip 240 miles long and 10 miles wide from just east of Dallas to Vernon Parish, Louisiana—equivalent in area to the state of Rhode Island. 400,000 people lived in the area. It was estimated 20 percent or more of Columbia could have survived reentry. How could NASA find all of the pieces spread across 2,400 square miles—much of it forests, swamps, and other inaccessible terrains?

Even the Department of Defense and Federal Emergency Management Agency (FEMA) would be hard pressed to come up with enough resources. One logical choice was the Bureau of Land Management's firefighters, who are normally hired only during the summer fire season. They were physical people experienced at working outdoors. Those who didn't have other commitments could be hired to search for the debris. In addition, FEMA put out a call for volunteers, especially ones with useful skills.

The public enthusiastically reported potential pieces of debris—in many cases too enthusiastically. NASA received 1,459 reports from people who believed they'd found a piece from Columbia, including 37 states Columbia did not pass over during its reentry—as well as Canada, Jamaica, and the Bahamas!

At least one person was astonished to see on the news, the day after the accident, a heavily charred but very recognizable piece. CIBX manager John Cassanto immediately recognized that it was part of his experiment.

The search involved underwater divers, airplanes, and—most important—a lot of shoe leather. Each day, on average, there were 5,600 searchers in the field. Most firefighters are American Indians. There were 256 teams of American Indians involved in the search. (This book's website has a list of all of the participating tribes.) Besides trained searchers, many NASA workers felt the need to go out and help search, even if it was just for a couple of days. They wanted to be part of the recovery process.

Each search team consisted of about 20 firefighters and volunteers, somebody from the Environmental Protection Agency (EPA) to check debris for potential hazards, and a NASA person to help identify the debris. Many of the searchers wore T-shirts with the motto "Their mission has become our mission."

The debris ranged from coin-size pieces to a 400-pound 12-foot-long piece of Columbia's fuselage. The pieces were transported to Barksdale AFB in Louisiana and entered into a database. Columbia had been at Barksdale previously, during ferry flights to Florida from Edwards AFB in California and White Sands, NM—riding on the top of NASA's modified 747. The debris was shipped by the truckload from Barksdale to KSC.

The search operation had a hard deadline — the spring growing season would make it far more difficult to spot pieces, and the firefighters would only be available until May before they were needed for their normal jobs. A map of the debris zone was laid out, and a schedule was developed based on the available resources: how much could be searched each day, and when the searches would have to end.

NASA Administrator Sean O'Keefe noted there were no turf battles between the diverse organizations, most of whom didn't have any experience working with each other. O'Keefe said, "The unprecedented debris recovery operation involved over 14,000 people representing over 130 different agencies and volunteer groups, contractors and private organizations, state and local governments — you name it, everybody was involved. Well beyond the expectations of what we ever could have imagined, with that many disparate groups who had never worked together before, this has proven to be, in my judgment, the greatest inter-agency, inter-government, inter-personal activity we could have conducted anywhere — any time. It has been absolutely flawless, and it is for that reason there has been absolutely no controversy to speak of at all. It has been scantily covered [in the news]. If there was any controversy, we would have been reading about it on the front page."

Astronaut John Herrington echoed O'Keefe's comments: "The people who were there had this real motivation and dedication to what they were doing. They did it. When you asked for something, it would happen. There wasn't any, 'We can't do that because of this.' You needed something — it would happen. "

Herrington also noted the passion of the searchers, "People have a real deep appreciation for the human aspect of human spaceflight. It's the people involved in it. If anything came out of this tragedy, [it is that] you realize there's a huge support for flying humans in space from the general public. That, if anything, makes you feel you're doing this for a reason. People appreciate this — they realize what it means to them. I appreciate the fact there were people there going way beyond what was expected to recover both the crew and the hardware to figure what happened. You'll never have any idea unless you're out in East Texas, what a fabulous thing the people out there did. From the minute it happened and debris started coming down in Texas, to all the people who came from around the country to participate in the recovery. I was really, really impressed with the people I met — Texas forestry service, U.S. forest service, FEMA, EPA, you name it. It was really good, coming together with really talented people doing really important things."

FILM AND VIDEOTAPE SURVIVAL

It was astonishing how much film and videotape survived. Columbia had 337 videotapes, and 137 rolls of 35-mm. and 70-mm. film onboard. 28 of the videotapes and 21 rolls of film were recoverable.

The most important for the investigation was the handheld 35 mm. film and videotape Mike and Dave shot of the ET after it separated from Columbia, and the automated cameras within Columbia's belly. They would have provided hard evidence for how much bipod foam had fallen from the ET. But those film cartridges and videotape were never found.

The recovered video and film which was not damaged by the heat are a fascinating behind-the-scenes look at life in space. Below are two of the photos NASA chose not to release.

The first usable videotape was the one Laurel shot during the reentry. As Laurel shot the video, it wound on the cartridge's pickup reel. The camcorder would have been exposed to intense heat, but its body would have partially insulated the tape inside. As the heat crept inside it would have done the most harm to the outer layers, with the last video she shot.

Many have questioned whether or not there could be more video, including Laurel's husband Jon. He said, "The thing I have always wondered about is was there more that was recovered that they just wouldn't show us. I'm not sure I have that answer."

He said, "I have found things out long after the fact that have very much disturbed me. Suffice to say that information was not supplied to [the families]. Does he think it's possible NASA is still withholding information from the families? Clark said, "I think, based on the fact that it took so long for some information to get to me, that it's possible." If information is being intentionally withheld, it's because some high-level manager felt it would cause the families emotional harm.

The recovered videos show life on Columbia; little clips the crew videotaped. Some of the video was transmitted during the mission and was familiar. But the most interesting was the raw footage the crew shot. Dave was clearly videotaping segments for his movie; there are clips where he shot the same scene multiple times to get it just right.

A rare photo shows the crew's exercise shoes for the ARMS experiment.

This photo of Laurel working inside the Biopack glovebox has stains on its negative caused by the accident.

The only 70-mm. film NASA released was a group shot of all seven crewmembers. The 35-mm. film included 92 photos of the crew working on their experiments, photos of the failed water separator, and Earth views.

This book's website has many of the recovered photos and video clips, including ones that are not available on NASA's Internet sites.

FALLING DEBRIS AND MIRACLES

One of the biggest myths was that it was a "miracle" nobody was hurt or killed by debris. It's been repeated many times: NASA was incredibly lucky nobody was killed. It's often said it's fortunate the accident happened on a Saturday morning, when most people were still at home, with the implication that if it happened on a weekday with more people outside, the results would have been different. Why was NASA so lucky?

A little common sense shows it isn't surprising at all—in fact it would have been surprising if anybody was hit by debris. The simple fact is people are small, buildings are large. That's pretty obvious, but something many people don't think about when they believe it was a miracle nobody was hurt.

There were 84,124 pieces of debris recovered during the primary search operations. Dividing 84,124 pieces by the 2,400-square-mile debris area works out to just 35 pieces per square mile. Even if you assume every person under Columbia's flight path was a large adult lying down outdoors, the odds of any individual getting hit were less than 1 in 41,000. And even if somebody was hit by a piece of debris, it would not necessarily result in an injury. Getting hit by a lightweight cloth patch falling from a height of 38 miles feels identical to getting hit by a lightweight cloth patch falling from a height of 10 feet.

What the statistic *does* show is how difficult it was to find the debris. On average, each square mile searched produced only 35 pieces of debris. That's why it took such an intense effort to recover everything.

ONE ASTRONAUT'S ACTIVITIES IN THE FIELD

Hundreds of NASA personnel joined the recovery effort in the field, including many astronauts. Astronaut John Herrington was assigned to coordinate the aerial searches from the Lufkin Command Center. Herrington had performed similar tasks in the Navy before coming to NASA.

The Texas Forest Service had up to 36 helicopters, 10 light aircraft, a handful of aircraft with special sensors, the civil air patrol, and even ultralight powered paragliders.

Helicopters could cover four times as much land as search teams on the ground, but they could only spot larger pieces or things that were unusual looking, which would warrant sending a ground team to inspect the area more closely. The helicopters were sent primarily to areas where winds could have blown lighter debris away from the main search corridor.

Herrington was the air ops officer on March 27 when a Bell 407 helicopter crashed in the Angelina National Forest in San Augustine County, Texas. Charles Krenek, an employee of the Texas Forest Service, and pilot Jules "Buzz" Mier were killed in the crash. Matt Tschacher, a U.S. Forest Service employee, and KSC workers Richard Lange and Ronnie Dale were injured. Herrington said, "It just added to the feeling of how bad it was already. Two people had perished in the process, in the recovery effort. For me that was the hardest time in the whole ordeal."

As part of the search and recovery operations, the astronauts spent much of their time meeting volunteers. Herrington, the first enrolled Native American astronaut, said, "A good portion [of the searchers] were fire crews, and a good portion of people who fight fires in this country are Native Americans. I could show them the respect and honor they deserved because they did a phenomenal job. It was amazing. I heard 89 percent of the people walking the debris line were Native Americans. What was neat about it was they treated it with a reverence that a lot of people didn't expect. These were not inanimate objects they were finding, these were very, very living pieces they had a connection to. Here's a group of folks who for centuries had a very intimate connection to the sky. They are a part of what's going to return us to flight, they're now part of the space program and what we're doing."

Many of the volunteers, especially the American Indians, wanted to meet Herrington. He said, "I just tried to make sure I would talk to every camp—I went to Nacogdoches, Hemphill, Palestine, and Corsicana—the four major camps out there. I wanted to go out and see people."

The two members of the debris recovery team who were killed in a helicopter crash—Jules "Buzz" Mier (second from left) and Charles Krenek (on the right)—pose with three other recovery team members. Photo courtesy of the CAIB.

A WILD GOOSE CHASE

Each of the reentry videos submitted by the public was studied for possible anomalies, and 24 unexpected events were seen, including unexpected brightenings or objects falling off Columbia. Each of those anomalies was assigned a number. Based on the analysis of the videos and radar records, search parties were sent to Caliente, Nevada, to look for debris #6. But that search was in the wrong place.

Skywatchers John Sanford in Springville, California and Jay Lawson in Sparks, Nevada both videotaped debris #6. Space scientist Rob Matson, an experienced satellite observer, explained, "John's video showed a distinct piece of fairly significant debris slowly separating from Columbia. As soon as I saw that I realized this is not just a "knot" in the plasma, it's a physical object of some significant size. I went through the video frame-by-frame and measured the separation distance from Columbia. The debris he saw in his video was the same as the Sparks, Nevada [video] from a totally different perspective." Lawson's video from Sparks showed Columbia flying in front of the planet Venus. (The video iss available on this book's website). Knowing Lawson's location and Columbia's path pinpointed the exact time the debris fell away.

The purpose for all of this work was to determine the object's "ballistic coefficient". The ballistic coefficient is an engineer's method for measuring an object's drag. Imagine tossing a piece of litter from a car. The litter and car are traveling at the same speed when the litter is tossed, but the car continues to drive down the highway while the littler quickly slows down and falls to the ground. If the litter is a small dense object like a fishing weight it will quickly drop, but if the litter is a piece of cardboard it will flop around before hitting the ground. The cardboard has a large area with a small mass (a high area-to-mass ratio), while the weight has a small area and a higher mass (a low area-to-mass ratio).

Matson explained, "Knowing the correct field of view [in the video image] is critical to getting an accurate ballistic coefficient. I plotted all of the positions of the debris; time tagged everything, got the range to Columbia. Eventually you can come up with an equation with the curve of the separation versus time." Matson calculated a ballistic coefficient of 0.049 square meters per kilogram, a fairly dense object. Matson's calculation was fairly close to NASA's value of .058. Next Matson looked at the direction and intensity of upper level winds when the accident took place. Then Matson noted, "When you get the ballistic coefficient it's just a mater of flying it to the ground." In other words — it's rocket science.

Matson's analysis resulted in a location of 37.44 degrees north and 114.00 degrees west. Matson said, "It puts it just over the Nevada-Utah border into Utah. Looking on a map it's in the Cougar Canyon area a little to the southeast of Beaver Dam State Park." NASA's calculations, which also included radar data, differed by 50 miles. Matson said he

spent hundreds of hours working on the problem because, "I couldn't figure for the life of me why [NASA was] searching where they were searching—it makes no sense." Most astonishing Matson did all of his calculations from publicly available information, including Columbia's reentry
timeline on NASA's Internet site and NOAA weather data.

BALLISTIC COEFFICIENTS, TERMINAL VELOCITY, AND ROCKET SCIENCE

Much of Columbia was burned to ash during its catastrophic reentry.
When you expose materials to 3000° F, not much will survive. But some
did, and in surprisingly undamaged condition.

Debris moving extremely rapidly, even in the thin air of the upper
atmosphere, encounters friction as it rubs against the air molecules, and
the friction creates heat. Slower debris encountered less friction, and less
heat. In addition, some materials conduct heat very slowly, so while the
outer layers were exposed to intense reentry heat, the inner portions
encountered less heat.

How fast any piece was falling when it hit the ground was determined by its ballistics coefficient –a combination of the mass and surface
area. At extreme altitudes there's less pressure and less drag so objects can
fall faster and most pieces would go supersonic. The largest pieces created sonic booms that were strong enough to be heard on the ground. As
the pieces fell through lower altitudes, where the atmosphere is thicker,
the pieces slowed down.

Physics professor Bart Lipofsky said, "Terminal velocity is the maximum speed anything can achieve when falling through an atmosphere."
Instinctively, most people think the higher an object falls from, the faster
it will be going when it hits the ground—but they're wrong. Lipofsky said,
"Take two sheets of paper, hold them out at arms length and release them
at the same time. Both will hit the ground about the same time. Crumple
up one of the sheets of paper and repeat the experiment." No matter what
the shape of a piece of paper, even if it's streamlined like a paper airplane,
its terminal velocity will be pretty small.

Light, fluffy items, such as pieces of cloth, hit the ground as gently as
a snowflake. But even the densest aerodynamically shaped objects would
not have been traveling faster than a couple of hundred miles per hour
when they hit the ground. Lipofsky noted, "For an object entering the
atmosphere from outer space the fringes of the atmosphere will slow it
down to terminal velocity far above the Earth's surface."

Nevertheless many refused to believe this. One reporter insisted,
"Columbia's pieces were falling faster when they hit the ground because
of its high altitude. I've taken physics, I know what I'm talking about."
Lipofsky said, "Maybe next time when he takes physics he should actually open his book and read it because he's completely wrong."

When Columbia came apart, everything inside — from stickers to the massive turbopumps — fell and accelerated until they reached their own terminal velocities. In the rarified upper atmosphere, they would have continued to accelerate to very high speeds, but once they entered the lower atmosphere they would slow down to ordinary terminal velocities.

What was incredible was how many items made it through the carnage of Columbia breaking up, the reentry heat, and the 38-mile fall, and survived relatively intact. Items like cloth patches, notebooks, film, videotapes, and even computer hard drives survived.

Part of it was just luck — similar to how flimsy items survive after a hurricane has decimated an area. One piece may have survived while a hundred identical pieces burned up. Cloth patches were packaged in extremely tight bundles. During reentry the outer patches probably burned up completely and the bundles fell apart. The inner patches would have received the least heat and fluttered the rest of the way to the ground.

FINISHING THE SEARCH FOR DEBRIS

The three-month $302 million search for Columbia's debris was completed in early May.

Debris collection head David Whittle said, "Thanks to the efforts we've had here in Lufkin and on the Texas corridor and over 100 federal and state agencies, we've recovered very close to 40 percent of the shuttle. It's a major major recovery. It's about twice as much as predicted that we would recover. On February 1, I stepped off the airplane at Barksdale AFB to start the first part of this search, what has turned out to be the largest search of this nature in the United States, in the history of the U.S., perhaps the world."

The original estimate of 20 percent was based on reentries of various satellites not intended to survive reentry. Columbia's thermal protection system did prevent many pieces from burning up, even though it was pushed beyond its limits. Columbia also has some extremely cold materials, like liquid hydrogen and oxygen, that helped protect items from the intense reentry heat and increase the chances they would survive. But the most important reason so much was recovered was the intense search.

FEMA's Scott Wells said, "We've had air, ground, and water operations. We've had ice storms, I think we had 40 days and 40 nights of rain back in February." — Well only in Texas could you have 40 days and 40 nights of anything in a 28 day month!

No matter how the numbers are given, the search was staggering. Over 30,000 people participated. Whittle said, "We had people walking within 10 feet of each other on a 210-mile strip, four miles wide, and they didn't miss very much. There was an incredible amount of pride in finding things and being complete. We have physically covered, with people

A group of searchers pace a field, looking for debris from Columbia.

walking, over 700,000 acres. We have searched over 1.6 million acres with our air assets. We've mapped 23 miles of the bottom of Lake Toledo Bend and Lake Nacogdoches. The U.S. Navy Supervisor of Salvage was a major player in our underwater operations, and they dove on over 3100 targets in Toledo Bend and over 326 targets in Lake Nacogdoches. The days I was out there, the water temperature was 47 degrees. The visibility underwater was about inches." The underwater searches yielded only one piece from Columbia.

Wells gave a comparison, "Just the ground operation was the equivalent of one person walking from the Earth to the moon, seven laps around the moon, then back to the Earth and one lap around the Earth and still have two thousand more miles to go."

With that intense a search it isn't surprising other lost items would be discovered. A body was found, a man who had wandered away from a nursing care facility about a year before. Also found was a pickup truck involved in a homicide investigation.

Whittle readily acknowledged they didn't find everything, "There are still pieces out there. We're still getting 10 to 16 calls a day from the public. Should somebody go out in their backyard and find something that gave us an interest we may go out to do a limited search in that area."

The recovery teams have asked anybody who finds something they believe is from Columbia to call 1-866-446-6603. Whittle said, "Depending on the debris, if it's something small there's a good chance we'll just ask you to FedEx it. If it turns out to be something larger or something toxic they'll send someone out to respond to it. They'll decontaminate it."

Flyers were distributed to sporting goods stores in the areas and Texas hunters when they renew their licenses.

CHAPTER 45   OTHER EVIDENCE

Besides debris, the investigation had evidence from several other sources. The first information was the telemetry received in Mission Control during Columbia's reentry.

How and when the sensors went bad provided some of the earliest clues. The almost simultaneous cutoff of a variety of unrelated sensors was a mystery. Jeff Kling said, "[They] trickled off. Normally when a sensor goes bad, it goes open circuit, and it's immediate." Engineers thought—what could cause a sensor or its wiring to fail that way? Could the wire have been torn apart? Could heat have melted a wire's insulation until it failed? Engineers determined if wires were overheated, they would fail in the same way Columbia's sensors failed. What all the failed sensors had in common was their wiring all went by the left wheel well.

After the accident, engineers reexamined every bit of telemetry they received during the reentry for anything abnormal. Starting at 8:52:17 A.M. EST there was a gradual temperature rise in the left wing, in the main landing gear. It was a subtle increase over time, so gradual that nobody noticed it was unusual during the reentry.

The Orbiter Experiment (OEX) telemetry recorder was found in excellent condition, and its tape had a gold mine of data. The 1970s-era

A chart shows which sensors had failed at a particular time (8:58:39 A.M.) during the reentry. At this point, Columbia was at an altitude of 39 miles and traveling at 13,391 miles per hour, and it was 17 minutes away from the landing at the Kennedy Space Center. Contact with Mission Control would be lost in just 53 seconds.

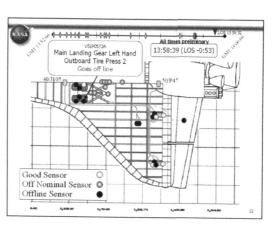

383

storage system monitored hundreds of sensors and proved that the damage occurred during the launch and showed a heat spike in the left wing during the reentry.

Many people asked an interesting question—Did Columbia fly differently from the other shuttles? Some commanders had reported Columbia felt "different" to them. Columbia was heavier than the other shuttles because as the first operational craft it was designed very conservatively. Over time, based on the lessons learned while Columbia was built and its early test flights, engineers were able to determine places where weight could be saved by using a smaller bolt or thinner spar without impacting safety. Later shuttles had less "dry weight" and could lift more cargo.

I polled several astronauts who landed Columbia. My criterion was that they had to be the commander of a mission on Columbia and a commander on at least one of the other shuttles. Of the 19 commanders who met my criterion, I interviewed 6: Joe Engle, John Blaha, Jim Wetherbee, John Casper, Tom Henricks, and Ken Bowersox.

They agreed there's no difference in Columbia's flying characteristics. Blaha said, "If you put a bag over my head I couldn't have told the difference between Columbia and Atlantis." Wetherbee, the most experienced commander, having landed Columbia, Discovery, Atlantis, and Endeavour a total of five times, said: "I don't know why anybody would say anything like that [that there were handling differences between Columbia and the other shuttles]." Ken Bowersox was less willing to commit. He noted in his limited experience (landing Columbia with a heavy payload and Discovery with a lighter return) he couldn't say for sure whether or not there was any difference. On Bowersox's STS-73 mission Columbia did fly differently. A shim between a pair of tiles had slipped out of place, resulting in a rougher descent through the atmosphere.

The CAIB also noted some unusual coincidences—including the high percentage of bipod foam losses on Columbia's flights, and the fact that Columbia flew almost all of the flights in 39-degree orbits. The later was easy to explain, 39-degree orbits are only required for scientific flights, most that were automatically assigned to Columbia because of its lower performance.

More unusual was Columbia's track record—it took more tries to get Columbia off the ground. Three Columbia missions account for a large percentage of its total scrubs: STS 61-C in 1986, STS-35 in 1990, and STS-73 in 1995. These missions had a total of 16 launch delays; almost one-fifth of all the delays in the entire shuttle program. Only three non-Columbia missions needed more than four launch attempts to get off the ground. As far as anybody has been able to determine, there was no common thread among Columbia's different delays—other than bad luck.

All together, Columbia spent 1,208 days on the launch pad—200 more than any other shuttle. Columbia's pad time did raise a concern because damage due to exposure to KSC's salt-air ocean environment might accumulate slowly over time.

Because of its memorable delays, Columbia got the reputation as the shuttle that had the most problems on the ground, but almost always performed flawlessly in space. Launch director Mike Leinbach said after the STS-107 launch, "Typically, Columbia's a little difficult to get off the pad and then works outstandingly on orbit. Today it was very easy, we launched on time."

MYSTERY OBJECT

Satellite tracking data is normally released as new objects are detected by ground-based radars and cataloged. After the accident, a mystery object was discovered in the raw radar data. The Air Force Strategic Command (USSTRATCOM) went through its radar data to see if there was anything unusual that went unnoticed during the mission. General Duane Deal said, "This was the most intensive examination of orbital data in the history of Space Command."

USSTRATCOM has a worldwide network of satellite tracking stations. There is interference from other transmissions on similar frequencies, flying overhead, and natural phenomena. Computers have to filter through the noise to find real space objects. Over 8,000 objects in low-altitude orbits, as small as a 4-inch bolt, are tracked by USSTRATCOM. For example, Columbia itself was booked as the third launch of 2003, and the 27,647th object tracked in the history of space travel.

The analysis resulted in the discovery of the flight day 2 "mystery object." A radar at Eglin AFB in Florida detected the object about a day after launch during its normal operations. During STS-107, the object slipped through the system because the computers believed it was electronic noise. But the intense search of the raw data permitted orbital mechanics experts to "connect the dots" between the radar observations and determine the object's orbit.

Natural gravitational forces pulled the mystery object away from Columbia. Based on the strength of the radar returns it was .3 by .4 meters (about 140 square inches). The object made a natural reentry on January 20 over the South Pacific.

Based on that sketchy information, satellite tracker Ted Molczan was able to calculate an approximate orbit. For it to have reentered so quickly it would have to have an area-to-mass ratio of about 0.049 square meters per kilogram, giving a clue to its composition.

Shuttle tiles were too light to match, but something made from RCC would fit. By the end of February I had enough confidence in Molczan's analysis and my own research to say it was likely the mystery object was made out of RCC. Molczan's analysis is available on the Internet at http://www.satobs.org/columbia/STS107mysteryobject.html

What's fascinating is that it has the same area-to-mass as Rob Matson's debris #6 object, which came off during Columbia's reentry. That implies both the mystery object and debris #6 could be made from the same material.

On April 9, the mystery object's orbital data was released. It's formally known as "STS-107 DEB" (for debris), and assigned serial number 27,713. It matched the analysis performed seven weeks earlier by Molczan.

So what was the Mystery Object? The most likely candidate is either a piece of an RCC panel or T-seal. In this scenario, the impact from the foam shattered the RCC, then Columbia went through 16 orbits—traveling from daylight to darkness and back. The constant heating and cooling in each orbit helped loosen the piece, and the maneuver to align the IMUs 23 hours after launch dislodged the piece and it came loose.

FOAM GUN TESTS

The Southwest Research Institute in San Antonio, Texas, has a giant compressed-air gun to test the strength of aircraft components against debris. The projectiles range from birds to shrapnel. There are urban legends about "chicken guns" and how they're used to test aircraft, but they actually exist, and the "bullet" is always a thawed dead bird, complete with feathers and head. The most vulnerable aircraft parts, like windshields and engines, are tested with "chicken guns." SWRI's gun shot blocks of bipod foam at a simulated shuttle wing.

Where the foam hit was an important part of determining what happened. During STS-107, engineers believed the bipod foam had hit the bottom of the wing, close to the main landing gear doors. After the accident, additional analysis with supercomputers moved the most probable impact point to RCC panel #6, and later to RCC #8.

The first foam gun tests were at the bottom of the wing, which is covered with black tiles. The tests confirmed the conclusions the engineers came to during STS-107; the bipod foam against the black tiles produced little damage.

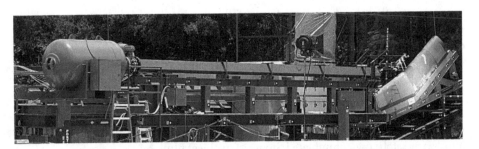

The compressed-gas "chicken gun" used to fire foam blocks at a simulated shuttle wing.

What was more challenging was the RCC on the leading edge of the wing. A test rig was built with RCC panels #5 through #10. The prototype shuttle Enterprise, built before Columbia, had some RCCs. Most of Enterprise's leading edge panels were fiberglass since they would never be exposed to the actual reentry heat. But a few of Enterprise's panels were real RCC.

The tests could not duplicate what actually happened. The gun was on the ground, while Columbia's damage took place at a far higher altitude, in a vacuum. The foam that hit Columbia was irregularly shaped and tumbling, the foam gun could only shoot box-shaped pieces directly at the target. The engineers who set up the tests could control the size of the foam "bullets," their velocity, and the angle the foam hit the target. They adjusted those values to try to compensate for the differences between their tests and what actually happened to Columbia.

The precise values for the impact were uncertain. The only information on the size of the bipod foam came from the computer-enhanced launch camera views. The velocity could be determined a bit more precisely, but it too had some uncertainty. Hubbard noted, "The variation in velocities we could have used is as low as 650 and as high as 825 feet per second. The size of the foam was estimated from 850 cubic inches to 1,600 cubic inches. The density of the foam varies from 2 pounds per cubic foot to 2.6 pounds per cubic foot." In other words, the amount of energy from the foam strike could have been as little as 2,220 to as much as 8,730 foot-pounds—4 times as much. That's the difference between a compact car hitting a wall at 5 miles per hour and a school bus hitting the wall at the same speed.

A suitcase-size foam bullet produced a 5.5 inch crack in Discovery's RCC #6, a 30-flight veteran.

The final test was far more impressive. The gun was aimed at RCC #8, a 26-flight Atlantis veteran, and the results were astonishing—a jagged 16-by-10-inch hole—big enough to put a person's head through.

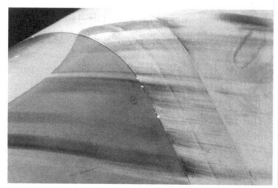

An early foam gun test produced relatively minimal damage—a 5.5-inch crack. The photos show pieces of foam imbedded in a seam of the wing and, on the right, a close-up of a 1-inch crack

Engineer Dan Bell measures the hole created by the 1.67-pound foam block traveling at 500 m.p.h.

Many news media show only the last test—an extremely dramatic large hole—and ignore the earlier tests, where far less damage took place. They use Scott Hubbard's quote, "We have found the smoking gun. The test today demonstrates that this is, in fact, the most probable cause creating the breach that lead to the accident." Rarely do the media use fellow CAIB member James Hallock's more important comment: "If it were as large as the one that we just saw—that's so large that you're gonna be bringing lots of energy in. And I would think (Columbia's breakup) would've happened much sooner. In other words, Columbia would not have made it to the state of Texas."

There's absolutely no doubt the giant hole was about 50 percent larger than what actually happened.

Hallock added, "I personally believe that we are talking about something that definitely has to be in the 6- to 10-inch [range], because that sort of agrees with all of the calculations."

How could the two foam gun tests produce such different results? One key factor was the RCC's highly variable strength. Since there was no requirement for the panels to withstand any damage, their strength was not controlled during their construction. Hubbard explained, "The variation in freshly made RCC breaking strength is 70 percent." It's quite possible the first test was one of the stronger panels, and the last test was a weaker panel. If the tests were reversed, the results could have been dramatically different.

Hubbard noted because of all of the unknowns, "It's all in the same ballpark. And we demonstrated the important thing, which was the clear connection between foam and a breach of about the approximate size." In other words the size of the hole doesn't count, just that the tests proved the bipod foam could create a hole large enough to account for what happened to Columbia.

Quantitatively, the foam gun tests did not provide very useful information. Qualitatively they showed the foam could damage the RCC, something that was evident to those who knew the RCC's strength and the amount of energy a 1.67-pound piece of anything traveling at 500 m.p.h. has.

THE STORY TOLD BY THE DEBRIS

Columbia's pieces were reassembled at KSC. They clearly showed the left wing was much more heavily damaged than the right wing. Even missing pieces helped tell the story. Much of the left wing was never found, not because it was missed during the search operations, but because it burned up.

The State of Florida built a 50,000 square foot $4.1 million hangar, hoping that future reusable rockets would want to use it. But the reusable rockets never appeared, and the only use for the hangar was as an overpriced warehouse. The hangar was laid out in a grid with a template of Columbia's surfaces including the wings, rudder, body, and flap. As pieces arrived, they were placed in the appropriate locations.

Items from the crew cabin were placed in a separate room, out of consideration for the astronauts' families. The seats, notebook computers, clothing, and even uneaten food were considered crew compartment contents.

Gregory Kovacs, a Stanford University professor working for the CAIB, noted, "It is like putting together a multi thousand piece 3-D jigsaw puzzle on a 2-D surface." As the jigsaw puzzle came together, the debris told its story. Even non-engineers could see debris from the left side was damaged far more than the right side. Clearly the left side was where things happened, while most of the damage to the right side was caused by the reentry heating or hitting the ground.

A set of Plexiglas supports were built so the recovered left wing RCC fragments could be studied in the correct orientation. As the pieces were mounted, it became clear, even to non-engineers, that the greatest damage had occurred around RCC #8.

Gehman acknowledged, "The debris was more important than I thought it was going to be."

The closed room where debris from the crew cabin was kept separately out of respect for the families. Photo by Philip Chien.

Reconstruction chair Mike Leinbach shows the left wing RCC Plexiglas displays. Photo by Philip Chien.

CHAPTER 46 COMPLETING THE INVESTIGATION

While searching for clues to the direct cause of the Columbia accident, the CAIB concentrated on the underlying causes: What was the environment within NASA that permitted the accident to happen? Why wasn't whatever went wrong caught in time? Why didn't NASA consider falling foam to be a more serious safety concern? Did managers and workers do everything they could to maximize safety and minimize risks? What was the morale of the workers? And even—What were the compromises in the shuttle's design and Senate support that contributed to the accident? The debris and engineering tests helped establish *how* the accident happened. The investigation into NASA's culture showed *why* it happened.

As the evidence came together, it all pointed in the same direction. The burn patterns on the recovered debris, the telemetry, the OEX recorder, the orbital characteristics of the mystery object, and the foam gun tests all pointed to the same conclusion. RCC #8 was damaged during the launch, which led to Columbia's destruction during reentry. The bipod foam that hit the wing had enough energy to create the breach. No additional contributing factors were needed.

The evidence was overwhelming—the foam was the direct cause of Columbia's destruction. A brand-new RCC was just as vulnerable as a 20-year-old RCC that had been through 25 or more missions. Safety cutbacks did not affect how the bipod was inspected. Quality control during the bipod foam's manufacture had nothing to do with it—the bipod foam could have come off just as easily on any mission. There was nothing unusual about Columbia or its age that made it more vulnerable.

The CAIB was open during its investigation, and the general conclusions were known well in advance. The final report, polished by professional writers, was edited to a reasonable length.

The technical aspects were easy—investigator Scott Hubbard said, "In four words 'the foam did it.' I'll point out one thing about the statement. We do not include the words, 'probably, likely, most probable.' All of this exhaustive work that we've done, all the discussion and the testing, have led us to a simple statement that the foam resulted in the breach that led to the loss of the orbiter."

But the far more important underlying issue was how the falling foam issue was treated as unimportant throughout the shuttle program. Gehman emphasized, "We determined NASA is not a learning organization, they do not learn from their mistakes." Gehman noted, "The fact that this piece of foam that came off on STS-107 was much, much larger than NASA's previous experience is, of course, important—because it gets into the question of why didn't that alarm the engineers? I mean, that's kind of basic to our investigation."

NASA's complacency about the continuing problems with lost bipod foam on previous flights, and in particular its avoidance of the issue after the STS-112 near miss, were blamed. Also blamed was NASA's culture, where people didn't speak out when they should have, and flawed decisions by the MMT. The board had dozens of recommendations for changes.

The CAIB was cautious with its wording. Gehman said, "We very carefully used a sentence with two negatives—'The shuttle is not inherently unsafe'—to send a signal: it's not safety, it's risk, but it's not unsafe."

While it has mistakes, the CAIB report is an excellent document. More than one person has noted if the Warren Commission, which investigated the assassination of President Kennedy, was as thorough, there wouldn't be a cottage industry of conspiracists. The CAIB's report is on this book's website.

Gehman noted, in a big-picture, "The real regretful part is not the NASA part. *The real regretful part is the inability of the United States government to understand the life expectancy of a generation of vehicles of this type—they have life expectancies. And we've come to the end of the life expectancy of the shuttle, by any measure. We have failed as a nation to provide its replacement.*"

Some of the NASA managers involved in the decision making before and during STS-107 have chosen to leave—most significantly, Marshall Spaceflight Center External Tank manager Jerry Smelser and space shuttle program manager Ron Dittemore. Dittemore and his bosses claimed that the retirement had been planned all along, and he was not asked to resign or step aside because of the accident. But insiders say they heard no talk about any plans for Dittemore to retire or discussions about possible successors before the surprise announcement. Others have been moved laterally to different positions. Perhaps most surprising, many of the high level STS-107 managers retained their positions or received positions with greater responsibilities.

20/20 HINDSIGHT

After the accident, critics pointed fingers and said NASA ignored the warnings from outside safety panels and others who expressed concerns and said a shuttle accident was imminent. There are always people concerned about safety who express their concerns—as they should. In April 2002, eight months before the accident, Richard Blomberg, former chairman of NASA's Aerospace Safety Advisory Panel, said, "I have never been as concerned for space shuttle safety as I am right now." Blomberg blamed the elimination of planned safety upgrades, a reduced workforce due to hiring freezes, and aging infrastructure. Certainly he was right to be concerned—many safety positions and inspectors were eliminated during cutbacks in the 1990s.

NASA was under heavy budget pressure in the late 1980s and the 1990s, and its spending power was reduced. ISS had major cost overruns, so funds were transferred from the shuttle's budget to ISS's. NASA had to fly additional shuttle missions and cargo because of Russia's financial problems, including Rick Husband's STS-96 flight. When push came to shove, NASA cut back workers, including many safety workers and quality inspectors. NASA administrator Daniel Goldin claimed the cutbacks would not reduce the safety level, but the workers in the field wondered how that claim could be made.

The Congressional committees that approved NASA's budget were also in the habit of taking out a couple of million dollars here and there for pet projects within their home districts—a total of $167 million in 2003. Among others the pork barreling included $15.5 million for scientific research in Fairmont, West Virginia; $7.6 million for hydrogen research at Florida State University; $2.25 million for a life science building at Brown University in Rhode Island; $1.8 million for the Gulf of Maine laboratory; and $1.35 million for the expansion of an Earth science hall at the Maryland Science Center. Each of these pieces of pork was in a Congressional district represented by one of the members of the congressional committee which oversees NASA's budget.

One engineer said technicians and engineers were being overworked, inspections were being missed, and there were too many meetings and presentations that just wasted time while workers struggled with obsolete equipment. While all of that was true and needed to be corrected it wasn't the cause of the Columbia accident.

For all the justifiable concerns about 1990s cutbacks in personnel and specifically safety personnel, the brutal fact remains: *The bipod foam that fell off the ET and hit Columbia's left wing could have happened just as easily on any shuttle flight from STS-1 to STS-107.* On the average, on 1 out of 11 flights, a chunk of foam would fall off the left bipod. It simply wasn't cutbacks to NASA's budget, overworked technicians, laid-off safety inspectors, mistakes that were missed by overworked personnel, or the aging infrastructure that caused the accident. Pure and simple, it was a flawed design that had been in place from the inception of the shuttle program three decades earlier.

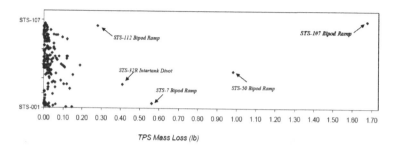

A misleading version of this graph—which leaves out the most important data point, STS-107—appears in one of the appendixes to the CAIB report. On STS-107, 1.67 pounds of foam was lost, far more than was lost on any previous mission. Corrected graph by Philip Chien.

For years people said they were worried about an impending accident. But their crystal balls were inevitably foggy when it came to predicting what would cause an accident. Nobody who claimed a shuttle accident was imminent mentioned any concerns about a piece of foam fatally damaging a shuttle, even though four previous bipod foam loss cases were well documented.

Admiral Hal Gehman said to those who claimed they expressed warnings before the Columbia accident, "If you're so smart, tell us what the cause of the next accident is going to be ahead of time so we can avoid it."

Bipod foam falling off and hitting a critical place on the shuttle was just a random occurrence waiting to happen. The odds a piece would hit a critical location were fairly low. NASA's engineers thought they were insignificant.

One of the most damning portions of the CAIB report is in the description of how NASA reacted to the STS-112 incident, where the bipod foam hit a Solid Rocket Booster—fortunately without damaging anything critical. Instead of fixing the problem, NASA missed the warning and decided to continue to fly the shuttle as-is while the issue was studied.

It's interesting to speculate what would have happened if quality inspector David Strait didn't see the tiny cracks in the flowliners in June 2002 and the accident still occurred on STS-107. STS-107 would have flown in July 2002—before STS-112. The most recent known loss of bipod foam would have been STS-50, in July 1992—a decade earlier. Engineers believed the bipod foam issue was understood and would not happen again. Would the CAIB have been as harsh if the Columbia accident had occurred without the STS-112 incident? Gehman said, "We viewed [STS-112] as just another data point on a straight line, we did not view it as a warning or red flag. It was much more significant to us as confirming evidence that the bureaucracy was not serious about safety."

Even within NASA, many engineers were "blind" to the fact a high-velocity object packs a large punch. A couple of days after the accident, manager Ron Dittemore said, "It just does not make sense to us that a piece of debris could be the root cause for the loss of Columbia and her crew. There's got to be another reason." But there wasn't. A 1.67-pound piece of foam traveling over 500 m.p.h. was the direct technical cause of the accident.

It's difficult to accept that once the foam hit Columbia's wing the crew was doomed, but those are the hard facts. Certainly it's regrettable Mission Control couldn't have had enough information to even attempt an Apollo 13-style rescue or repair, but there just wasn't enough information for anybody to truly come to the conclusion Columbia was fatally damaged until after the accident.

Could the legendary Apollo-era managers have done a better job than today's shuttle engineers? Absolutely not. Only 20 missions were flown during the Apollo moon program, including the spinoff Skylab and Apollo-Soyuz missions, without any inflight deaths. Had the shuttle program stopped after the 24th flight, it too would have had a 100 percent success rate. STS-107 was the 113th shuttle flight, and it followed 87 successful missions in a row. The more times you fly, the more times you're exposed to risk and potential failures. It's also important to remember that two of the best known of the legendary Apollo flight directors, Gene Kranz and Gerald Griffin, were high-level NASA managers during the early days of the shuttle program. It was under their watch that foam first fell off the ET and became an acceptable risk.

While the STS-107 MMT made mistakes and they weren't as good as they could have been and they should have done better, nobody had anything but the best intentions. Nobody involved with STS-107 was negligent or wanted to cause the crew or shuttle any harm. At most all that can be said is they were human and could make mistakes.

NASA'S CULTURE

The CAIB conclusion most difficult to quantify was "culture" — whether workers felt informing managers about problems would hurt their jobs. There are a couple of cases where workers did speak up and were publicly praised. David Strait, the inspector who discovered the flowliner cracks, was mentioned many times in glowing terms by NASA administrator Sean O'Keefe. Another worker accidentally bent a metal pipe on a payload and instead of ignoring it or hiding it told his boss he made a mistake and was praised for doing so.

One engineer halted two separate shuttle launches. Eagle-eyed Jorge Rivera spotted a 5-inch pin sitting on the ET before the STS-92 launch. The launch was scrubbed, and a technician went out on a catwalk to retrieve it. On another occasion, Rivera examined a photo of an ET falling away from the shuttle and saw a bolt that had not retracted completely. The next launch was put on hold while engineers studied the situation and compared the photo to photos from earlier missions, and said it was acceptable. In that case it could be argued Rivera had cost taxpayer thousands of dollars because his observation unnecessarily put a launch on hold for a day. But that's the wrong interpretation— he saw something which looked wrong and spoke up. It's rare to be in an industry in which, when you spot something out of place that causes your employer thousands of dollars, you get applauded, and especially when what you thought might be a problem turned out to be a non-issue. But that's exactly what the space program must do to encourage safety.

After one scrub, a frustrated VIP said, "Why can't NASA do things right?" O'Keefe chastised the individual, declaring, "They did everything right. They decided not to launch because things weren't perfect."

But how these examples, and there are many others, trickle down to the technician in the field is unclear. The attitude of workers in almost all organizations is to keep your mouth shut so you won't get in trouble. Certainly, every shuttle technician has to be thinking, at least on a subconscious level, if I bring up a minor issue and especially if it turns out to be nothing, my boss may think I'm a troublemaker. And I'm not absolutely sure it's a problem anyway. I'm better off if I keep these thoughts to myself and don't say anything, especially if it's something I'm not specifically responsible for.

Nobody stood up at STS-113's preflight Flight Readiness Review and said, "This is ridiculous. The bipod foam fell off and damaged an SRB on STS-112. If it happens again, maybe we won't be so lucky and it will hit something more critical." None of the engineers who saw the video of the bipod foam hitting Columbia's wing and was extremely concerned went to a member of the MMT and said, "We need more hard information to determine whether or not this is safe."

NASA must encourage its engineers and technicians to speak out whenever they have any concerns, without any fears of retribution or backlash from their bosses. In fact, NASA has to make sure its managers *reward* engineers and technicians who bring up concerns even if they turn out to be non-issues. But even after all of NASA's high-visibility efforts to improve its "culture" after the Columbia accident, at least one group of engineers felt more comfortable anonymously sending their concerns to *The New York Times* instead of to their managers. Others still keep their mouths shut when they have concerns or believe things should be done better to avoid being labeled "troublemakers" by their bosses.

On the opposite side, there will always be critics, the media, and Congress, who will complain whenever there's another delay or a scheduled milestone can't be accomplished—completely missing the point that delays because somebody's brought up a potential safety issue are good delays. How NASA's going to get its rank and file workforce to speak up whenever they think things may be wrong is always going to be one of its most difficult challenges. That will always be NASA's biggest "cultural" problem.

COMPARISONS BETWEEN THE COLUMBIA AND CHALLENGER ACCIDENTS

There are some similarities between the Columbia and Challenger accidents, but they aren't as strong as some have claimed. In both cases there were warning signs ahead of time that a component was being abused— the SRB's O-Rings for Challenger and ET's bipod foam for Columbia. In both cases there was pressure to keep the shuttle manifest on schedule, although the pressure was far higher before the Challenger accident. In both cases managers made the wrong decisions to go ahead. But in the case of Challenger the raising of the critical issue and wrong decision came up the night before launch, when everybody's anxious—launch fever, as it's called by insiders. In Columbia's case, the discussion of the critical issue and the wrong decision to fly took place three months before launch, at an ordinary management review.

In both cases the space shuttle orbiter was not the cause of the accident, and in both cases it was a propulsion element at fault due to limitations in the original design. For Challenger the O-Rings became brittle at low temperatures, for Columbia the bipod foam's design was not strong enough. But in the case of Challenger the O-Rings were pushed beyond the temperature range where they could be safely used, and in the case of Columbia the bipod foam was flying under allowable conditions.

Contrary to popular myth, NASA did not forget all of the lessons from the Challenger accident. But some complacency did set in after 17 years and over 80 successful missions in a row. Many believed everything was working fine on the shuttle, so it was less of a test vehicle and more of an operational vehicle, and the only concerns were how well the vehicle and the support systems were aging over time, and budget pressures which cut back safety workers.

There was one change to shuttle processing which did affect safety negatively, and it was a conscious decision: to make more money. The entire aviation and aerospace industries have a simple definition for Foreign Object Debris (FOD—pronounced "fod"): anything that doesn't belong around or on flight hardware. KSC redefined FOD in January 2001, adding the new category "processing debris," meaning out-of-place items found during the routine processing of the flight hardware. Processing debris was considered acceptable. Why change the industry standard by adding a non-standard and potentially risky definition? Any FOD found in flight hardware after final closeout inspections would harm prime shuttle contractor United Space Alliance's performance awards, but "processing debris" did not affect the bonus payments. The KSC unique definition pretending that "processing debris" was acceptable was eliminated after the Columbia accident.

Dr. Sally Ride, America's first woman in space, was the only person to serve on both the Challenger and Columbia accident investigations. In April 2003 I asked her to compare the similarities and differences. Ride said, "The investigative work is actually rather similar—but harder in this case [Columbia]. The times are different, NASA is different, the accidents were different, the details are different, but a lot of the questions we're asking are the same and we'll just see what we come up with here." Dr. Ride's entire answer is on this book's website.

A FINAL LOOK AT THE DEBRIS

The press had a final opportunity to see Columbia's debris in the reconstruction hangar on September 11, 2003.

Especially touching, somebody had left a couple of long-stem flowers by the window frames. And at least a sense of humor wasn't totally absent—on the table with the left-wing tiles, somebody had left a small stuffed space alien doll and that brought a smile to my face.

Anything I recognized was especially painful. Items like propellant tanks, tires, windows, and left wing RCC panels were especially poignant. The window 10 frame was especially emotional to me. Window 10 is the port aft flight deck window that looks into the cargo bay. This was the window Willie looked though as he saw the incredible colors of the atmosphere and sunrise he described to Laurel in a videotape that survived. During Columbia's 28 missions, 126 individuals had looked through that window. Some had made multiple flights on Columbia, and four had the opportunity to fly on Columbia three times.

 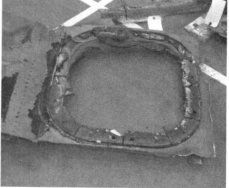

The port aft flight deck window (Window 10) as it looked during the mission and after the accident.
On-orbit photo by the Columbia crew, recovered debris photo by Philip Chien.

Only scraps of metal remained from the tunnel which connected
Spacehab to Columbia's crew cabin. Photo by Philip Chien.   One of Columbia's landing gear. Photo by Philip Chien.

Very little of the tunnel between the middeck and Spacehab survived, just a couple of circular pieces of metal with large portions missing
which used to be the hatches.

A year before, Scott Thurston had told me about the crow's feet that
were added to the thrusters. He showed me a thruster with the crow's feet
still firmly attached and doing their job of preventing the nuts from turning.
Seeing Columbia's debris left the profound feeling that it was something
special, and worth the lives of the seven astronauts who set out to accomplish their goals — it was far more than just twisted pieces of metal sitting on
the floor.

CHAPTER 47   KOOKS AND MYTHS

Within minutes after the accident, the first mistakes appeared. Benjamin Laster of Kemp, Texas, called CNN and said he saw an airplane close to Columbia. Anchor Miles O'Brien warned, "This is no reflection on Mr. Laster or anybody else we're going to be hearing from. But frequently, we get witness accounts at this juncture of any incident, plane crash or whatever it may be, that do not bear themselves out. The space shuttle was at 200,000 feet, so even if there was a commercial airliner or any sort of airplane in that area, it would have been significantly lower." At least that was an innocent statement by an eyewitness who just described what he thought he saw.

With Ilan Ramon's presence, there was immediate speculation some terrorist group shot down Columbia. But no terrorist group and very few governments have weapons capable of hitting a target flying at Mach 18 at an altitude of 38 miles.

Other early speculation included structural failure, corrosion, old worn bolts, faulty computers, or damage from a piece of space junk.

There were plenty of theories about sabotage, describing how a disgruntled worker could have placed explosives onboard. The CAIB report noted, "In its review of willful damage scenarios mentioned in the press or submitted to the investigation, the Board could not find any that were plausible. Most demonstrated a basic lack of knowledge of shuttle processing and the physics of explosives, altitude, and thermodynamics, as well as the processes of maintenance, documentation, and employee screening."

"A basic lack of knowledge" was a common characteristic of many theories.

Many reporters and NASA public affairs officers didn't know or care about the difference between the black tiles and gray leading edge RCC panels and used the terms interchangeably. Many didn't realize the handmade bipod foam was far larger than the pieces of acreage foam sprayed on the tank's cylindrical surfaces that had fallen on almost every mission.

One reporter said salt air had damaged the RCC so much it was brittle and couldn't handle as much of a strike as undamaged RCC. The foam gun tests proved brand-new RCC was just as suspect to damage as the panels on Columbia which had been flying for 21 years.

Many talked about the Columbia "explosion." Columbia was ripped apart by aerodynamic forces once its overheated and weakened left wing tore apart. No explosion caused the accident. It's interesting to note that even with all of the rough forces in the breakup and intense heat of reentry, small explosive bolts and other pyrotechnic devices were recovered intact: their charges were not activated by the reentry heat. People in Texas did hear booms, but what they actually heard were sonic booms created as larger pieces fell.

Many misinterpreted NASA's statement that the bottom of the wing and the areas of the wing closest to the fuselage were not visible from within the crew cabin. They thought NASA claimed it was impossible to see any portion of the wing.

Numerologists had a field day when they noticed the accident took place on the first day of the second month of the third year — or on a date of supposed astrological importance. Or it was the Chinese New Year. Or a New Moon. Or whatever other coincidences they could find or create.

And then there were some really bizarre stories.

THE WARPED WING

One of the strangest stories came out of Israel two days after the accident. The television news program "Erev Hadash" declared, "On the fifth day of his journey into space, Ilan Ramon spoke with Prime Minister Ariel Sharon through a video link. He presented the Prime Minister his view out of the windows of the Columbia shuttle, and right there on the surface of the wing on the left side you could see a long crack and a dent. Eleven days after, it was that very same wing that broke off the shuttle and finally brought it to its destruction." The story included a video clip, which sort

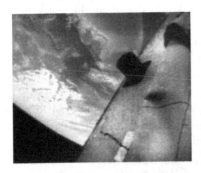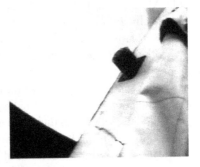

Israeli television showed the left image, which they claimed was a crack in Columbia's wing. The right image is the actual view which was transmitted on NASA TV the day before. What looked like a "crack" is actually the natural folds in the thermal blankets which cover the front wall of the payload bay. Note the cylinder in the middle of the photo, part of the payload bay door latch-closing mechanism.

Dave Brown studies the latch which closes the payload bay doors several months before Columbia's launch.

of looks like an off-white surface in an unusual shape. But that isn't the shuttle's wing—it's just a close-up view of the forward part of Columbia's cargo bay which was transmitted the day before the Sharon interview. What looked "warped" were thermal blankets. The black "knob" in the middle is part of the latch used to close the payload bay doors.

The black knob in the Israeli television image is very visible in photos taken before launch. A photo during the "Crew Equipment Interface Test" where the astronauts inspected Columbia shows Dave Brown kneeling on a work platform and staring at the knob at the back of Columbia's bay. The reason Dave was so interested was that the knob would be very important if he had to do an emergency spacewalk to close the payload bay doors manually.

DAVE BROWN'S E-MAILS AND SENATOR ALLEN

One of the most astonishing revelations about STS-107 came out in Congress on February 4, three days after the accident.

Virginia senator George Allen called Dave Brown's brother Doug, one of his constituents, at the JSC memorial service. Doug was still in an extremely emotional state after the memorial. Doug noted much later, "I probably shouldn't have been speaking to anybody when I was in such an emotional state." But Allen was insistent—he absolutely wanted to talk to Dave Brown's brother.

Later that day, Allen announced in Congress that Dave had expressed concerns about the shuttle's left wing in his e-mails to his brother. Allen's comments are on page S1856 of the Congressional Record. A transcript of the entire Senate session is on this book's website.

Allen said he was on a first-name basis with Doug Brown, and he went on to describe his conversation, starting with what Doug told him:

> [Dave] talked about NASA, about how they cared about, for
> example, specifically, one of the culprits or suspected culprits in
> this tragedy, which was that piece of foam that hit the left wing.
>   His brother—and he communicated with him by e-mail when
> he was up in space—had actually taken photographs of that wing
> because they were concerned about it.
>   I said: Did those photographs get back?
>   He said: No, they didn't send those photographs back. But
> that will be part of the investigation, at least his oral description
> of the situation.

Astonishing if true—it would indicate NASA was lying when it said the astronauts had no knowledge they were doomed. But the e-mail Doug Brown mentioned to Senator Allen was just Dave talking about the wonders in space, not expressing safety concerns. That e-mail is in Appendix C of this book, courtesy of Doug Brown.

Doug Brown also told Allen he had watched the NASA TV video showing Dave looking out of the overhead window to videotape the tank—the ordinary activities performed after every shuttle launch. Allen apparently misinterpreted Doug's comments to indicate Dave Brown had told him he saw something out of the ordinary. Doug explained, "It was an unfortunate miscommunication. It happened as a result of speaking to people after a very emotional memorial."

At first Doug Brown hoped the situation would just blow over, but it didn't. NASA public affairs made it worse by noting e-mails from the astronauts to their family and friends were considered private unless the recipient chose to reveal their contents; they did not put out any clarifying statements about what Dave could have seen—if anything—or what he might have meant in the e-mails. Since there was a lot of confusion and people continued to repeat Senator Allen's statement that Dave had told his brother he was concerned about the left wing, Doug Brown realized he had to do something to stop the rumors from spreading even further.

On February 10, NASA issued a press release with Doug Brown's statement explaining what he told Allen. "I wanted to clarify a couple of facts reported recently regarding my brother, Dave Brown. Dave sent several personal e-mails during the mission, but at no time did he write about any concerns with damage to the left wing of the orbiter or any other safety issues. As they reached orbit, Dave took his planned photos of the External Tank separation, which is standard procedure. These are the photos I discussed with Senator Allen."

Senator Allen has misinterpreted the remarks by a grieving family member about the ordinary post-launch video taken on every mission into a remark that the astronauts knew about potential shuttle problems and had expressed their concerns.

AN INTERNET VIRUS DESTROYED COLUMBIA

The Russian newspaper Pravda reported, "it was the first time that a spacecraft got an Internet address of its own which provided it with connection with the Earth through the satellite. Such an experiment was held for the first time; it is no wonder when the catastrophe occurred, it was almost immediately reported. It could be somehow connected with the Internet experiment carried out during the flight. It will take some time to find out whether such suggestions are true or absurd." There were rumors the virus came from either North Korea or China.

The Low Power Transceiver did have its own IP address, but it was on a secure closed Intranet—not the publicly accessible Internet. LPT used a Linux operating system computer with its own radio. It was not connected to the shuttle's computers. LPT communicated via TDRSS and ground stations. In effect, the experiment ran on its own, and Columbia was just its platform in the sky. LPT was an incredible success, returning all of its data before Columbia's reentry. Even if a virus could have somehow uploaded itself inside LPT and taken over the operating system, there would be no way for it to make it into Columbia's operational computers. There just wasn't any connection.

ON-ORBIT PHOTOS OF THE SHUTTLE'S WING

One of the more unusual conspiracy claims was titled "Shuttlegate" by its author. He claimed, "I thought they had no images of the wing. Apparently the astronauts decided to go into the cargo bay and take pictures over the side of the wing to see the damaged portion of the wing, which is not visible from this angle. NASA took all the photo images offline and hid them—now they say they never existed."

Er—no. NASA said they had no photos of the BOTTOM of Columbia's wing. The Shuttlegate photo was taken from within the shuttle's crew cabin with a fisheye lens that gives it an unusual feel. Clearly the astronaut

A perfectly ordinary photo of Columbia's cargo bay, wings, and the Earth. But misinterpreted by somebody who thought that NASA was trying to hide something.

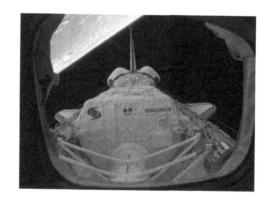

who took the photo was going for an artistic effect showing Spacehab within Columbia's cargo bay with the Earth in the background.

Spacehab had put in a low-priority request for the astronauts to take a photo of the Spacehab module. Manager Pete Paceley even showed the photo during the status briefing on January 30! Not only is NASA not trying to hide this photo—it's available for anybody who wants it—for free. It's on this book's website and NASA's Flight Day 7 photo collection.

The photo does show how little of the wing is visible from inside the crew cabin. It is possible for somebody inside Columbia to take photos of the outermost portion of the wing. But unlikely because it's pretty boring, and you've got this magnificent gorgeous Earth to look at instead.

FOAMSICLES AND THE STRENGTH OF THE FOAM

Many people refused to believe a block of foam you can puncture with your finger could destroy the shuttle. Their theory was the foam was like a sponge, absorbing water that changed to ice because of the supercold propellants inside the tank. These hypothetical "foamsicles," if they existed, would be far stronger than ordinary foam.

The New York Times claimed Columbia spent two weeks longer than usual at the launch pad, where it was exposed to four times the usual amount of rain. The problem is the phrase "than usual." The schedule from rollout to launch requires a minimum of two weeks, but it's extremely flexible depending on holidays and other scheduled activities. Columbia spent 39 days on the launch pad for STS-107—exactly what was planned, to the minute, when it rolled out on December 9. There wasn't a single delay. The average time a shuttle spends on the launch pad is 38.2 days, so STS-107 was almost exactly average. (There were 33 missions that spent more time on the pad, as much as 5.3 months in one case.)

STS-107's ET was exposed to rain, just like every other mission. Weather records show 12.78 inches of rain during Columbia's stay on the launch pad, versus an average of 5.45 inches for all launches — not the "four times" claimed by the Times. There was nothing extraordinary about rain and STS-107.

Could any amount of rain be absorbed by the foam and turn into ice? No. The ET's foam is formulated as closed cell foam and designed to repel water. (A Styrofoam picnic cooler is an example of closed cell foam; it doesn't absorb water. A sponge, in contrast is open cell.) Water is weight, and if the foam did absorb even a tiny amount of water, the vast amount of water-soaked foam would reduce the amount of payload that could be carried into space. CAIB head Hal Gehman noted, "In all the testing we did, we were unable to get this foam to absorb much moisture. It doesn't matter how much it rains on it." No water, no ice. No ice, no foamsicles.

The reason a piece of foam could damage Columbia is simple—kinetic energy. Just like a pencil penetrating a tree during a tornado anything with a large velocity carries a large amount of energy. The 1.67 pound piece that struck Columbia's wing at over 500 m.p.h. carried as much punch as a compact car hitting a brick wall head-on at 9.4 m.p.h..

ENVIRONMENTAL POLICIES DOOMED THE SHUTTLE

According to some claims, EPA regulations that eliminated Freon caused the foam to fall off.

CFC-11 Freon was used to apply the Spray On Foam Insulation (SOFI) to the ET, and the formula was changed because of EPA regulations. The new method did result in more foam falling off and hitting the shuttle, most notably STS-87, which had 308 damaged tiles, but that was not the type of foam which doomed Columbia.

In the mid-1990s, the EPA banned CFC-11 Freon. NASA has many waivers from the EPA for critical items. In each case a commercial supplier is licensed to produce the limited quantities NASA needs, but it's incredibly expensive to manufacture the relatively small quantities just for one customer. Lockheed-Martin went through a major effort to find a more environmentally friendly propellant. (It wasn't something they wanted to do, but a necessity.) They selected HCFC 141b (Dichlorofluoroethane). HCFC 141b is only used to spray acreage foam –applied to the large cylindrical surfaces with a giant robotic sprayer.

The bipod foam was BX-250 foam, which was excluded from the EPA rules. Technicians built the bipod by hand, layer by layer, and carved it into shape. The manufacturing process for the bipod and its chemical composition did not change and still used CFC-11. No changes to environmental regulations caused the Columbia accident.

RELIGION AND PROPHECY

Religious fanatics tried to tie in the fact that Ramon was Jewish, or that he had a Torah onboard, to the accident. By coincidence, much of the debris fell in Palestine, Texas, but it's pronounced "Pal es teen."

One commentator speculated, "Could the explosion over Palestine, Texas, be God's way of drawing world attention to the matter of Palestine? Could the Columbia shuttle's name and tragic destiny be a prophetic wake-up call to the world in general and to the United States of America in particular? Are American efforts to re-divide the Biblical land of Israel leading it toward a fatal crash? Is it coincidence, or is there a Divine Message in the fact that the shuttle broke up over Palestine—a U.S. town few people in the world knew about?"

The answer to all of the above is no.

TOP TOP SECRET SPY SATELLITE PHOTOS SHOW COLUMBIA WAS DOOMED!

There are some incredible images that show a shuttle exploding. Supposedly an Israeli spy satellite took them. But it isn't Columbia—it's a clip from the fantasy movie "Armageddon" and somebody's idea of a sick joke.

> The following pictures are photos taken of the Columbia explosion from an Israeli satellite in space. They were sent to me by a friend who works for the government. He received them from a friend of his from the justice dept. He wrote the following in his e-mail.
> Just in case you get a bit woozy looking at these, I apologize up front for the photos below. They are sequential pictures of the Shuttle Explosion, taken from an Israeli Satellite in space. They were taken....obviously, just as the Shuttle began to explode over the Earth. A friend of mine in the Justice Department sent them to me.
> TRULY AMAZING—AND SO SAD!

The only sad part is what was in the mind of the person who came up with the absurd and tasteless hoax in the first place.

Israel's EROS-A spy/Earth resources satellite did take an image of Columbia—but it was the day before launch while Columbia sat on its launch pad. The image is similar to other commercial Earth sensing satellites and has far lower resolution than American military spy satellites.

NASA WON'T PERMIT SHUTTLES TO REENTER OVER THE UNITED STATES ANYMORE

With all of the debris falling in East Texas many wondered whether NASA would ever allow shuttles to fly over populated areas again and whether or not that would put an end to shuttle landings in Florida.

There are three landing sites –KSC in Florida, Edwards AFB in California, and White Sands in New Mexico. White Sands is normally considered a backup backup site and has only been used once, the STS-3 mission in 1982. But what many people don't realize is many of the reentry paths to Edwards go over heavily populated areas in the Los Angeles basin.

A photo of Columbia from an Israeli "spy satellite"–taken the day before launch, while Columbia was still on its launch pad. Photo courtesy of Imagesat.

CAIB chair Hal Gehman said, "When NASA calculates its entry path they do not take into account what's under it. We felt in the future, all things being equal—they still have to keep the shuttle safe—we do need to think about what's under it."

Almost every shuttle mission since 1999 goes to the ISS. When reentering the shuttle can take a northwest to southeast path over the continental United States or a southwest to northeast path which passes over Mexico or Central America, Cuba, and Florida before landing at KSC. NASA selects the later, not because of risks to the public, but because of a variety of reasons including high altitude ice particles in high latitude clouds and the crew's sleep cycle. The astronauts have to be awake for launch, landing, docking, and undocking. That drives the rest of the mission's events and the best landing times occur on landing passes from the southwest. Some of those reentry paths do go over heavily populated areas, including Mexico City, Orlando, and Sarasota.

When it won't interfere with the mission's objectives Mission Control will plan future reentries to avoid populated areas, but only when it doesn't affect anything else.

NASA has analyzed the risks to the population underneath the shuttle's flight path during reentries and if there are concerns about a future shuttle before reentry it will be sent to White Sands where the risks to the public are the lowest.

. . . AND MYTHS FROM NASA

And sad to say, NASA also spread plenty of misinformation. The press kit says STS-107 used a superlightweight External Tank because the public affairs person who put in that page couldn't be bothered with checking and finding out STS-107 had an older lightweight tank.

NASA's website says the Kirtland image of Columbia's reentry was taken "as it passed by the Starfire Optical Range, Directed Energy Directorate, Air Force Research Laboratory, Kirtland Air Force Base, New Mexico." It doesn't mention that it was taken with a small consumer telescope, not the giant adaptive-optics Starfire telescope.

When NASA head Sean O'Keefe testified before Congress he compared the lost bipod foam to a Styrofoam picnic cooler falling off of a pickup truck. That would only be true if by "falling" you meant propelled at over 500 m.p.h..

Many within NASA, including O'Keefe and engineers, said many times it was a miracle nobody was hurt or killed by debris. Given that there were only 35 pieces collected per square mile it would have been surprising if somebody was hurt. Certainly it's fortunate nobody was hurt—but hardly a miracle.

CHAPTER 48

WAS A RESCUE
POSSIBLE?

Like Monday morning quarterbacks, Civil War reenactors, and Titanic fans, people naturally wonder whether something could have been done differently. Should NASA's engineers have realized Columbia was doomed? And what could have been done to save the crew? The question asked again and again is, Could Columbia have made it over to ISS, where the crew could stay until a rescue shuttle was sent up?

The answer to that often-asked question is not only no, but emphatically and absolutely no. Columbia was in an orbit that traveled 39 degrees north and south of the equator. The International Space Station is in a 51.6-degree orbit. Even if the two spacecraft traveled within a couple of miles of each other, they would be going in different directions at full speed. The change in velocity required to move from Columbia's orbit to the ISS orbit is an astounding 3,780 m.p.h.. Columbia had enough propellant to change its speed by just 305 m.p.h., enough for the small on-orbit maneuvers and the deorbit burn. For Columbia to change its orbit to the ISS's orbit would take over 115,060 pounds of propellant—more than twice what the shuttle could lift into orbit. The laws of rocket science are absolute—moving from one orbit to another requires lots of propellant.

During the in-flight analysis of the foam strike, engineer Rodney Rocha noted there were some limited things that could have been done to reduce the amount of heating if the wing was damaged. He e-mailed colleagues three possibilities: "[1] On-orbit thermal conditioning for the major structure (but is in contradiction with tire pressure temperature cold limits), [2] limiting high cross-range deorbit entries, [3] constraining right or left hand turns during the Heading Alignment Circle...." Columbia's reentry already had an extremely low cross-range (just 15 miles out of a possible 1,000 miles) and Columbia didn't survive long enough to get to the Heading Alignment Circle Rocha mentioned.

The other factor Rocha mentioned was to "cold soak" the left wing, reducing its temperature as much as possible before reentry to give it a bit more margin. Essentially you do the same thing by removing cold beverages from the refrigerator just before leaving home, so they stay cool for as long as possible. But as Rocha noted, it would not be possible to cold soak the wing while also making sure the tires were not cooled below acceptable limits. Columbia's wing had a temperature of 6 degrees

Fahrenheit on its lower surface when it reentered. If the wing was cold soaked longer, it could have been another 10 degrees cooler. In an extreme case, cold soaking over a period of two days would result in a 65-degree reduction. That would have helped, but by an insignificant amount—delaying the destruction for about 37 seconds.

While mistakes were made in determining the extent of the damage, there's no plausible scenario where the outcome would have been different. Fantasy scenarios have been proposed, where a request for a spy satellite photo of Columbia the day after launch is granted immediately and there's a suitable opportunity for a spy satellite to take a photo immediately, the photo has enough resolution to give firm evidence Columbia's been mortally wounded, and a daring rescue plan is devised. But those scenarios assume (a) more concrete evidence from the launch videos that Columbia was damaged, (b) spy satellite photos with enough detail to determine the nature and extent of the damage, and (c) all of this happening very early in the mission, before much of the supplies, like life support, power, and food, were used up.

If NASA did have knowledge Columbia was damaged, the answer was simple—NASA Administrator Sean O'Keefe told Congress, "We'd move heaven and Earth and use every resource we had to try to save the astronauts." O'Keefe said he would have been willing to give the orders to go ahead with a rescue mission, realizing any rescue mission would involve putting another shuttle and crew at risk. It's important to remember that with any rescue scenario, whatever damaged Columbia could also damage the rescue craft, since there would have been no time to make fixes before its launch. A problem with a rescue mission could result in a 100 percent fatality becoming a 200 percent fatality.

As a thought experiment the CAIB instructed NASA to create a way to save Columbia's crew based on the assumption there's undeniable evidence Columbia's doomed four days after launch. The CAIB gave these instructions not to blame NASA for what they didn't do, but for insight into how managers acted. If you believe nothing can be done, you will go to less effort than if you believe it is possible to do something. CAIB head Hal Gehman explained, "We did this not because we felt that there was a chance to save the crew, but to prove that the philosophy 'there's nothing we can do about it so let's not dwell on it' was wrong." For STS-107, the evidence shows there's nothing which could have been done in time, even if everybody made the correct decisions when analyzing the foam strike. But there could be a situation with less damage, where there would be more margin for a possible rescue or repair.

Based on the CAIB's rules, NASA came up with a repair scenario and a rescue plan. The plans are described in Volume II of the CAIB report, Appendix D.13—STS-107 In-Flight Options Assessment. In the repair scenario, Brown and Anderson would make a close-up inspection on a spacewalk. They could stuff the hole with whatever heat absorbent materials were available—

most significant would be heavy tools and a bag of water. The water would freeze over time, and then get covered with Teflon tape. It's assumed the purpose for these repairs is to get Columbia to a low enough altitude for the crew to bailout without any plans to save the spacecraft itself. But it's doubtful the effort would have been enough to keep the wing intact through reentry. In effect, this plan falls in the category of it's better at least to do *something*, even if it's got only a miniscule chance of success, than to do nothing at all.

But how about a rescue? Could another shuttle have been sent up in time with a rescue crew? The study concluded this scenario was just barely possible. Shuttle Atlantis was being prepared for an early March launch. Could it be prepared before Columbia's supplies ran out? Gehman said, "I've got no idea if it would have been successful or not. It's very, very risky—but not impossible."

Columbia used lithium hydroxide (LiOH) canisters to remove carbon dioxide exhaled by the crew. That was the limiting factor for how long the Columbia astronauts could live. The crew transmitted a video clip of Willie swapping an used LiOH can. There were enough spare LiOH cans through February 5 with normal usage. Less exercise and minimizing activity results in less carbon dioxide, which would have permitted the LiOH cans to be used longer. Allowing the carbon dioxide level to double would result in shortness of breath, fatigue, and headaches—literally the crew would be half suffocating. Combined, those techniques would stretch the LiOH supply to February 15. Again, it must be emphasized Mission Control would have to have extremely firm evidence very early in the mission that Columbia was doomed to make the decision to tell the astronauts to stop exercising and minimize their activities.

Columbia's crew would shut down everything that wasn't absolutely necessary. The rats are humanely killed. No exercise or unnecessary activity is allowed, and everybody's on short rations. Their job for the next three weeks is simple—stay alive and conserve supplies as much as possible.

Workers would accelerate preparing Atlantis, working shifts around the clock. Unnecessary tests and checks would be eliminated—time is of

As a LiOH canister was saturated with carbon dioxide a crewmember would swap it out with a fresh canister. Here Willie McCool is seen exchanging canisters on flight day 11.

the essence. But all of the needed tests are performed—no cutting corners, since a botched rescue is worse than no rescue at all. The astronauts and their backups begin training. They're all very experienced, so the training will go quickly for "ordinary" shuttle tasks. Training the rescue astronauts don't need are ignored.

The rescue mission launches on February 10. Rules that apply for a normal launch, like marginal weather conditions, may be waived. Atlantis launches for a rapid rendezvous with Columbia. Over a series of spacewalks, the seven Columbia astronauts would transfer to Atlantis.

There would certainly be a big party once all of Columbia's crew and Atlantis's spacewalkers were safely inside Atlantis. But what would certainly be on everybody's mind is whether or not Atlantis had the same flaw as Columbia, and would its reentry be successful?

Gehman said, "If you could have reduced the processing time satisfactorily and if you could go through the launch countdown and prelaunch preparations without a mechanical problem and if the weather was suitable and if the rendezvous was successful and the two orbiters could maintain station on each other and if the multitude of EVAs worked all right, you could have done this. It is possible. There are no showstoppers. But it all turns out to be extraordinarily [difficult]." That's a lot of "ifs" and "ands."

While there are a thousand things that could have gone wrong, nothing in it violates the laws of physics or the capabilities of the shuttle. It would require new techniques and of the coordination of activities that had never been accomplished together. It would also require violating important flight rules. For example, the airlock hatch is always left open during spacewalks, so the spacewalkers can get back inside. The rescue would require Atlantis's hatch to be closed each time a pair of Columbia's astronauts entered. During those periods, Atlantis's spacewalkers outside would be at far more risk.

And if this amazing rescue scenario did happen, there's one thing that would absolutely be certain. The Hollywood version would be even more exciting.

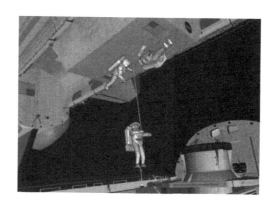

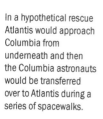

In a hypothetical rescue Atlantis would approach Columbia from underneath and then the Columbia astronauts would be transferred over to Atlantis during a series of spacewalks.

CHAPTER 49

MEMORIALS

More than one astronaut has joked, "If I die in a space accident they'll name a high school after me." That's an understatement. There have been scores of memorials for Columbia's astronauts.

The formal memorial was held at the Johnson Space Center on February 4, three days after the accident. Additional memorials were held that week at the National Cathedral in Washington, D.C., Lufkin, Texas, and the Kennedy Space Center. At each ceremony, politicians and religious figures talked about generic biographies, and people who knew the astronauts talked about them as human beings. Chief astronaut Kent Rominger and astronaut Jim Halsell spoke from the heart, sharing anecdotes about each of the crew. Their remarks are on this book's website.

The family members have attended scores of memorial ceremonies. VIP trips to Israel and India had been planned for the crew. The families of the crewmembers were asked to take those trips in their place, and they got the thanks from the many people who had some connection with the mission, or felt some connection to the astronauts or desired to learn more about them.

Mike Anderson's high school classmate Mike McKinley said, "We named the Math-Science wing at Cheney high school the 'Michael P. Anderson Math Science Wing.' We had a bronze likeness of him made. There's a showcase with pictures and the Cheney high school pendant he flew on STS-89 that he presented to the school. We started the Michael P. Anderson memorial scholarship for an outstanding math or science student that is planning on pursuing a career in one of those fields."

On April 1, 2003, seven of the astronaut's children and representatives, threw out the official first pitches for the Houston Astros 2003 baseball season. The children were Matthew Husband, Cameron McCool, Kaycee Anderson, Iain Clark, and David (Yiftah) Ramon. Also there were K.C.'s husband Jean-Pierre Harrison and astronaut Joan Higginbotham, who represented Dave Brown.

The Astros wore the STS-107 patch on their uniforms for the 2003 season. The logo was also painted on the home team's and visitor's dugout walls. Not surprisingly, the Astros recognition of Columbia's crew only lasted that season; by the next year the STS-107 logos on the dugouts were painted over.

Each year Jews around the world commemorate Yom HaShoah, the Holocaust Commemoration Day. Six candles are lit in a solemn cere-mony—each candle representing a million Jews killed. Normally, Holo-caust survivors are asked to light the candles. One of the biggest Yom HaShoah ceremonies is in Cleveland, Ohio, also the home of NASA's Glenn Research Center. Mist scientist Suleyman Gokoglu is a Turkish-American Muslim. He told friends in Cleveland's Jewish community about Ilan Ramon's involvement with Mist, and they asked him if Ilan would be willing to light one of the six candles in April 2003. After the accident, the community decided to add a seventh candle that year in memory of the Columbia astronauts, and especially Ilan. They asked Gokoglu to light that candle and speak about Columbia's crew.

The first spacecraft NASA launched after the accident was GALEX—the Galaxy Evolution Explorer. GALEX is an ultraviolet tele-scope, like the Hubble Space Telescope. But Hubble's super-telephoto lens can only look at an extremely small portion of the sky—as if looking through a soda straw. GALEX is wide-field telescope looking at large regions of brighter objects. It would take Hubble years to photograph all of the stars in a single GALEX image.

The GALEX team wanted to dedicate their "first light" image to the Columbia astronauts. They selected the constellation Hercules because it was over Columbia when communications was lost with Mission Con-trol. GALEX took the "Columbia Dedication Field" images on May 21 and 22 and showed the constellation in a way it's never been seen before.

One astronaut made a personal dedication to Columbia. American astronaut Ed Lu and Russian cosmonaut Yuri Malenchenko were the first humans to launch into space after the accident. Their launch was moved from the shuttle to a Russian Soyuz. Lu wore the STS-107 patch on his Sokol spacesuit.

Mist scientist Suleyman Gokoglu lights a candle at Cleveland's Yom HaShoah service in honor of the Columbia astronauts. Photo courtesy of Dr. Suleyman Gokoglu.

A fountain in Laurel's hometown, Racine, Wisconsin, has been named the Laurel Salton Clark Memorial Fountain. The city is planning a permanent display with a sundial and information on Clark's life.

On August 6, 2003, the International Astronomical Union (IAU) announced asteroids 51,823 through 51,829 were named after Columbia's crew. Amateur astronomers with decent telescopes and CCD cameras can view the Columbia asteroids under good lighting conditions, but they're 1,600 times fainter than the faintest stars visible to the naked eye.

Coronado High School, in Lubbock, Texas, named the field where Willie ran during his junior and high school years the "Willie McCool Track and Field".

Rick's hometown, Amarillo, Texas, renamed its main airport "Rick Husband International Airport," with a 7-foot bronze statue in the terminal's main concourse. Another statue was placed downtown.

The Astronaut David Brown Memorial Endowment was established at William and Mary College, Dave's alma mater, to provide funds for promising gymnasts.

While Kalpana Chawla was a naturalized U.S. citizen, she's had many memorials in her native land. India renamed its first dedicated meteorology satellite Kalpana-1 in her honor. The satellite is stationed at 74 degrees east and views about a third of the world.

K.C. is praised as a role model for Indian students, especially girls, who want to achieve greatness. Taken literally that would imply India is encouraging promising students to leave India to achieve their life's goals elsewhere!

The astronauts' names were added to the Astronaut Memorial at KSC in July 2003, but not unveiled until October 28, at the dedication ceremony attended by the astronauts' families. The "Space Mirror" is a 42-foot-high 50-foot-wide polished black granite surface.. Already on the memorial are the astronauts from the Apollo 1 and Challenger accidents, an X-15 accident, and several aircraft accidents. Ilan Ramon became the first non-American honored. Dr. Jon Clark spoke for the families:

> We truly all are [members of the] Columbia families because we've all shared the loss of the crew and the first space shuttle. It would be hard to capture the thoughts and spirits of the crew."
> It's not a memorial to the past, but a testimony to the future. And we must decide whether we are a space fearing or space faring [nation] as we step into the next phase of returning to flight and beyond. This memorial has many blank spots and they will not go unfilled, because the destiny of mankind will come at some cost. And therefore do not ask for who the mirror shines, it shines for you.

Legendary 40-year veteran astronaut John Young said,

> We all know that when you're involved in great endeavors you're involved in great risks. The crew knew that and they accepted

those risks. Their mission was the most complex multi-disciplinary research mission that NASA has ever flown. Columbia's crew members will forever be heroes to us because they found the best within themselves and shared it with their families, their colleagues, their community, and their nation.

Many of the European experiments were controlled from the Florida Institute of Technology in Melbourne. FIT was building seven new student residences, to be named the Crane Creek Apartments. Student Rebecca Mazzone suggested naming a building after each of the astronauts and the group of buildings the "Columbia Village."

The Spirit rover landed on Mars on January 3, 2004. Its landing site was designated the Columbia Memorial Station. A dedication was placed on the antenna used to transmit photos and scientific data to Earth. NASA accidentally released a spacecraft photo with the Columbia logo before Spirit's launch in June 2003, but quickly pulled the photo from its public web sites. A Mars project insider explained, "We wanted to dedicate our mission to the Columbia astronauts. But we were afraid if it failed it might be perceived as being disrespectful."

Seven hills on the Martian horizon were named the Columbia Hills. The hills were used to triangulate Spirit's location and determine exactly where it landed. Within a couple of days, project scientist Steve Squyres announced the long-range plan—go to the nearby Bonneville crater and, after exploring its rim, turn right and head for the Columbia Hills: "I can't think of anything more exciting to do in terms of exploring." Squyres noted at the time that he couldn't promise Spirit would last long enough to reach the Columbia Hills, but he was confident whatever they found along the way would be interesting.

Spirit was built with a goal to last 90 Martian days, a milestone it reached in April 2004. Squyres noted, "That's just when the warranty runs out. We're in good shape, and the engineers tell us it should last a lot longer." Spirit reached the Columbia foothills on June 15. Squyres said,

The names of the Columbia astronauts on the Astronaut Memorial.

The Columbia Village dorms at Florida Tech in Melbourne Florida. Photo by Philip Chien.

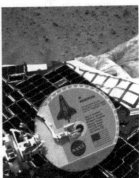

The Mars Exploration Rover with the STS-107 logo on its high-gain antenna before launch, and a closeup photo of the antenna on Mars.

"I think it's a very appropriate thing to memorialize the [Columbia] crew on the Martian surface. I hope we do Rick Husband's family proud by our exploration of the hill named after him."

Spirit and its sister rover Opportunity were both running strong over a year later. In August 2005, Spirit reached the summit of Husband Hill, 90 meters above the base of the Columbia Hills. Squyres said, "I can remember when we landed, seeing Husband Hill off in the distance and thinking, 'Man, wouldn't it be cool.' And to actually be on top right now—it feels really good. I think it's an amazing engineering accomplishment."

The first anniversary of the accident was marked by memorial ceremonies at NASA centers, in towns in Texas where debris was recovered, and in many of the astronauts' hometowns.

Well before STS-107, the National Football League had selected Houston for Super Bowl XXXVIII on February 1, 2004. Managers quickly realized the Super Bowl would coincide with the anniversary, and they decided to include a tribute to NASA and the Columbia crew in their pregame ceremonies. The Columbia families and the crew for the next shuttle mission, STS-114, were given VIP seats. K.C.'s husband Jean-Pierre and Jonathan Clark chose to go to the memorials in East Texas instead, wanting to thank the thousands of people who helped recover Columbia's debris. Clark also visited where his wife's body was recovered.

The day after the anniversary, the Columbia Memorial was dedicated at Arlington national cemetery.

A park at Polly Ranch in Friendswood, Texas, where Dave Brown lived, was dedicated on what would have been his 49th birthday, April 16, 2005.

Dave's brother Doug made his own tribute to the crew. Dave had planned to edit the video clips he shot during the crew's training and mission into a movie about the crew and their mission. Doug said, "We spent every Christmas talking about the movie—we didn't talk about NASA and the science, we talked about him making the film. The excitement for Dave was making the film. For me the movie is to fulfill what he

A spectacular panorama from the summit of Husband Hill, 90 meters above the surrounding Martian plains. McCool and Ramon Hills are visible in the distance, slightly to the left of center.

Columbia Memorial Station on Mars.

would have finished, and the way Dave would have done it." Doug Brown was put in contact with producers Howard Swartz and Mark Marabella. "Astronaut Diaries" premiered on the Discovery Science channel on May 14, 2005. It shows a tender behind-the-scenes look at the crew during their training and includes family members talking about the astronauts. The producers noted that they had far more footage available than what they could put into an hour—they could have easily made a show twice as long.

The first shuttle to fly after the Columbia accident was Discovery, on the STS-114 mission in July 2005. During the mission, the combined Discovery and space station crews talked about the sacrifices humans make for exploration. The tribute is on this book's website.

THE FAMILIES AFTER THE ACCIDENT

The families of the Columbia astronauts are adapting to life without their loved ones, like any other family does after a death. But unlike most families, they had to mourn in the public eye.—something that only happens when major celebrities or royalty die tragically.

The Columbia family with the closest NASA connection is certainly the Clarks. Laurel's husband Jon is a NASA flight surgeon, responsible for the health of the astronaut crews he works with. His dedication to the space program has not wavered in the slightest, but he has become an outspoken critic about how some within NASA have taken a "business as usual" attitude and are resistant to safety improvements. Clark said, "In the aftermath of how we're trying to change, I see some overt changes, but deep down inside I still see some of the same things going on. Not the least of which, [NASA] Headquarters is not always doing the right thing." Clark is writing his own book about Columbia and his wife Laurel.

Evelyn Husband had described herself as a stay-at-home mother before the accident. She's become a speaker, talking primarily to Christian women's groups about her faith and her relationship with Rick. She told me, "My children always go with me. I feel I've already lost Rick, they're not going to lose me."

As originally planned, Rona Ramon and her children moved back to Israel. Son Assaf still wants to enter the Israeli Air Force and become a fighter pilot like his father. Lani McCool and her children moved to Anacortes, Washington, to the house she and Willie called home. The other families have chosen to stay in Houston, where they've grown roots in the NASA community.

Some of the families, understandably, are starting to tire about constantly being asked to attend memorials or for interviews. They're frustrated having to answer the same questions again and again, especially when they're constantly asked whether or not they blame NASA or some individual within NASA for their loved one's death. While they still feel the loss, they want to move forward with their lives. Others are still willing to do interviews and attend memorials but have to pick –and choose because it is so disruptive to their lives.

For Jon Clark, taking his nine-year-old son to the memorials has resulted in additional stress. He said, "It has of late taken a huge emotional toll on myself and my son. I don't take him to these things anymore. I showed this video, and it just tore him apart."

Each of the family members is adjusting at their own rate. All of them miss their astronauts. Some acknowledge that they knew that there were risks involved and accepted those risks and are glad that they were able to accomplish something extremely special. Others still mourn and haven't moved forward with their lives as quickly.

Like the Challenger 51-L crew, the STS-107 Columbia astronauts have been honored with countless memorials. It's a pity the main reason they're remembered is because they died, and the other astronauts and missions are not remembered as well because they were fortunate enough to complete their missions.

Of all the memorials dedicated in their honor, the Columbia crew would have considered one to be more important than all of the others put together—the science that was accomplished on their mission. The crew had spent years dedicated to learning everything they could about how to operate their experiments so they could maximize the amount of science they would bring home.

CHAPTER 50

LEGACY:
THE SCIENCE
NOT LOST

Columbia's seven astronauts willingly put their lives at risk because they felt the mission's benefits were worth it.

One of the most offensive things said about the Columbia accident is the mission wasn't worth the risk to the crew's lives. The astronauts felt it was worth the risks, and that's why they flew. Microgravity science is not as glamorous as a mission to service the Hubble space telescope or an ISS assembly flight. But the science is certainly just as important in the long run.

Microgravity is pure science — science in a state-of-the-art laboratory that can't be replicated in any laboratory on Earth. Much of it was science where a human was the test subject. The pure science on STS-107 is the kind of science that leads to unexpected discoveries. Nobody can predict what will come out of science in a laboratory — that's why science is done. Most scientists would be disappointed if an experiment came out the way they predicted: The more interesting science occurs when things turn out differently from what was expected, and investigators try to figure out why the results were different.

Many of the experiments which got a lot of visibility or media attention were student experiments flying on a "space available" basis. They were not the reason for flying the mission. In fact, no single experiment was the reason to fly STS-107. But combined, the experiments formed an incredible suite of microgravity, Earth observations, and technology development investigations.

Columbia's crew spent 16 days working on 85 science experiments. After the accident, the initial reaction from the scientists was shock and grief for the astronauts and their families. By comparison their experiments no longer seemed important. Fellow astronaut and scientist Peggy Whitson was the principal investigator for the kidney stone experiment. She said, "We lost seven friends that day. Whether or not I lost seven subjects is trivial." SOFBALL scientist Paul Ronney said, "When I look at the notebook of graphs and the stack of CD-ROMs we produced from the downlinked data, I really start to feel guilty about what happened. After all, SOFBALL was one of the most crew-intensive experiments on the flight. They were flying largely to do my experiment. It's as though seven USC graduate students were killed in an accident in my laboratory."

Many of the scientists who had data radioed during the mission wondered whether doing research with it would be offensive or disrespectful to the memory of the astronauts. But then they quickly realized that attitude was exactly the opposite of what Columbia's crew would have wanted—the astronauts had willingly and gladly risked their lives because they felt the science was worthwhile. The mentality changed—do the science, as much science as possible—and dedicate it to the astronauts. The astronauts' families have encouraged this attitude. Mission Scientist John Charles said, "If we had any doubts, the families told us directly—not only do we want you to work on [the science], we want to be briefed as well."

How much science was accomplished on STS-107 or could still be performed after the accident is difficult to quantify. The crew accomplished every task they were asked. They were able to troubleshoot experiments that didn't work properly, and in many cases make repairs. Many of the experiments radioed some or most of their data during the mission, permitting scientists to start their research immediately. In a couple of cases, it was possible to provide feedback to the astronauts so they could make changes and improve the results. In other cases, especially with the life science experiments, the scientists had to wait until they got their samples back. In some extraordinary cases, there were amazing scientific results that wouldn't have been possible had Columbia landed safely—tiny animals and small off-the-shelf electronics units survived through all of the heat and rough forces.

For many of the educational experiments, the primary objective was to get students interested in science, so that goal was accomplished even before the launch took place; getting back the results after the mission was just an additional (but certainly planned) bonus. In many cases, the scientists and engineers wanted their hardware back so they could evaluate how well it stood up to the stresses of flying in space, but it wasn't a necessity for completing the science.

Shortly after the accident, NASA estimated approximately a third of Columbia's science could still be accomplished, primarily data radioed to the ground during the flight. In May 2003 Spacehab played back all of the data from the flight and was able to collect some additional information which was missed during the mission.

The student S*T*A*R*S experiments are a good example of the entire spread of results. Some accomplished their complete objectives, others had partial results, and one returned very little useful science.

The "ant farm" returned excellent video of the ants as they created their tunnels. There was never any plan for the students to get the hardware back. At most, they would get a bowl with the ants and gel material so they could show off their "space ants." Nothing was lost from that experiment.

The silkworm experiment also sent back excellent video. There never was any intent or desire to even return the silkworms, cocoons, or moths to the students, so the students also got their complete data set.

The "chemical garden" had a minor problem. One of the eight actuators failed to push its crystal in position but other than that the experiment worked well. The loss of the one crystal was unimportant—the other crystals grew well and produced exactly what the students wanted. Video of the crystals growing tentacle-like filaments in random directions was transmitted to the ground. In contrast, the control experiment on the ground had filaments which grew upward, away from gravity. The students said they would like to get the crystals back to put on display, but it's doubtful the delicate crystals would have survived the shock from Columbia's landing, plus the trip to Israel. So the students got all of the data they needed to do science, but not the "bonus" of getting their crystals back.

The bees experiment returned video of the bees as they ate into their balsa wood "home." The students wanted to weigh the wood block after the mission to determine how much the bees had chewed in comparison with the ground bees, so that portion was no longer possible. Not getting the experiment back was a partial science loss. In addition they had a non-scientific purpose to get their experiment back—to send the wood hive on a nationwide tour.

The spider experiment transmitted video of the spiders in action as they wove their webs. The astronauts collected web material to return to the students for postflight analysis. The students got video of the spider in space, but did not get the opportunity to test the web's strength.

The S*T*A*R*S experiment which lost the most was the medaka fish. It featured fish eggs, snails, and plants in a closed-cycle ecosystem. Video was broadcast during the mission, but graduate student Maki Niihori needed to get the experiment back to complete her research.

All of the SEM student experiments were lost, so the students couldn't compare the samples that had flown with their control samples. A much smaller number of replacement samples were squeezed into a Russian Progress spacecraft and launched to the International Space Station in December 2004; they returned to Earth in July 2005 on the STS-114 mission.

The SOFBALL experiment transmitted some video and radiometer (thermal) telemetry. Scientist Paul Ronney said, "[My] notebook and those CD-ROMs are their legacy."

During the mission Ronney sent the crew a preliminary summary of SOFBALL's results. He noted, "When the Gods want to punish you, they answer your prayers. It will take me years to analyze all of the data obtained on STS-107. Flame balls live by the old stage performer motto—"leave 'em wanting more..." Ronney says the flame ball data transmitted to the ground will keep him busy for quite some time.

Ronney decided to continue the crew's tradition of naming the flame balls with the ones that weren't named during the mission. A pair of flame balls that flew around each other in spirals were named Crick and Watson, after the scientists who discovered the double helix struc-

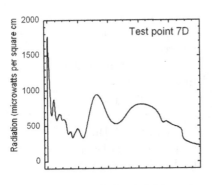

A graph showing the intensity of a flame ball nicknamed "Dave" as it changes over time. The up-down oscillations had been predicted theoretically, but never seen until this flame ball was generated on STS-107. Courtesy of Dr. Paul Ronney.

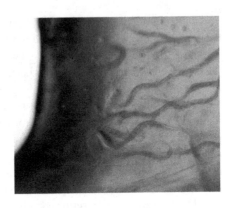

A magnified view of the C. elegans worms that survived the Columbia accident.

ture of DNA. A third in that group was named Franklin after Rosalind Franklin, who grew the crystals for those DNA discoveries. A full list of flame ball names is on this book's website.

Many payloads were recovered in decent shape, even after being subjected to the harsh forces of Columbia's breakup, the heat of reentry, and impact on the Earth's surface. In many cases, the scientists were amazed their recovered experiments could still yield decent science. Mission scientist John Charles said, "It is still stunning to think about the fact these things went from Mach 20 to zero in the most unplanned and hopefully non-reproducible way."

In a couple of cases there were extraordinary results. Worms survived through the carnage, proving multicellular organisms can survive reentry with very little shielding and possibly giving additional weight to the theory life could have been brought to Earth by a meteoroid. At least a couple of electronics boxes continued to record data after they hit the Earth!

It took about two months before the CAIB finished its paperwork and determined none of the payloads could have caused the accident. The scientists were anxious—did their payloads survive? Was it still possible to do any useful science? Even if an experiment was intact when it hit the ground, it would have spent anywhere from a day to several weeks exposed to the Texas environment. There was certainly reason for hope—very ordinary items had made it through intact

It seems incredible—like a bad science fiction movie titled "Worms from outer space." Hundreds of tiny *Caenorhabditis elegans* worms were the only multi-celled animals to survive. Technicians opened the containers and found hundreds of tiny living worms. Each worm is about the size of a comma in a newspaper—just visible to the naked eye.

The purpose for the experiment was to compare worms grown with the standard NGM (Nemotide Growth Medium) worm food and a new

CeMM (C. elegans Maintenance Media) formula. The CeMM food is a synthetic mixture that doesn't spoil for over a month, an important factor for longer-duration experiments

The smaller an animal is, the less sensitive it is to gravity. If you halve each of an animal's dimensions, the result is one which weighs only an eighth as much. The tiny worms are almost completely insensitive to gravity. The rough ride the worms got through reentry wouldn't affect them—but the heat could have cooked them.

Dr. Catharine Conley said, "We were surprised they recovered it, we weren't sure the canisters would survive [reentry]. But given that they survived we started pestering people to say if the canisters survived the worms were probably alive"

The worms' life cycle is about 7 to 10 days so by the end of the mission there would be worms born in flight. There was enough NGM food to last a week. When the worms run out of food they enter a dormant stage. If things had gone according to plan the NGM fed worms would be dormant second generation worms after landing. The CeMM food would last longer and presumably those worms would still be alive and reproducing

In the recovered NGM canisters the worms ran out of food as expected and were dormant. They included the original worms loaded before launch and their children. In the CeMM canisters there was enough food so the worms kept reproducing and going through their normal life cycles—even after the canisters hit the ground. While the exteriors of the canisters had been exposed to temperatures as high as 983 degrees the worms had escaped the heat and survived.

The other BRIC experiment grew moss. Ohio State University plant biologist Fred Sack said his reaction was, "Oh, that's really interesting and amazing, I can't believe they found the canister. It almost felt irrelevant—why take this seriously, it's a tragic event. It is amazing they found it, but it's unlikely it will be anything meaningful [in terms of science]. It fell from such a huge height, subject to vibrations, heat spikes. We didn't expect anything to come of it, we didn't have much hopes for it."

BRIC's clumps of moss culture were mostly destroyed, but the individual cells survived. Sack said, "When I saw the cultures showed some intact cells, it was just extraordinary. Just seeing the hardware was recovered with all of the pock marks on it, and still seeing some meaningful biological samples, it was almost a disconnect; it's hard to imagine how it survived all that."

Some of the moss survived and grew in spirals again. Sack said, "Most of the cultures were totally messed up. We were able to photograph some cells. A little mound of moss was mechanically destroyed, so we couldn't tell how the cells were positioned in space. Sitting a long time in the fixative, remarkably, seems to have preserved cellular structure just like the first flight."

The worm and moss experiments used off-the-shelf sensors to record temperature changes. The computer mouse-size HOBO has a battery, a

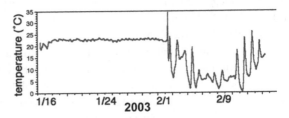

One of the off-the-shelf HOBO temperature recorders that survived the accident and continued to work.

An amazing graph shows the HOBO temperature curve—very stable during the mission, rapidly rising during the reentry, and then, after taking a couple of days to recover, recording the normal day-night temperature changes where the canister fell in a field in Texas. Photo and graph courtesy of Dr. Fred Sack.

small computer, a temperature sensor, and a memory chip. The HOBO saves temperature measurements to the chip every couple of minutes. In at least a couple of cases, HOBOs collected data in space, through part of the reentry until they reached their maximum temperature settings, hit the ground, and then continued to record the temperature until their batteries ran out!

One of the biggest surprises was the survival of delicate protein crystals. Dr. Larry DeLucas, director of the University of Alabama Birmingham's Center for Biophysical Sciences & Engineering, said, "I couldn't believe that we had anything that was able to survive. I was absolutely flabbergasted. The bottles were intact with the solution in them, and we could see sparkly things in there." The sparkles were interferon crystals, grown for drug company Schering-Plough. Scientists note that the crystals were of extremely high quality, and have improved knowledge of how interferon works. Schering-Plough is patenting the recovered STS-107 crystals.

The CIBX team saw their experiment on the news the day after the accident. What was incredibly frustrating was knowing it had time-sensitive urokinase crystals. If they could get access to those crystals within three to four days, then they could be examined and possibly hold the information for an anti-cancer drug. But if they had to wait any longer, the crystals would dissolve, becoming worthless. Unfortunately they had to wait. When the CIBX team got to Florida they wondered if the payload was in as good shape as it appeared on the news? Were the seals intact? Were the samples in good condition? Six of the eleven experimenters believed they could still get good science. CIBX manager John Cassanto said, "We saw the seals had held after three months. That's a long time for a seal to hold. There's four bolts to get out. But we physically had to lever it out with a chisel and hammer. We've got samples; we've got science. It's an awesome experience—to have this survive that long and not be burned up is almost a miraculous thing."

NASA sponsored the urokinase anti-cancer drug and microencapsulated antibiotics. Dr. Dennis Morrison got to examine his experiments,

The CIBX payload made it to the ground intact, landing in the driveway of a gas station in Nacogdoches Texas. Photo courtesy of the Daily Sentinel.

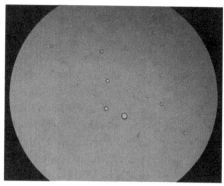

Closeup photos of the microencapsulated drugs which survived the Columbia accident. Photos courtesy of Dr. Dennis Morrison.

and as anticipated the urokinase crystals dissolved. The antibiotics survived, although some had degraded over time.

The space experiments helped determine the best way to manufacture microcapsules on the ground. Morrison noted they're focusing on using the microcapsules to deliver drugs injected after cryo-surgery for prostate tumors which can't be removed completely by surgery or cryo-freezing. The team has completed animal tests and is preparing to begin human clinical trials.

Another recovered CIBX experiment was the Planetary Society's "Experiment for Peace." Planetary Society head Louis Friedman said, "Our hopes were dashed when the Columbia accident occurred. That we're actually recovering scientific results where the students will participate, is something that transcends just the individual science." The two students didn't see their experiment opened up in person—they were busy studying for their finals.

The imitation meteorite with the bacteria film was put under scanning electron microscopes to determine if the bacteria attached themselves to the meteorite. The samples were mixed with each other, and no meaningful scientific results could be obtained. But of course the objective of promoting peace among Israelis and Palestinians was achieved with Israeli Yuval Landau and Palestinian Tariq Adwan's friendship and cooperation.

Mist transmitted 90 percent of its data. Dr. Angel Abbud-Madrid said, "One test after another, we were downlinking everything, so we could decide [the parameters] for the next one. The last two or three tests we were running so close together [that we didn't have time to downlink] Those final two tests were lost."

Some of the recovered Combustion Module hardware included the hard drives, Hi-8 videotapes, and memory cards. The manufacturer was able to recover the data from one of the memory cards. It had one of the Mist tests which was not downlinked during the mission—and the most valuable to the researchers. Abbud-Madrid explained, "It turned out it was a very critical test. The very last few tests is when you're closing in on the data you want to get, and that was the specific test point we got [recovered]. It gave us the those last few pieces of information we needed." The scientists don't have the video for that run, but the memory card data includes the critical temperature readings for how the flame front interacted with the mist. Abbud-Madrid said, "We wanted to see that very last velocity at the very end—when it slowed down almost to a halt."

The VCD experiment received some telemetry, along with the crew's verbal descriptions of how their samples looked. But they couldn't examine their purified water samples and lost the rest of their telemetry. Overall, they accomplished more than 60 percent of their experiment's goals.

MEIDEX transmitted video of dust and sprite observations over several orbits. But most of the video, recorded for analysis after the mission, was lost. Some of MEIDEX's data, recorded on digital videotapes within the crew cabin, survived, and copies were made for the scientists.

The Critical Viscosity of Xenon experiment transmitted most of its five tests. All of CVX's data was stored on an onboard hard drive. The hard drive was an off-the-shelf 340 Mbyte 2.5" disk, something used in a mid-1990s laptop computer. After the heavily damaged drive was recovered a commercial data-recovery company was able to restore almost all of the data that had not been transmitted to the ground. Somehow, an off-the-shelf computer hard drive survived a 38 mile fall to the Earth!

Even more incredible than the hard drive was CVX's test cell. It was recovered intact and held its pressure. If there was even the tiniest leak, the Xenon gas inside would have escaped. Engineers hooked up the cell

The recovered memory cards from the Mist experiment. Courtesy of Dr. Angel Abbud-Madrid.

The CVX hard drive was found but heavily damaged. A commercial company which retrieves data from hard drives was able to recover most of the lost data. Photo courtesy of Dr. Robert Berg.

to a gauge and verified the pressure was the same as when they prepared the experiment.

European scientists have expressed frustration that their requests to get the Biopack hardware back for analysis have not been approved by NASA. NASA's big concern is how the hardware will be kept under control, the nightmare situation being the theft or sale to a private collector outside of the U.S., where the U.S. government has no power to get it back.

A report on much of the science accomplished on STS-107 was released in January 2005, and is included on this book's website. While NASA has overstated the potential benefits of microgravity science, the potential *is* there—even if it isn't as great as NASA's optimistic claims make it sound. Who knows—it's still possible a new sprinkler nozzle for putting out fires with minimal water damage, a new drug, or other major scientific discovery can come out of the science performed on STS-107. And that would be the greatest memorial to the STS-107 astronauts.

And then there is the legacy of Columbia herself. Many engineers didn't want a repeat of what happened after the Challenger accident. After its investigation was completed, Challenger's pieces were sealed inside an old missile silo. The engineers hoped Columbia could be put to better use—some situation where it could continue to be useful to the engineering community.

They came up with using Columbia's pieces to analyze how materials act under extremely stressful situations like reentry. The only other accident involving the breakup of a vehicle returning from space was an X-15 rocket plane accident in 1967. The team was able to convince NASA's upper level management it was the right thing to do.

Columbia reconstruction team head Michael Leinbach said, "Columbia will continue to be a research vehicle even if it's not in the manner it was originally intended. We're going to study Columbia for years to come. We're going to store Columbia, we're going to catalog it, we're going to make it available for folks who want to study hypersonic reentry or spacecraft design, flight crew systems. We're going to learn from Columbia—this is the legacy we're going to leave the STS-107 crew and their families."

Unused space in the Vehicle Assembly Building at KSC, where Columbia was mated to its External Tank and Solid Rocket Boosters 10 months earlier, was selected as the permanent storage location. The area was painted, and fire sprinklers and security systems were installed. A display documents the Columbia recovery operation and what's stored inside. But the display is only visible to high-level VIPs who have special permission to enter that area. The Columbia repository is a memorial and a research library. But unlike Challenger's burial in an abandoned missile silo, the Columbia library certainly isn't a tomb.

Space shuttle Discovery launched on the first post-Columbia mission on July 26, 2005, 905 days after the accident. Coincidentally, the launch took place at 10:39 A.M. EDT, the same time as Columbia's launch on its final flight.

STS-114 was the culmination of a $1.4 billion effort to improve the shuttle, most notably the External Tank. The bipod foam was replaced with an electrical heater to prevent ice from forming. Marshall Spaceflight Center External Tank manager Sandy Coleman promised that no foam larger than a marshmallow would fall off of the improved tank.

In the 147-page press kit's description of all of the improvements to the shuttle, KSC's acceptance of the industry standard definition for FOD (Foreign Object Debris) is presented as a positive. In a spin doctoring attempt it's described how new FOD procedures improve safety, and ignores that FOD rules existed until two years before the Columbia accident when the rules were reduced in a conscious move to make more bonus money for the contractor.

Over 100 tracking cameras viewed Discovery's launch. The E208 camera in Cocoa Beach, the one that had been "soft focused" on STS-107, was replaced with a state-of-the-art setup. Cameras were also mounted on Discovery's External Tank and Solid Rocket Boosters, and

The bipod fitting on STS-114, on the right, shows the most significant external change—there is no longer any foam on the bipod fitting.

two aircraft with high-definition cameras offered the unique perspective of a shuttle flying toward the viewer.

Everybody was shocked to see something fall off the ET shortly after the boosters separated. Fortunately, it didn't hit Discovery. It was a piece of the Liquid Hydrogen PAL (Protuberance Air Load) ramp, an aerodynamic cover over the ET's wiring. That 0.9-pound piece was the equivalent of 54 marshmallows. It was the third largest known ET debris, with only the STS-112 and STS-107 bipod foams larger.

The entire PAL foam weighs 21.8 pounds. That's an extremely large piece, and potential source of lethal debris. In two previous cases (STS-4 in 1982, and STS-7 in 1983) umbilical well cameras detected lost PAL foam. As with the bipod, there were many cases where adequate photography wasn't available. NASA records say 62 flights had adequate PAL ramp photography. It's likely PAL foam was lost on some of the 49 missions without adequate photography and went unnoticed because it didn't damage the shuttle.

Manager Wayne Hale acknowledged that if the PAL foam had fallen off earlier, while Discovery was still within the denser atmosphere, it could have hit the shuttle."

In September 2003, Coleman noted that PAL foam was lost in the past and would require improvements. Later the decision was made not to change the PAL foam on the STS-114 tank because engineers believed it wouldn't be a problem. NASA thought the PAL ramps were okay, and improvements could be deferred.

Shuttle flights were put on hold again until it could determined why the foam was continuing to fail. Certainly the PAL foam incident was a major embarrassment for NASA. After saying for two years that no large pieces of foam would fall off of any External Tanks, just that happened on the very next flight—in a dramatically visible fashion. Smaller amounts of foam were also lost from four other areas. But those losses need to be taken in context— they were far less than any other shuttle mission, and Discovery's tiles and RCC came back in remarkably clean shape—with very little damage.

The media and members of Congress who thought STS-107 wasn't important followed the STS-114 mission in detail, but concentrated only on the changes since the Columbia accident and Discovery's reentry health. Many media called it a "problem plagued mission" even though overall it went quite well.

Discovery docked to the International Space Station and transferred 5,000 pounds of cargo. One key item was the Human Research Facility-2 (HRF-2), a more advanced version of the ARMS experiment the STS-107 payload crew dedicated so much of their time to. 6,000 pounds of cargo was returned to Earth—completed experiments, non-working hardware, Russian autopilot units for reuse on future missions, and even trash. The crew made three spacewalks to make repairs to the space station, installed new scientific experiments, added a storage platform for spare parts, tested tile and RCC repair techniques, and even made a simple repair to the shuttle's belly. All of the major objectives were accomplished, and many tasks were added while the mission was underway.

Many STS-107 Mission Control team members also worked STS-114. Entry flight director LeRoy Cain, MMACS officer Jeff Kling, and others sat in the same seats. Flight directors Paul Hill and Cathy Koerner, who handled photo analysis from the public, were two of STS-114's flight directors. Flight director Kelly Beck led a team that analyzed Discovery's thermal protection system. A couple of months before the mission, NASA announced a new class of flight directors, including STS-107 entry FDO Richard Jones, and entry GNC Michael Sarafin.

Bad weather in Florida resulted in Discovery landing at the backup site at Edwards AFB in California. As Discovery passed over Oxnard, its sonic booms were heard by amateur astronomer Brian Webb. He was an eyewitness to Columbia's reentry two and a half years earlier and an "ear-witness" to Discovery's reentry. Webb said, "I heard a loud BOOM-BOOM that shook the house. The window was partially open and I saw the curtain move. Dogs in the neighborhood started barking, and one of my neighbors walked outside, apparently wondering what the explosion was."

After the landing commander Eileen Collins said, "Today is a very happy day for us—but it's also a very bittersweet day for us too as we remember the Columbia crew and we think about their families. The Columbia crew believed in their mission, and we are continuing their mission—that's very important to us, that that mission of space exploration goes on."

While monitoring the STS-114 mission something totally ordinary which occurs multiple times on every mission creeped me out. Communications was temporarily lost with Discovery while it was in orbit. Capcom Julie Payette in Mission Control called "Discovery, comm check" and didn't get an answer. She tried a second time. Intellectually I knew that noisy communications was no big deal and there almost certainly nothing wrong. But in my heart I remembered Charlie Hobaugh's ordinary call "Columbia, Houston, UHF comm check." After Payette's second attempt Eileen Collins responded and I breathed a sigh of relief. The phrase "Comm check" may be a standard part of NASA's communications protocol, but there's no way it's ever going to sound as ordinary in the future.

IRONIES

Had the upper atmosphere winds been slightly different, or had Columbia's solid rocket boosters burned just slightly differently, the bipod foam could have missed the wing entirely, or hit the wing at a different angle with a non-fatal blow. If the foam had come off earlier, it wouldn't have slowed down so quickly, and even if it did strike the wing, it wouldn't have hit it with as much force. If the foam came off a little later, at a higher altitude, then there would-n't have been enough atmosphere to change its speed relative to the shuttle. Only under a very narrow range of circumstances would a piece of debris become deadly. It just happened that NASA's luck ran out on January 16, 2003, and Columbia's wing was in the way of a stray foam bullet.

STS-107 didn't need the shuttle's 50-foot robot arm, so it wasn't installed. But if the arm was onboard, it could have been used to inspect the left wing. If the damage occurred elsewhere, it could have been out of the robot arm's reach, but Columbia's damage was in a place where the robot arm could see it. If the damage was extensive enough to be visible on the video from the robot arm's camera, it would have given engineers posi-tive evidence Columbia's wing was badly damaged. In that hypothetical case, the rescue scenario in Chapter 48 might have actually happened.

If STS-107 was a mission to ISS, then it would have been easy to use the space station's robot arm to examine the place the engineers believed the foam had hit and determine the amount of damage. It would be extremely easy to ask the astronauts to inspect the wing for damage on an already planned ISS spacewalk before their other planned tasks. The ISS would have had the additional benefit of a shelter where the crew could wait for rescue, or even conceivably wait for an emergency repair kit to be designed on the ground and launched up to them.

If the shuttle had been damaged on almost any other mission in the previous six years, the reentry path would have been from a completely dif-ferent direction. Most shuttles take a reentry path over the Gulf of Mexico on the way to Florida, or over Pacific Ocean if landing at the backup site in California. All of the debris, including the bodies, OEX recorder, and dam-aged RCC, would have ended up in the deep ocean. (In contrast the shuttle Challenger's debris landed on the relatively shallow continental shelf where it could be recovered.) The CAIB would have been limited to the launch video of the foam striking the wing, the telemetry radioed to the ground during reentry, the tracking data from the Mystery Object, and the foam gun tests. At best, the CAIB could have determined that maybe the foam strike could have caused the accident, but not why. They would have to note there were many other possibilities they couldn't eliminate. CAIB chair Hal Gehman acknowledged, "If we did not have the debris, if we did-n't have the OEX recorder, I don't know what we would have concluded. I don't know what we would have been able to prove without it."

If the E208 camera in Cocoa Beach, Florida was in focus it would have given better quality images of the bipod foam as it hit Columbia's wing. That would have given the analysis team more accurate informa-tion on the size of the foam and where it hit. After the accident super-computers were used to improve the resolution, and the enhanced video was dramatic—showing more detail than the fuzzy video views the engi-neers had during the mission. This book's website has the enhanced video clip. If the tracking camera was in focus, it's very possible engineers would have come to the conclusion that Columbia was severely damaged during the mission.

If a fatal accident had to happen on STS-107 it happened in the least bad manner. The astronauts got to spend their entire mission enjoying themselves in the most productive work they had ever done in their lives. They had worked most of their lives to reach that goal. They had no knowledge Columbia was crippled and their deaths were mercifully swift. Their families got to see them enjoying themselves and share in their joy and didn't know about any problems until after it happened—there was no waiting in terror during reentry. Columbia's reentry path over the U.S. made it far easier to recover the debris, than if it had fallen on a foreign country or over an ocean. Because so much debris was recovered it was possible to determine—without any doubt—what had happened, and even possible to still accomplish much of the science that had been feared lost. Certainly it will always be regrettable the accident took place, but it could have been far worse.

If NASA did know that Columbia was damaged, the agency would have done everything it could to repair the damage or attempt a rescue. But then Mission Control would have been put in a difficult situation— informing the crew Columbia probably wouldn't survive reentry. Maybe in this particular case, ignorance was bliss.

Was STS-107 worth all of the effort? Many people have asked me whether or not I felt the science on the mission was worthwhile and I've always unequivocally said yes—because it was worth it. There's no doubt about that. Pure science in a laboratory may not lead to a new product or the hypothetical spin-offs mentioned in NASA's fancy color brochures, but pure science always leads to knowledge, and isn't knowledge what exploration is all about?

The STS-114 launch as seen from next to the E208 tracking station in Cocoa Beach, Florida. Photo credit: Philip Chien.

APPENDIX A:
CHARITIES

Many people have asked how they can help the families of Columbia's astronauts. As with any charitable contributions, you should contribute directly to the source that will benefit if you want to make sure your money's going to the right place. The families released a statement, "The families of the Space Shuttle Columbia crew are deeply grateful for the generous outpouring of support and affection we have received from around the world over the past weeks. Many people have asked how they can honor the STS-107 crew and assist our families during this difficult time. There are several charitable funds that have been established on our behalf."

A NASA contractor contributed the following website:

www.columbiashuttlefund.com

It describes three categories for contributions: funds that help out all of the families, funds that help specific crewmembers, and funds that contribute to the education of the children.

There are several funds for the education of the astronauts' children. Five of the Columbia astronauts had children (K.C. and Dave didn't) ranging in age from 5 to 22. After the Challenger accident, NASA received an overwhelming response from people who wanted to help. Legally, NASA could not accept contributions from the public, so a separate non-profit organization was set up. After the last of the Challenger astronaut children became adults, the decision was made to keep the organization in place just in case there were future space accidents. More information is available on the organization's website:

www.spaceshuttlekidsfund.org

Contributions can be sent to:

The Space Shuttle Children's Trust Fund
P.O. Box 34600
Washington, DC 20043-4600

Some of the families have no need for charity, but recognize the public's desire to give as part of the grieving process. They've passed on the contributions they've received to their astronaut's favorite charity.

Dave Brown's favorite charity was "For Inspiration and Recognition of Science and Technology" (FIRST), an educational group where students build robots. Here Dave poses with students and their robot at a competition at the Kennedy Space Center.

APPENDIX B:
COMPANION CD-ROM

The companion CD-ROM for *Columbia — Final Voyage* is available for purchase separately from this book. The CD can be ordered directly from the book's website, http://www.sts107.info, or by mail with the form printed on the next page.

The CD includes over 1,000 pictures, audio and video clips, NASA technical documents, transcripts, and additional information about the STS-107 mission, as listed in the menu below.

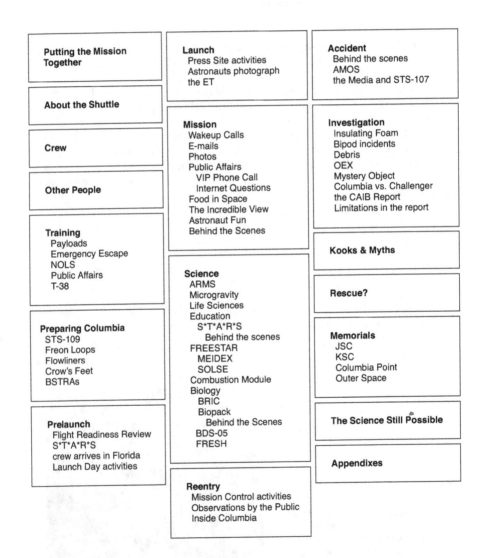

Putting the Mission Together

About the Shuttle

Crew

Other People

Training
Payloads
Emergency Escape
NOLS
Public Affairs
T-38

Preparing Columbia
STS-109
Freon Loops
Flowliners
Crow's Feet
BSTRAs

Prelaunch
Flight Readiness Review
S*T*A*R*S
crew arrives in Florida
Launch Day activities

Launch
Press Site activities
Astronauts photograph
the ET

Mission
Wakeup Calls
E-mails
Photos
Public Affairs
VIP Phone Call
Internet Questions
Food in Space
The Incredible View
Astronaut Fun
Behind the Scenes

Science
ARMS
Microgravity
Life Sciences
Education
S*T*A*R*S
Behind the scenes
FREESTAR
MEIDEX
SOLSE
Combustion Module
Biology
BRIC
Biopack
Behind the Scenes
BDS-05
FRESH

Reentry
Mission Control activities
Observations by the Public
Inside Columbia

Accident
Behind the scenes
AMOS
the Media and STS-107

Investigation
Insulating Foam
Bipod incidents
Debris
OEX
Mystery Object
Columbia vs. Challenger
the CAIB Report
Limitations in the report

Kooks & Myths

Rescue?

Memorials
JSC
KSC
Columbia Point
Outer Space

The Science Still Possible

Appendixes

Columbia—Final Voyage CD-ROM ORDER FORM

Shipping Information

Name

Address

Address

City   State   Zip

Phone number

email address

Purchasers of the book receive a $5 discount
for the companion CD-ROM. If you order the book
online enter the code "Springer-1" into the order form.

It's okay to photocopy this order form if you don't
want to remove the page from the book.

Product Selections

		Quantity		Subtotal
STS-107 multimedia companion CD-ROM	$10.00	☐		☐
		☐		☐
		☐		☐
		☐		☐
		☐		☐
		☐		☐
		☐		☐
			Subtotal	☐

Payment Information

Make checks payable to:
Space Coast T-shirts

Florida addresses add 6% sales tax ☐

P.O. Box 541759
Merritt Island, Florida 32954

Shipping and handling ☐
Shipping and Handling is $4 per item via Priority mail.

Phone: (321) 452-5682
Mon.-Sat. 10AM to 5PM Eastern time.
Email: spacecoast_tshirts@earthlink.net

Grand total ☐

Credit card number

Expiration date

3-digit ID code (located on the back
of the card in the signature box)

Discover   Visa   Mastercard
☐   ☐   ☐

Signature

Contact us for quantity orders.
We regret that we can only ship orders to U.S. addresses.

Or order online at http://www.sts107.info

APPENDIX C:
E-MAILS WITH COLUMBIA

The space shuttle has excellent communications, far more than earlier spacecraft or even the shuttle a decade ago. Off-the-shelf notebook computers are used for e-mail, and there's an inkjet printer onboard. The computers run ordinary commercial software. The astronauts can send an e-mail to anyone, but only certain people, selected by Mission Control and the astronauts, have the capability to send them e-mails in space.

The STS-107 astronauts got a daily "mail" delivery, a package of messages from Mission Control. This "execute package" includes flight plan updates, the status of fuel quantities and emergency landing sites, predictions for dust storms over MEIDEX's regions of interest, updates on the status of the payloads, interesting places on Earth to photograph, and even news from around the world. The cover page includes a cartoon or other humorous item, usually something appropriate for the day and often inside jokes that can only be understood by people who are working on the mission. The humor tends to lean toward Gary Larson's "Farside" or Scott Adams's "Dilbert." Here's a couple of the messages the crew received during the mission—operational as well as humorous.

The Red Team was sent this list on flight day 5:

```
Top 5 reasons for doing PhAB4
5)  You love being referred to as "The Count"
4)  You love the smell of alcohol swabs in the morning
3)  You now win every game of "Operation"
2)  Two words: blue gloves
1)  Constantly saying "This won't hurt a bit."
```

On Flight day 16 this message was sent to the Blue team::

```
Top ten calls that probably won't be heard during this
flight:
10. The Sleep PI has requested that you sleep in tomorrow
    night to obtain enhanced science.
9.  The Red Cross says they have achieved their quota for the
    month, so no more blood draws are required.
8.  You have a "go" to release the STARS bee into Astrocul-
    ture to begin the pollination portion of the experiment.
7.  We've stowed some T-bones for you to cook up in the CM-2
    chamber when you get a chance.
6.  CM-2 is mystified by Mist.
5.  We have completed 150% of all science objectives so you
    can take the rest of the flight off!
4.  You have a "go" to remove the Dove bars from EOR/F—enjoy!
3.  We've just been notified that Domino's Pizza delivers to
    orbit, so you can configure PDIP for Pepperoni.
2.  Pool party in Lake Butler!
1.  We copy, everything is "Okey-dokey"!
```

On landing day, the messages included this:

> All of the Mission Control teams want to pass along they have
> enjoyed working with you and appreciate your patience, pro-
> fessionalism, and dedication. The FREESTAR team is quiet for
> the first time all mission (many folks are out buying the
> champagne and Kosher goodies to celebrate). The first toast
> will surely go to the entire crew for making this flight such
> a monumental success!

One limerick summed up the mission's progress and lack of problems quite well:

> There once was a long science mission
> Whose plan went without much revision
> The Execute Pack
> Was filled front and back
> With poetry of questionable precision

Okay, so the execute package humor wasn't always funny. What do you expect from engineers? This book's website includes all of the STS-107 "execute packages."

The astronauts also sent many professional and personal e-mails. Professionally they sent messages to Mission Control and the payloads team to thank them for their hard work. Personally they e-mailed family and close friends to tell them about the wonders of space travel and how much they missed them and wished they could be there with them.

The crew carried photos of their families with them in space and emailed photos of those photos to their families to give them unique souvenirs of their mission.

Dave's brother Doug decided to compile his own top ten list of reasons for flying in space, with a humorous twist, and solicited Dave's family for their thoughts. He got more than he bargained for and sent 14 responses, some serious but mostly humorous. Several of the thoughts came from kids, including Dave's nephew and niece, who had their own pre-teen priorities for what's important in life. The e-mail was sent to Columbia five days after launch.

> From: Doug Brown
> Sent: Tuesday, January 21, 2003 1:27 PM
> To: 'MS1'
> Subject: Top 10 Reasons to go to space by your relatives
>
> 14 Eric Ploeg: Because you can
> 13 Eric Ploeg: To seek out new places for humans to survive
> 12 Eric Ploeg: There are still places and things no one has
>   seen or done
> 11 Alison Brown: You can dance like no one's looking; except
>   for those with the cool NASA channel, then the whole
>   world is watching
> 10 Eric Ploeg: To impress the opposite sex
> 9   Eric Ploeg: To get a better view of the earth
> 8   Paula Brown Miller: Government paid travel and enough
>   frequent flier miles for another NASA trip
> 7   Doug Brown: You can go to work, sleep on your feet and
>   not get fired
> 6   Danny Brown: You can flick your buggers farther

5   Casey Brown: Your spit balls will fly straight
4   Sandy Brown: George Bush announces "leave earth—pay no
     taxes"
3   Gerry Webber: Excused absence from the marathon Brown
     family reunion at the Wakulla
2   Marilyn Brown: It's only eight and a half minutes to
     weight loss
1   Eric Ploeg: It's the best ride there is ... better than
     anything at Disney

Here's a personal e-mail Dave Brown sent to his family and friends the day before landing, courtesy of Dave's brother, Doug Brown:

Subject: Flight Day 16

Friends,
It's hard to believe but I'm coming up on 16 days in space
and we land tomorrow.
I can tell you a few things:
Floating is great—at two weeks it really started to become
natural. I move much more slowly as there really isn't a
hurry. If you go to fast then stopping can be quite awkward.
At first, we were still handing each other things, but now we
pass them with just a little push.
We lose stuff all the time. I'm kind of prone to this on
Earth, but it's much worse here as I can now put things on the
walls and ceiling too. It's hard to remember that you have to
look everywhere when you lose something, not just down.
The views of the Earth are really beautiful. If you've ever
seen a space Imax movie that's really what it looks like.
What really amazes me is to see large geographic features
with my own eyes. Today, I saw all of Northern Libya, the
Sinai Peninsula, the whole country of Israel, and then the
Red Sea. I wish I'd had more time just to sit and look out
the window with a map but our science program kept us very
busy in the lab most of the time.
The science has been great and we've accomplished a lot. I
could write more but about it but that would take hours.
My crewmates are like my family—it will be hard to leave them
after being so close for 2 1/2 years.
My most moving moment was reading a letter Ilan brought from a
Holocaust survivor talking about his seven year old daughter
who did not survive. I was stunned such a beautiful planet
could harbor such bad things. It makes me want to enjoy every
bit of the Earth for how great it really is.
I will make one more observation—if I'd been born in space I
know I would desire to visit the beautiful Earth more than
I've ever yearned to visit to space. It is a wonderful planet.

Dave

MS1
NOTE: This is private/personal mail and not for release to media.

Laurel sent an e-mail to the payload team and other support personnel, thanking them for all of their hard work on the mission:

> From: MS4
> Sent: Monday, January 27, 2003 1:37 AM
> Subject: Hey from 150 NM above the Earth
>
> Hello all,
> I planned all along to write several notes to share a little of our mission perspective from space. Unfortunately, I've been too busy for much e-mail at all. Things are going terrific up here, mostly because of all the hard work and dedication of all of you on the ground. Some of the MCC team are quite talented poets. We have had wonderful laughs and smiles over the verses in the Execute Pkg. We'll have to make a book of STS-107 poems after the flights. The experiments all seem to be doing well, esp now that the module is cooler. It is really fun to see the plants and tissues and other visual evidence of all of your efforts. I can't wait to see how all the results turn out. The first few days are pretty much a blur. I am very thankful for all the tremendous training and preparation we had over the last 2.5 years. The views of our planet are beyond words and not nearly frequent enough for me. I wish each and every one of you could float up here with us for a few minutes. I continue to be overwhelmed at the smooth integration of this whole operation. A million thanks to all of you down there who are making it work!!!
> All my best, Laurel
>
> MS4
> NOTE: This is private/personal mail and not for release to media.

Laurel Clark uses a notebook computer on Columbia's flight deck.

APPENDIX D:
THE STS-107 CREW LOGO

Each shuttle crew designs its own logo. Some of the more artistically talented astronauts draw their own; others use professional artists. The logo has to be approved by NASA Headquarters, and in at least one case an astronaut had to do some fancy talking to convince Headquarters to let the crew include their inside jokes on their official logo.

Normally one of the rookie astronauts takes responsibility for designing the crew logo; and Laurel Clark took that responsibility for STS-107. Johnson Space Center graphic artist Terry Johnson worked with Clark and created the final design that the crew approved.

A variety of concepts were presented to Laurel and she chose the ones she liked, offering input into what should be changed or added. This book's website has all of the early versions of the logo. The earliest logos just have placeholders for the names of the pilots because they were created before they were assigned to the crew.

The final design is one of the better shuttle logos. It's simple and tells the mission's story. It's the first shuttle patch in the shape of the space shuttle.

Many people were confused by the "μg" on the logo. It's actually the Greek letter "mu" which is used by scientists as an abbreviation for "micro", indicating a millionth, so "μg" means "microgravity," the theme for many of the mission's objectives.

THE 18 DELAYS TO STS-107

The STS-107 mission had an unenviable 18 delays, totaling two and a half years. Many of the delays were the result of being "bumped" by higher priority missions to support the International Space Station and Hubble Space Telescope. But there were also slips due to delays involving missions to other satellites, and delays due to a short circuit in Columbia's wiring, among other reasons. Altogether, the delays added up to 32 months from the original contracted launch date. Surprisingly, very few of the delays were specifically related to STS-107, and there were no delays after Columbia finally left its hangar in November 2002. Here is a list of the delays, with several other milestones:

- *December 1997: Baseline contract, launch date May 11, 2000.*
1. NASA Technical Direction (TD) dated August 8, 1998, to slip launch to No Earlier Than (NET) September, 2000.
2. NASA TD dated April 27, 1999, to slip launch to NET December 1, 2000.
3. NASA TD dated September 23, 1999, to change launch to November 30, 2000.
4. NASA TD dated November 16, 1999, to slip launch to NET January 11, 2001.
5. NASA TD dated March 28, 2000, to slip launch to NET February 22, 2001.
6. NASA TD dated May 3, 2000, to slip launch to NET April 9, 2001.
7. NASA TD dated July 7, 2000, to slip launch to NET June 14, 2001.
- *July 25, 2000: Dave Brown, Kalpana Chawla, Mike Anderson, Laurel Clark, and Ilan Ramon were officially assigned to the mission.*
- *October 27, 2000: Rick Husband and Willie McCool were added to complete the crew.*
8. NASA TD dated November 29, 2000, to slip launch to NET August 2, 2001.
9. NASA TD dated February 14, 2001, to slip launch to NET October 25, 2001.
10. NASA TD dated March 23, 2001, to slip launch NET April 4, 2002
11. NASA TD dated May 11, 2001, to slip launch to May 23, 2002
12. NASA TD dated October 29, 2001, to slip launch to June 27, 2002
13. NASA TD dated January 28, 2002, to slip launch to NET July 11, 2002
14. NASA TD dated April 22, 2002, to slip launch to July 19, 2002
- *May 24, 2002: Spacehab and FREESTAR are installed in Columbia's cargo bay.*
- *June 24, 2002: Columbia's launch is put on hold due to flowliner cracks.*
15. NASA TD dated June 26, 2002, to slip launch to NET mid-August 2002.
16. NASA TD dated July 18, 2002, to slip launch to NET September 26, 2002.
17. NASA TD dated August 16, 2002, to slip launch to NET November 29, 2002.
18. NASA TD dated September 17, 2002, to slip launch to NET January 16, 2003.
- *November 18, 2002: Columbia is rolled over from its hangar to the Vehicle Assembly Building.*
- *December 9, 2002: Columbia is rolled out to its seaside launch pad.*
- *January 16, 2003: Columbia launches.*

APPENDIX F:
SPECIAL THANKS

When I realized something really bad had happened to Columbia it hit me hard. At first I didn't want to accept it, but as the clocks ticked away and events didn't happen I had to accept the inevitable conclusion Columbia was lost. More important than that—seven people who I knew, people who I had talked to just a couple of days before, had died.

Several friends, colleagues, and astronauts encouraged me to write a book about Columbia, since I could write a fair and honest book about the people and the mission, not just the multitude of books that were sure to be written about the accident.

For the longest time I didn't want to touch my collection of audio tapes with my personal interviews with the astronauts. I thought it would be emotionally difficult to listen to Rick's laugh when he found something I said amusing, Mike's quiet voice, the excitement whenever Dave talked, Willie's optimism about everything, K.C.'s lovely Indian accent, or the others. I thought listening to my interviews with them would be extremely difficult. But it wasn't—it was uplifting. It brought great pleasure to hear the voices of people I knew, talking about their passion for what they wanted to do, how important it was to them personally, and even the jokes and laughs we shared during those interviews.

Many of the people interviewed for this book were initially reluctant to talk about Columbia because of bad experiences with the press after the accident or bad memories of the accident, but brought out their box of memorabilia and remembered the good times. The many people interviewed ranged from close personal friends of the astronauts to ones with no connection to the mission other than seeing Columbia during its reentry. All of my interview subjects were extremely cooperative and wanted to find out more about the mission and the people. To all of them I give my sincere thanks for sharing your memories and thoughts. I hope I've done them justice in this book.

This book would not have been as detailed as it is without the cooperation of the engineers, scientists, and managers in the space program I've worked with for over two decades. Of course there's no way I can thank Rick Husband, Willie McCool, Dave Brown, Kalpana Chawla, Mike Anderson, Laurel Clark, or Ilan Ramon. Their many interviews brought me far more than just the quotes seen in black and white in this book—they shared their knowledge with me, and their passion about how they felt about spaceflight and their mission.

The STS-107 crew poses for the press during their dress rehearsal in December 2002. Photo by Philip Chien.

Extremely important thanks go to family members who shared their time with me talking about a very difficult subject and many of the personal photos, e-mails, and insights about their loved ones, especially—Doug Brown, Jon Clark, Evelyn Husband, Lani McCool, and Dorothy & Paul Brown.

The folks at Spacehab, especially Pete Paceley and Kimberly Campbell, were incredibly cooperative providing me with information and behind-the-scenes access during the crew's training and during STS-107.

At NASA and its contractors special thanks go to—Kay Grinter, Dr. John Charles, Tom Dixon, Mike Gentry, Margaret Persinger, Linda Mullen, Mark Rupert, Jeff Kling, Paul Lockhart, Terry Wilcutt, Roz Hobgood, Paul Hill, Wayne Hale, Linda Ham, Milt Heflin, John Herrington, Jerry Elliott, Dr. Suleyman Gokoglu, Dr. Catharine Conley, Dr. Dennis Morrison, Scott Thurston, Dr. Howard Ross, Lucy Lytwynsky, John Stoll, Kandy Warren, and Mike Leinbach.

Other important people include—Christina Joell, Mary Lockhart, Beth Duke, Ted Molczan, Rob Matson, Dr. Paul Ronney, Dr. Angel Abbud-Madrid, Admiral Hal Gehman Jr., Al Saylor, Jeff "Goldy" Goldfinger, Cindy Swindells, Dr. Gary Wells, Dr. Fred Sack, Dr. Millie Hughes-Fulford, Dr. Jason Hatton, Martha Siepmann Wilson, Dave Santucci, Miles O'Brien, Tom Rose, Dustin Block, Rick Baldridge, Peter Goldie, Chris Valentine, Jay Lawson, Brian Webb, Dr. Yoav Yair, Tariq Adwan, Ed Galindo, Mike McKinley, Lawson Vankuren, Keith Flamer, Al Cantello, John Barootian, Tom Lewis, Michael Juley, John Cassanto, Dr. Robert Berg, P.R. Blackwell, Dr. Bart Lipofsky, Jon James, and Dr. Guillermo Sanabria.

Special thanks to Kathy Scobee Fulgham for her permission to publish her letter to the children of the Columbia astronauts on this book's website.

Very special thanks to those who provided photos for this book and the companion CD-ROM.

Additional thanks goes to everybody else who was interviewed before the STS-107 mission, during the mission, and afterwards or provided information about their connection to STS-107.

Special thanks to Tom Henricks, Jon Clark, Mark Kirkman, and Jason Hatton for reading a draft of my manuscript and providing excellent feedback.

Any mistakes that remain within the book are my responsibility.

APPENDIX G:
BIBLIOGRAPHY

The Internet is a wonderful resource for writers, but like any other reference source items should be taken with a grain of salt and double-checked. There are many excellent references for STS-107 and its crew, but far more which have many mistakes. The author's website —

http://www.sts107.info

—has links to these websites and other STS-107 Internet resources.

STS-SPECIFIC WEB SITES
STS-107 press kit
> http://www.shuttlepresskit.com/STS-107/STS-107_SPK.pdf

Space Research and You website outlines microgravity research at a beginner's level
> http://spaceresearch.nasa.gov/sts-107/

NASA's STS-107 crew page
> http://spaceflight.nasa.gov/shuttle/archives/sts-107/crew/index.html

The European Space Agency's webpage for its STS-107 activities
> http://www.esa.int/esaHS/SEM2GUZKQAD_research_2.html

An excellent article from NASA's science.nasa.gov website on STS-107's SOFBALL experiment—"A Flameball Named Kelly"
> http://science.nasa.gov/headlines/y2003/31jan_kelley.htm

Spacehab's student S*T*A*R*S experiments including all of the photos transmitted from Columbia during the mission.
> http://www.starsacademy.com/sts107/index.html

The Columbia Accident Investigation Board (CAIB). Note that NASA now has the archive for this website but has not changed the contents.
> http://caib.nasa.gov/

Satellite tracker Ted Molczan's excellent analysis for whether or not Columbia could have been photographed by a spy satellite
> http://www.satobs.org/columbia/KeyHolesattosat.html

Ted Molczan's analysis of the STS-107 Mystery Object
> http://www.satobs.org/columbia/STS107mysteryobject.html

NASA's History Office's Columbia memorial
> http://history.nasa.gov/columbia/index.html

Challenger Center for Space Exploration
> http://www.challenger.org/

OTHER REFERENCES
An excellent chronology of all of Columbia's missions
> http://science.ksc.nasa.gov/shuttle/resources/orbiters/columbia.html

Shuttle news reference manual—badly out of date and without the illustrations but still an excellent reference.
> http://science.ksc.nasa.gov/shuttle/technology/sts-newsref/stsref-toc.html

Shuttle Operations Data Book—excellent technical documents
> http://spaceflight.nasa.gov/shuttle/reference/sodb/

Archived shuttle statistics but badly out of date
> http://spaceflight.nasa.gov/shuttle/reference/green/

GLOSSARY

acreage The foam which covers the large cylindrical surfaces of the External Tank is referred to as "acreage foam." It's sprayed on by a robotic machine.

ARMS Advanced Respiratory Monitoring System. A European-sponsored human biology experiment.

area-to-mass A number which indicates an object's ballistic coefficient, or how much drag it has. A high area-to-mass ratio indicates a lightweight object while a low area-to-mass ratio indicates a dense object. Often abbreviated A/M. Units are assumed to be square meters per kilogram unless otherwise stated.

ballistic coefficient A scientific way of measuring an object's drag. (See "area-to-mass.")

bipod A two-legged version of a tripod, basically an upside-down "V" shape. The External Tank's forward attachment fitting for the shuttle is a bipod. The "feet" of the bipod had wedge-shaped pieces of foam.

BRIC Biological Research In Canisters; on STS-107, two biology experiments featuring moss and worms.

BSTRA Ball Strut Tie Rod Assembly—A ball and socket joint used to provide some flexibility to the shuttle's propellant lines.

CAIB Columbia Accident Investigation Board. The independent panel which investigated the Columbia accident

Capcom Capsule Communicator. The astronaut in Mission Control who communicates directly with the shuttle for all operational matters.

C-Band Roughly 4 to 6 Ghz. The radar dishes in Florida that track the shuttle for launch and landing are in this frequency range.

CDR Commander.

CEIT Crew Equipment Interface Test. A training exercise where the astronauts inspect everything they're going to use aboard the shuttle. In addition, the astronauts trained to perform a spacewalk do a "sharp edge" inspection where they examine the payload bay to see if there's anything which could snag or rip their spacesuits.

CIBX Commercial ITA Biomedical Experiments. Hundreds of small microgravity experiments activated on-orbit by the astronauts.

CIC Crew Interface Coordinator. A scientist or engineer who talks to the astronauts in space about their experiments and represents the entire payload team.

comm Communications.

Crater A simple computer program used to predict damage to the shuttle's tiles by pieces of foam or other debris.

downlink Transmissions from the shuttle to the Earth

ergometer A fancy bicycle-like device. It has a seat, handlebars, and pedals, but no wheels. The pedals are attached to sensors that calculate the astronaut's workload.

ESA European Space Agency.

escargot A nickname for the astronauts selected in 1995, including Rick Husband, Kalpana Chawla, and Mike Anderson.

ET External Tank.

EVA ExtraVehicular Activity. Spacewalk.

execute package The daily "mail" from Mission Control to the shuttle. It includes flight plan updates, maintenance procedures, interesting places on Earth to observe and photograph, a summary of the mission's progress, and some humor for the astronaut's enjoyment.

family escort Astronauts designate fellow astronauts as their family escorts to help their families before and during a shuttle mission. If there's an accident, then the family escorts become the "Casualty Assistance Call Officer" (CACO.)

FCR The Flight Control Room is what most people think of as Mission Control. It has about 20 consoles for the engineers who monitor the shuttle in real-time.

FIT Florida Institute of Technology, a/k/a Florida Tech—a university about 35 miles to the south of KSC which served as the control center for Biopack and Biobox.

flowliners Baffle-like holes that keep the propellant flowing smoothly through the shuttle's plumbing to its engines.

FREESTAR Fast Reaction Experiments Enabling Science, Technology, Applications and Research. A group of six experiments mounted on a truss inside Columbia's cargo bay. Some were automated, some were operated by the astronauts from within the shuttle's crew cabin.

FRR Flight Readiness Review. The management meeting about one week before launch that determines everything's ready to support the launch.

GOBBSS Growth of Bacterial Biofilm on inorganic Surfaces during Spaceflight. A small experiment sponsored by the Planetary Society to investigate whether or not bacteria could attach itself to a meteoroid.

IMU Inertial Measurement Unit—the gyros which determine the shuttle's location in space.

ISS International Space Station.

IVA Intravehicular Activity. Working inside the spacecraft. Also the astronaut inside who assists spacewalkers into their spacesuits.

JIS Joint Integrated Sim(ulation). A large simulation exercise involving Mission Control, the astronauts, and possibly other organizations like the payload teams. In a JIS, everything is simulated as closely as possible to an actual mission, with the sim-sup giving the team a variety of problems to resolve.

JSC Johnson Space Center. The home of NASA's human spaceflight operations, located in Houston, Texas.

KSC Kennedy Space Center. Where the shuttle launches and lands. Located in Florida, next to the city of Titusville, about 40 miles to the east of Orlando.

Ku-Band Roughly 12 to 14 Ghz. The shuttle's directional high-gain transmitter. This system can only be used when the payload bay doors are open on orbit and the antenna's extended over the side and has a clear view of a Tracking Data and Relay Satellite. It's used for high-bandwidth transmissions including large amounts of experiment telemetry and television.

LES Launch and Entry suit. The bright orange pressure suits used for launch and landing since 1988. Nicknamed "pumpkin suits" because of their color and bulk. A newer suit, the Advanced Crew Escape System, looks the same but has better capabilities.

LiOH Lithium Hydroxide. Oatmeal can-size canisters filled with Lithium Hydroxide crystals that absorb carbon dioxide.

loops Communications channels.

LPT Low Power Transceiver. An experimental radio transceiver for more efficient spacecraft communications.

LSP Laminar Soot Process. One of the three combustion experiments on STS-107.

MECO Main Engine Cut Off—8.5 minutes after launch, when the three main engines shut down.

MEIDEX Mediterranean Israeli Dust Experiment. An Israeli sponsored camera to image dust clouds and sprites from space.

MER Mission Evaluation Room. An offline Mission Control group that evaluates the mission's progress and any problems that are being addressed.

MET Mission Elapsed Time. The shuttle's clock which begins counting at launch.

microgravity Extremely low amounts of gravity experienced aboard the space shuttle in orbit. It's not zero gravity because the force of gravity can be measured, but it's extremely small—about one millionth of the normal gravity on Earth.

MMT Mission Management Team. A group of high-level engineers and managers who oversee the space shuttle's day-to-day operations.

MOD Mission Operation Directorate. The chief flight director or deputy which oversees real-time operations in Mission Control.

MS Mission Specialist.

OARE Orbital Acceleration Research Experiment. A sensor mounted within Columbia's cargo bay that measures minute changes to the microgravity levels.

OFK Official Flight Kit. Items of a souvenir nature flown for organizations with some connection to the mission.

OIG Orbital Information Group—the NASA website for public distribution of satellite tracking information.

OMDP Orbiter Maintenance and Down Period. The equivalent of a 30,000 mile checkup on you car—but in this case it's 50 million miles. Each shuttle is periodically taken out of service for an overhaul of its systems, detailed inspections, and upgrades. NASA has since renamed the OMDP the OMM (Orbiter Maintenance and Modification).

OPF Orbiter Processing Facility—the large hangars where the shuttles are serviced between missions.

OCA Orbiter Communications Adapter.

Osirak A French-built nuclear reactor in Baghdad, Iraq. It was destroyed by Israeli F-16 jets on June 7, 1981.

PAL Protuberance Air Load . An aerodynamic ramp on the side of the External Tank.

PDIP Payload Data Interface Panel. Switches which configure the shuttle's communications system.

PhAB4 Pronounced "Fab Four"—(Physiology and Biochemistry Team). A suite of four physiological experiments (experiments on the human body)

PLT Pilot.

POCC Payload Operations Control Center. The room where payloads are controlled from on the ground, engineers monitor Spacehab's systems, and decisions are made for payload operations in space.

PRSD Power Reactant and Storage Distribution. The large aluminum tanks that hold the liquid hydrogen and oxygen used to generate power and water.

PS Payload Specialist.

PSRD Prototype Synchrotron Radiation Detector. A payload originally intended for STS-107 but moved to the STS-108 mission.

PTO Protein Turnover experiment. One of the PhAB4 experiments which evaluates how the body consumes protein in microgravity.

RCC Reinforced Carbon-Carbon—the gray fiberglass-like material which protects the shuttle's hottest surfaces during the reentry.

RPM Revolutions Per Minute.

S*T*A*R*S Space Technology and Research Students. A commercial educational payload sponsored by Spacehab. It featured six student developed experiments.

sardines A nickname for the astronauts selected in 1996, including Willie McCool, Dave Brown, and Mike Anderson.

S-Band Roughly 2 to 4 Ghz. The frequencies used by the shuttle's low data rate transmitters. They can transmit voice and telemetry no matter what the shuttle's orientation and during almost all stages of the mission.

Sim Simulation.

Simsup Simulation supervisor.

SOFBALL Structure of Flame Balls At Low Lewis-number. One of the three combustion experiments on STS-107.

SOFI Pronounced "so fee." Spray-on Foam Insulation. The orange foam insulation that covers most of the External Tank. It's basically the same material as spray cans of foam insulation available in hardware stores.

SOLCON Solar Constant. A Belgian experiment which monitors the output from the Sun.

SOLSE Shuttle Ozone Limb Scattering Experiment.

Spacehab The commercial firm which built the Research Double Module where most of the STS-107 experiments were housed. The term "Spacehab" refers to both the company and the module. There are three versions of the Spacehab modules—the original single module used for either experiments or to carry cargo to the Mir or ISS, the Logistics double module used to carry cargo to Mir and ISS, and the Research Double Module (RDM), which premiered on STS-107.

Spacelab The European pressurized laboratory module used for science experiments aboard the shuttle from 1983 to 1998.

SRB Solid Rocket Booster. The large solid propellant boosters used for the first two minutes of the shuttle's flight. The two SRBs provide most of the thrust and noise during launch.

SSME Space Shuttle Main Engine. The liquid hydrogen liquid oxygen powered rocket engines that propel the shuttle from sea level to orbital altitude.

STS Space Transportation System. Almost always just used as an acronym, STS-107 designates a particular mission. The number is only partially related to the order the flights fly.

TCDT Terminal Countdown Demonstration Test. The dress rehearsal for the astronauts with the launch team.

TDRSS Tracking Data and Relay Satellite System. TDRSS consists of a group of Tracking, Data, and Relay satellites in orbit, which act like a switchboard in the sky. The shuttle and other satellites send their signals up to a TDRS instead of only when they fly over ground stations. The TDRS retransmits the data to a ground station in White Sands, New Mexico. Using TDRS permits Mission Control to keep in contact with the shuttle for over 85 percent of the time, as opposed to less than 15 percent for previous spacecraft that relied on ground stations. TDRSS is also used by a variety of scientific satellites and some classified satellites.

terminal velocity The maximum speed an object can travel as it's falling through the Earth's atmosphere. It's the speed where gravity and the drag on an object are equal. Dense aerodynamically shaped objects have the largest terminal velocities, a couple of hundred miles per hour.

TPS Thermal Protection System. The gray RCC panels, the black and white tiles, and white blankets which cover most of the shuttle's exterior to protect it from reentry heat.

uplink Transmissions from the Earth to the shuttle (e.g., Mission Control tells the shuttle "We've uplinked today's mail to you.")

VAB Vehicle Assembly Building. One of the world's largest buildings. Originally built for the Apollo program, the VAB is where the shuttle is mated to its External Tank and Solid Rocket Boosters. It also includes storage areas and is where Columbia's debris is permanently stored.

VCD Vapor Compression Distillation experiment. An experimental vacuum still designed to purify urine.

INDEX